LIVESTOCK

Judging, Selection and Evaluation

THIRD EDITION

Roger E. Hunsley

Executive Secretary
American Shorthorn Association
Omaha, Nebraska

W. Malcolm Beeson

Lynn Professor of Animal Nutrition
Department of Animal Sciences
Purdue University
West Lafayette, Indiana

Danville, Illinois   **I P P**   THE INTERSTATE Printers & Publishers, Inc.

LIVESTOCK

Judging, Selection and Evaluation

LIVESTOCK JUDGING, SELECTION AND EVALUATION

Third Edition

Library of Congress Catalog Card No. 87–82120

ISBN 0–8134–2773–8

1 2 3
4 5 6
7 8 9

Foreword

As improvement in domestic animals became increasingly necessary in the development of the crop – livestock economy, the significance of understanding those basic economic values that can be arrived at through judging was recognized. The judging of merit in livestock is as fundamental to selection as selection is fundamental to improvement. The competitive improvement efforts of sponsors of the various breeds of livestock has stimulated the search for economic efficiency. The U.S. show ring has served to accelerate competitive progress among breeds, to unify opinion on useful type and to point out various improvement opportunities.

As research contributes to a greater understanding of the general economic agreement sought by both the producer and the consumer regarding practical type, the use of livestock judging, as a basis for selection and improvement, will grow in importance.

Preface

The subject of livestock judging, selection and evaluation in this text is presented in a practical, direct manner. In addition, a large number of pictures and drawings have been used to illustrate modern types of market and breeding animals.

In Parts 1, 2 and 4, considerable discussion is devoted to the influence that conformation and finish have upon value in the beef, sheep and swine carcasses. In Part 2, the factors that are recognized in the judging of the unshorn fleece are emphasized. The skeleton and its influence upon the shape, muscle structure and general usefulness of the different types of livestock are also stressed.

An extensive list of descriptive terms is included for each type of livestock. These lists will enable you, the student, to enlarge your vocabulary of specifically applicable terms so that your discussions will be definite and concise. Moreover, the judging terminology should acquaint you with the current practical terms that are in use for judging beef cattle, sheep, dairy cattle, swine and stock horses.

By studying and applying the information within this text, you will have a sound background in knowing how to judge live animals as a basis for selecting and improving livestock.

The many breed associations, breeders and photographers who have so generously contributed to making this text possible are gratefully acknowledged.

Roger E. Hunsley
W. Malcolm Beeson

Contents

3. DAIRY CATTLE 217

4. SWINE 269

5. STOCK HORSES 355

General Introduction

Being successful in the production of livestock to a large extent depends on being able to select animals that are good feeders and breeders. It also depends on being able to recognize merit in breeding animals and to understand values when buying or selling. The ability to judge is essential to success in all these phases of livestock production.

Many students who have the desire to learn how to judge livestock do not have the opportunity to gain valuable experience through long periods of daily contact with livestock. Moreover, they may not have the time to acquire that experience. Skill in livestock judging can, of course, be obtained through experience without formal instruction. Although experience is a good teacher, it is expensive and time-consuming. As a substitute in part for this experience, well-organized courses in livestock selection allow students to profit more by the experience they do get than would have been possible without the courses. Proficiency is attained only after years of experience and constant practice. Visits to ranches or farms, stockyards, packing plants and breeders are valuable in that students are able to observe and evaluate large numbers of livestock.

The practical side of livestock selection must be stressed. *Whenever any point concerning any type of livestock ceases to be of practical value, it no longer has a major place in livestock selection.* At times, too much emphasis may be placed on some minor fancy of a breeder or group of breeders, in which case the student loses sight of the fundamental basis for selection. Aesthetic characteristics, which contribute to the style and beauty of an animal, should not be emphasized to the point of being a detriment to the person who is attempting to make a living from livestock production. The *practical,* the *profitable* and the *useful* type should be fundamental to livestock selection.

There are certain characteristics that students should cultivate if they expect to excel and be successful in selecting and evaluating livestock. These characteristics are as follows:

1. Having the desire to know livestock.
2. Being able to form a mental image of ideal types.
3. Having a keen sense of observation.
4. Being able to interpret and incorporate production data and performance records.
5. Being able to coordinate live animal selection with carcass evaluation.
6. Being able to analyze a class of animals logically.
7. Being able to think independently.
8. Being able to give effective reasons.

Having the desire to know livestock. — If you, the student, have a strong desire to know livestock and to know them well, the battle is half won. However, unless your desire reaches beyond the requirements of the classroom and a judging team, you may not be successful. This ambition should lead you to sacrifice time and energy for the sake of studying and working with good livestock. To progress toward perfection in livestock selection, you must become livestock-minded and become aware of the perplexing problems that confront breeders who are selecting and culling to improve their livestock. Observe the imperfections as well as the perfections of the breeds of livestock. Study groups in addition to individuals. Thus, livestock judging should become a real challenge, rather than a means of learning to place a class of animals in the order in which you expect the judge to place them.

Being able to form a mental image of ideal types. — Livestock judging involves the comparison of differences. If one animal is judged or appraised, it is compared to some standard — namely, the ideal of its kind. If two or more animals are judged, then each animal must be compared first with the ideal and then with the others in the class. This procedure precludes the necessity of having definite information relative to the details of type and their influence upon function. Without this information, you cannot observe correctly and think clearly. Moreover, you will not be able to reach accurate conclusions.

The ideal types of livestock that you formulate in your mind serve as a guide in your judging of livestock. Attempting to judge livestock without first establishing a fixed image of the ideal in your mind would be synonymous to driving a car without a steering wheel — both would soon end in tragedy. The ideal types are illustrated through the livestock pictures and drawings included in this book. Although pictures and drawings are at best a poor substitute for live animals, study them carefully and attempt to visualize each animal alive. However, clearly defined ideals of the various types of livestock come only through constant contact with types that are good representatives of the various breeds.

Having a keen sense of observation. — Being able to detect important differences in livestock quickly and accurately is the result of intensely analyzing the outstanding good and bad points of animals. Persistent practice in evaluating towards the ideal is a sure way of sharpening your mind to be able to detect differences. A livestock tender, who knows livestock well, realizes that each animal possesses a physical individuality that sets it apart from the rest of the group. These observations have been acquired through constant and habitual contact with that class of livestock. Therefore, if you, as a student of livestock judging, expect to attain this "keen eye," you must be *among* livestock and *think* about livestock. Looking at an animal without really analyzing its physical make-up is futile.

Being able to interpret and incorporate production data and performance records. — Production data and performance records can provide you with objective measures that should help you make a more realistic decision as to the value of the animal(s) in question. The more information and facts that are obtained, the more accurate the judgment will be, provided the information has been interpreted correctly and used properly. Select for traits that are highly heritable and can easily be measured by the producer. Use actual records and adjusted data for within herd comparisons only. Use estimated breeding values (EBV) or expected progeny differences (EPD) when you are comparing animals between herds. Using records should remove a great deal of guesswork from judging and should allow you to be more accurate in selecting high-performance, efficient-producing livestock that are of the desirable conformation and type.

Being able to coordinate live animal selection with carcass evaluation. — In selecting modern meat-type slaughter or market animals, you must always keep in mind the end product of red-meat animal selection — the carcass. The old saying that it takes a different type of animal to win a live show than it does to win a carcass show is no longer true. To be able to pick out the live animals that will hang up the best carcasses requires a great deal of time and repetition to train the eyes and the mind to detect the various points on the live animal that indicate excellence in the carcass. After evaluating animals alive, you should evaluate the end product of these same animals — their carcasses. This will tell you where you were right or wrong in your judgment.

Being able to analyze a class of animals logically. — Since judging is based on observations, appraising animals in terms of exact percentages is impossible. All parts of an animal do not have the same economic value, but since each part is interrelated with and dependent on all the other parts, in the last analysis, the animal must be considered as a whole, rather than as a large number of separate parts. To assign definite percentage values to each part of an animal often leads to confusion. (The numbers on the various drawings of animals and carcasses are only to help you learn the various parts.)

To be logical in your analysis, you should approach the task in an honest, open-minded manner. Knowing what constitutes merit in market and breeding animals is important. You must be able to appraise values of market animals from an economic standpoint but still recognize the desirable balance in development that breeders and producers prefer, even though some points, such as constitution, environmental adaptability, feeding qualities, soundness and breed type, may not be of much concern to the packer. You must be sincere in your efforts to learn the proportional values of the different characteristics of an animal, from the consumer's as well as the producer's point of view, so that you can properly balance the values involved in judging each class of animals. Be careful that your enthusiasm for perfection in certain points does not interfere with your judgment of total value. Do not allow prejudice or biased opinion to influence your placing of a class — if you cater to your emotions, you will basically disqualify yourself as a competent judge.

If you observe the animals accurately, weigh the facts and then conclude with logical placing, your "busts" (no correct placings) will be reduced to a minimum. Never follow silly hunches. *Judging is applied, sound reasoning.*

Few areas have as many new, varied and fascinating problems that call for the exercise of judgment as that of livestock selection. Thus, the student who is open- and fair-minded and who renders sound judgment in livestock judging will be open- and fair-minded and will use good judgment in other endeavors as well.

Being able to think independently. — You must be able to accept just criticism from your superiors and to make your mistakes steppingstones to right thinking instead of stumbling blocks to progress. Your judgment may be just as sound as that of the student who is learning with you, so be honest with yourself and rely on your own thinking. Only to the extent that you do this will you be able to improve your judging ability.

A livestock judge often encounters very trying circumstances. When large groups of uniformly high-class animals are judged, the decisions will be extremely close, and close decisions, regardless of the number of individuals involved, call for considerable emotional stability. The most useful assistance the judge can possibly have on these occasions is the ability to recognize values, a deep regard for honesty and discipline in independent thought and action.

Being able to give effective reasons. — Being able to give reasons is, no doubt, a very difficult task for the average student in livestock judging. It requires a large, well-selected vocabulary, concentration, detailed accuracy, orderly and logical comparisons, variation in the choice of terms and the ability to give the reasons in a persuasive, confident manner. Thus, being able to give effective reasons depends partly on natural talent but primarily on acquired knowledge and practice.

It is important to have a definite system of giving reasons and to be able to use it effectively. In brief, the form used in this text is first to state the outstanding good points of the top animal and then to tell why you placed this animal, even though it may have some faults, over the one below. Admit the outstanding good points of the lower-placed animal in the pair. Follow the same order of describing and comparing each pair of animals in the class, adjusting the relative degree of description and comparison to fit the class at hand.

Don't worry about what system you adopt. Just master some system and make it work! The following are some suggestions and guidelines that may be useful when you are giving oral reasons.

1. *Be accurate* in your statements.
2. *Compare* one animal to another. Example: "Smoother, more desirable top."
3. Use description sparingly. Describing something outstanding or some obvious fault may be appropriate.
4. Refer to the various animals by numbers, being careful not to mix up the numbers. This will help you to keep the animals straight and will give a clearer picture to the judge listening to you. Avoid excessive use of "he," "she," or "it."
5. Practice "seeing" the animals when you are talking about them.
6. Eliminate "better" in your reasons. Be more specific.
7. Always grant the "under" animal any credit that may be due.
8. With a close pair, give nearly as much credit to the "under" animal as you do to the animal you placed over it.
9. Do not overcriticize the bottom animal, even though it may be very poor. Give two or three main criticisms of it and then quit.
10. Stand in a straight, comfortable position — usually with feet spread apart and hands behind back.
11. Have confidence in your placing, yet do not assume a "cocky" attitude.
12. Talk in a clear, convincing manner. Make your main points impressive and emphatic.
13. Vary your tone of voice, but do not shout.
14. Look the judge squarely in the eye, or at least give that impression.
15. Present your points in a well-organized, logical sequence.
16. Employ a variety of terms; avoid repetition. Understand the meaning of all terms. Keep in mind the purpose of the animals in each class and then use appropriate terms.
17. For breeding classes, mention the name of the breed represented when you are talking about type and character.
18. Be sure to give set of reasons within time allowed (two minutes).
19. *Be yourself* when you are giving reasons.

1

BEEF CATTLE

Livestock judging teams and judging contests have their place in training young people to know how to select and improve breeds of livestock. However, competitive livestock judging is a means to an end and not an end in itself.

The most important principles to consider when you are selecting and evaluating beef cattle are discussed in this part and are illustrated with drawings and pictures. However, by no means does the information here provide adequate training in itself. You must avail yourself of every opportunity to observe and carefully judge beef cattle. Applying the principles given here to the live animal and the carcass that it produces is an indispensable factor in learning the art of selecting cattle. You can improve your ability to judge through constant contact with good cattle, through detailed study of live animal and carcass evaluation programs set up by constructive breeders and researchers.

It is essential, especially for the beginner, to become familiar with the various parts of the live beef animal (Fig. 1-1). Knowledge of the names and locations of these parts aids considerably in developing the ability to detect the differences in livestock. After continued practice, recognizing these parts becomes automatic, and more emphasis can be placed on the points discussed in the remainder of Part 1.

SCORING BEEF CATTLE

The customary procedure in judging livestock is to evaluate four placeable animals on their merits. This generally is too difficult for beginners, since they do not know just what is desired. The beef judging card, however, will aid in overcoming this difficulty. If you will place the animals on the basis of the points that are briefly described, you will soon fix in your mind what is desired, especially if your instructor checks the placings for correctness. After using the judging card for several classes, be prepared to judge the animal on the basis of a whole rather than on each individual point.

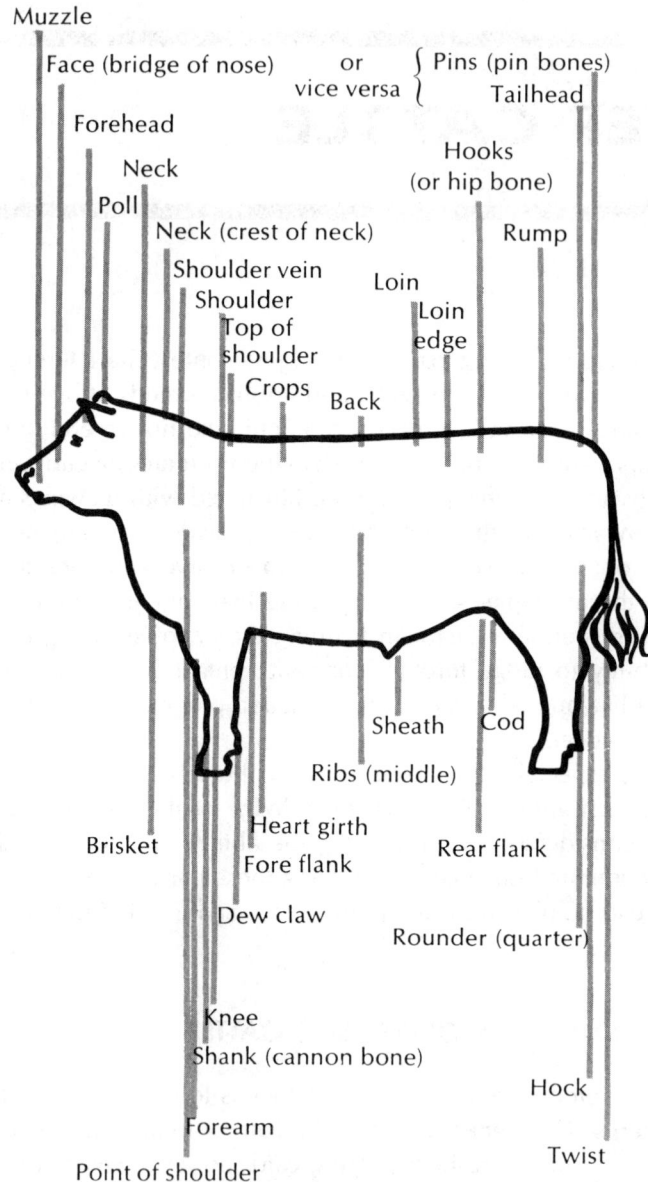

Muzzle

Face (bridge of nose)   or   { Pins (pin bones)

vice versa {   Tailhead

Forehead

Hooks
(or hip bone)

Neck

Poll

Rump

Neck (crest of neck)

Shoulder vein

Loin

Shoulder

Loin
edge

Top of
shoulder

Crops

Back

Sheath   Cod

Ribs (middle)

Heart girth

Brisket

Fore flank

Rear flank

Dew claw

Rounder (quarter)

Knee

Shank (cannon bone)

Hock

Forearm

Twist

Point of shoulder

Fig. 1-1. Parts of the beef steer.

BEEF JUDGING CARD

Points to Consider	Placing			
	1st	**2nd**	**3rd**	**4th**

General Appearance

Straight topline; thick and muscular throughout; desirable spring to rear rib; long-bodied, long-rumped; symmetrical and stylish; stands squarely on adequate substance of bone; trim middle, neat in the brisket and along the underline . . . .

Form

1. *Head* — Muzzle broad; nostrils large; eyes large and clear; face medium length, clean and slightly dished; forehead broad; poll definite and sharp; ears medium-sized and fine-textured; horns medium-sized and well-shaped . . . .

2. *Neck* — Average thickness, medium length, blends smoothly into shoulders; clean throat . . . .

3. *Shoulder* — Smooth, muscular (forearm also), minimum flesh covering, neat and smooth on top; shoulder vein smooth and full . . . .

4. *Breast* — Wide, full; brisket neat and trim with little dewlap .

5. *Forelegs* — Medium length, strong with adequate substance of bone; arm full and muscular; shank adequate size for weight of animal . . . .

6. *Chest* — Wide, average depth, girth adequate; crops neat and reasonably full . . . .

7. *Back* — Thick, muscular, straight, uniformly covered with minimum fleshing . . . .

8. *Loin* — Thick, deep, muscular, straight, uniformly covered with minimum fleshing . . . .

9. *Ribs* — Well-sprung, smoothly and uniformly covered with a minimum amount of firm flesh . . . .

10. *Flanks* — Slightly tucked up, neat, trim and firm . . . .

11. *Hips* (or hook bones) — Neatly laid in, smoothly and uniformly covered with a minimum amount of firm finish . . . . . .

12. *Rump* — Long, thick and muscular from the hooks to the pins; level and free from patchiness and excess finish . . . .

13. *Thighs* (or stifle muscle area) — Deep, thick, muscular, full . .

14. *Twist* — Medium depth and free of excessive finish . . . .

15. *Legs* — Straight, strong, placed out on the corners of the body; shanks should indicate average or above-average size or substance of bone . . . .

(Continued)

BEEF JUDGING CARD (Continued)

Points to Consider   Placing

	1st	2nd	3rd	4th

Finish

Smooth, uniform, firm covering, especially over ribs, back (rack) and loin; maximum of 0.6 inch of fat over 12th rib on 1,000-pound market animal; freedom from patchiness and rolls is desired; animal should be trim and free of waste in the brisket, along the underline, in the rear flank and at the base of the rear quarter. (Degree of finish is of great importance in market classes, but it is not so important in breeding classes as long as the ability to take on finish is indicated. However, excess finish in breeding classes is undesirable.) . . . .

Quality

Frame and finish smooth; hair fine and soft; hide thin and pliable; bone adequate size and clean-cut . . . .

Dressing Percentage

Smooth, firm finish; neat, trim middle; medium-weight hide. (Not considered in judging breeding cattle.) . . . .

Breed and Sex Character

(Applies only to breeding classes and will be discussed in the section dealing with the various breeds.) . . . .

FIRST STEPS IN EVALUATING LIVE BEEF CATTLE

A thorough understanding of the location and relative importance of each wholesale cut will give you a knowledge of the intrinsic importance of various parts of a beef animal, especially in judging slaughter cattle (see "Wholesale Cuts"). The locations of the various wholesale cuts are illustrated in Figs. 1-2 and 1-3. In Fig. 1-2, percentage values reflect the amount of total carcass weight typically represented by the various cuts in a choice slaughter steer. Fig. 1-3 shows the location of the wholesale and retail cuts in the beef carcass and the relative value of the different parts and the origin of some of the retail cuts commonly available in the retail store.

Figs. 1-4, 1-5, 1-6 and 1-7 point out the distinguishing characteristics of the modern beef animal (Animal A) and those of the old-fashioned beef animal (Animal B). The drawings, illustrations and explanations that follow pertain to the characteristics of bone structure and muscle development of this modern animal as well as the old-fashioned beef animal. The points and areas of the beef animal that can be used as indicators of muscling are illustrated and pointed out. However, some of the areas that are indicators of excessive finish are listed as follows. The point of the shoulder, the hip or hook bone and the backbone or vertebral column are all bones that have

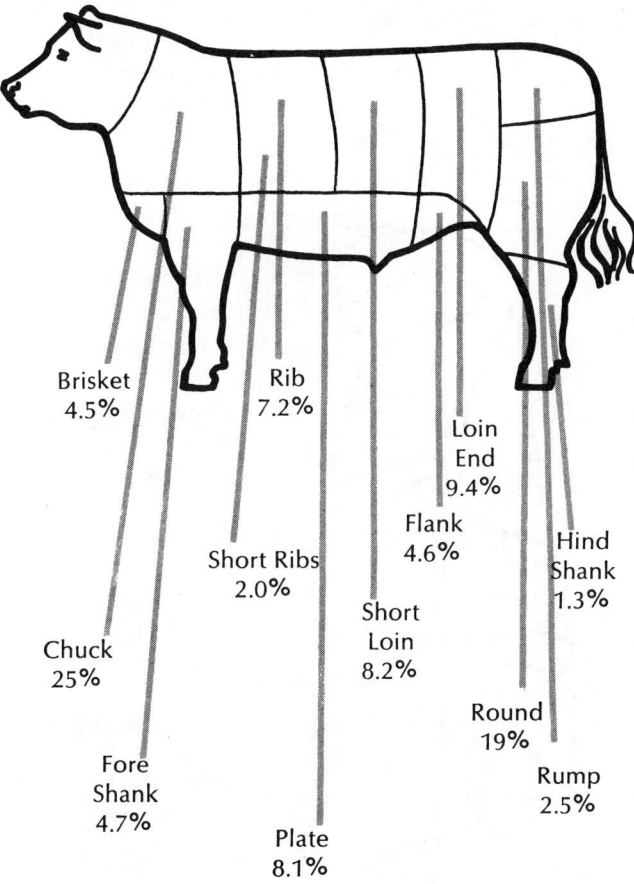

Brisket
4.5%

Rib
7.2%

Loin
End
9.4%

Flank
4.6%

Hind
Shank
1.3%

Short Ribs
2.0%

Short
Loin
8.2%

Chuck
25%

Round
19%

Fore
Shank
4.7%

Rump
2.5%

Plate
8.1%

Fig. 1-2. Location of the wholesale cuts commonly derived from a beef animal.

only a small amount of tendon and skin covering over them. As an animal fattens, a layer of fat forms between the tendon and the skin layer. When an animal is handled in these areas, everything under the hide is finish or fat. The tops of the shoulders, the fore-rib, the rear rib, the loin edge, the flank, the elbow pocket, the brisket and the cod are all areas that can be used to determine the amount of finish a beef animal is carrying at one time or another during the fattening period.

This information is based on research findings and practical experience and can be used to evaluate and appraise modern beef cattle — to help you select the big, stout, fast-growing, heavy-muscled beef cattle that carry a minimum amount of outside fat cover. These are the type of cattle we need to seek, find and breed. Only visual appraisal with the keenest eyes and the sharpest minds, along with performance records, can insure a modern beef cattle industry for years to come.

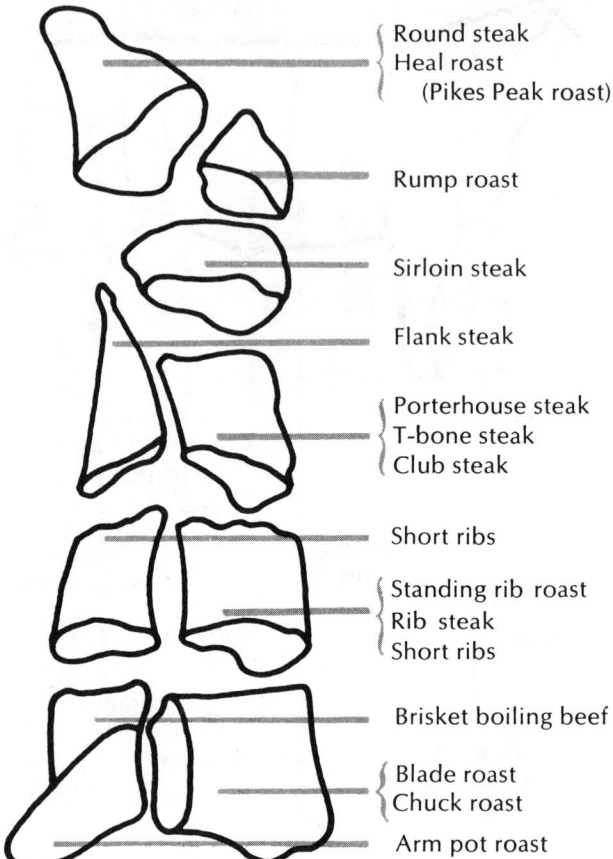

Round steak
Heal roast
(Pikes Peak roast)

Rump roast

Sirloin steak

Flank steak

Porterhouse steak
T-bone steak
Club steak

Short ribs

Standing rib roast
Rib steak
Short ribs

Brisket boiling beef

Blade roast
Chuck roast

Arm pot roast

Fig. 1-3. Origin of retail beef cuts indicated by the wholesale cuts from which the retail cuts would be trimmed. (Courtesy, Purdue University)

Figs. 1-4, 1-5, 1-6 and 1-7 are all of a steer, but the same principles of selection apply for breeding cattle. In judging breeding classes, you should also consider feminine and masculine traits and sound feet and legs.

Front View

Modern Animal A (Fig. 1-4) shows muscle development in the shoulder (1) and forearm (2). This area can be used as an area of muscle indication. There will be a certain amount of judgment involved in determining whether the shoulders are coarse and prominent due to the actual skeletal structure of the animal in question or if the shoulder and forearm indicate extreme muscle development.

The cannon bone area (3) is one of the most accurate and quickest indicators of bone size and substance of an animal. Research data show that heavier-boned animals have more muscle. In other words, there is a positive relationship between bone size or substance and muscling of cattle.

The brisket and dewlap area (4) is clean and trim and in no way indicates excessive finish.

Old-fashioned Animal B (Fig. 1-4) fails to exhibit the muscular appearance of Animal A. An animal of this kind may be as muscular as Animal A, but excessive finish will conceal extreme muscle development; however, in most cases, animals of this appearance (B) are usually light muscled and excessively finished. Animal B does not show an abundance of muscling in the shoulder (1) and forearm area (2). If it does have muscling, it is concealed by excessive finish. At any rate, the muscle

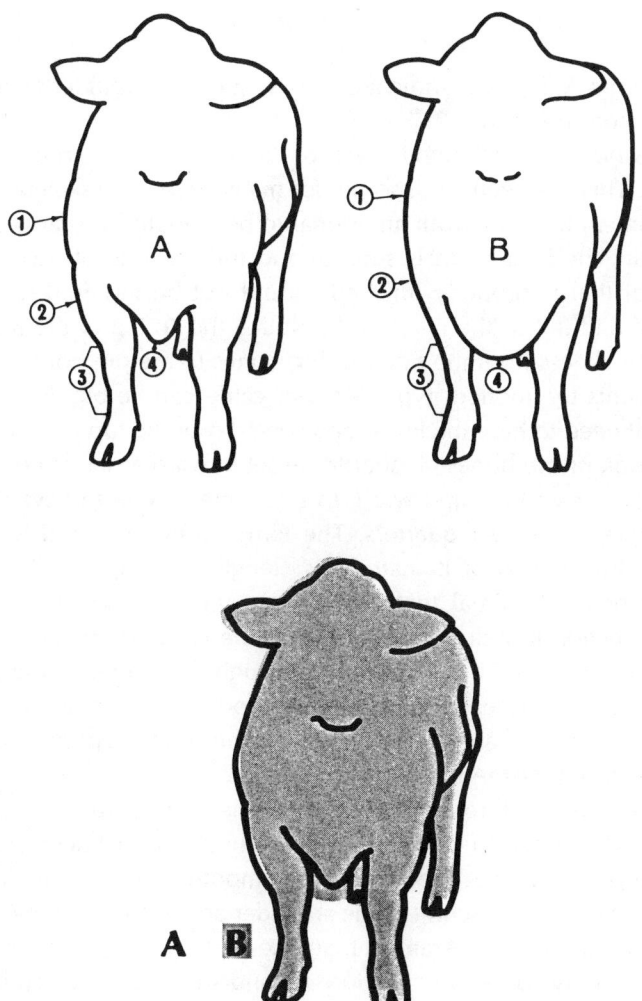

Fig. 1-4. Front view of the modern (A) and the old-fashioned (B) market beef steers, with superimposed comparison. (Courtesy, Purdue University)

seams are filled with fat to give the animal the smooth appearance that livestock evaluators have desired for many years. The animal is smooth only because it is relatively light muscled, and the muscle seams have been filled with fat to bring out the smooth effect.

Animal B is relatively light boned (3), which would tend to go along with the lack of muscling mentioned previously. The brisket and dewlap area (4) of Animal B is extremely full and rounded — this is nothing but fat-waste, as there is very little muscle found in this area of the carcass from a beef animal. Animals with excessive fat deposits in this area usually yield carcasses that are plenty fat both internally and externally.

Top View

Modern Animal A (Fig. 1-5) indicates muscling in the shoulder area (1) from this view as well as from the front.

Animal A is not full and smooth back of the shoulders (1) in the heart girth or fore-rib area (2). This has been a misnomer for many years. The skeletal structure and muscle development do not permit an animal to be smooth in its appearance in this area. Once again, the heart girth is smooth and full because of excessive finish. A muscular animal that is properly finished should not be required to be extremely neat shouldered and full in the heart girth. Notice the spring to the rear rib (3) and the length and thickness from the hook or hip bones to the pin bones (5) in drawing A, which represents the modern type. Modern cattle can be big, long and growthy, but they will still need to have thickness and spread over the top and through the rear quarters. The hook or hip bones (4) should be set forward as far as possible, and the pin bones should be set high and wide, to allow the animal to have the maximum amount of muscle in the rear quarters. The term commonly used to describe this area in either a live animal or its carcass is "length of rump," or "length from the hooks to the pins." An animal that has a long rump should also possess length through the rib section to help balance or correlate the body parts.

The thickest part of Animal A's body is through the center of the rear quarters. This is the greatest muscle mass in the animal's body, and if an animal is meaty and muscular, the rear quarter area should well be the thickest portion of the animal when viewed from either the top or the rear. Note the wedge shape of Animal A, with the narrower portion through the center of the rear quarter.

Old-fashioned Animal B (Fig. 1-5) possesses uniform thickness front to rear. An animal of this type can be described as being smooth and uniform. Animal B does not exhibit a great deal of muscling in the shoulder area (1) and is very smooth in the heart girth or fore-rib area (2). Animals that take on this general appearance usually carry excessive amounts of fat over the fore-rib and shoulder and also have heavy fat deposits in the elbow pocket.

There is no added expression or spring to the rear rib (3) of Animal B. The thickest part of Animal B is in the middle of its body, and this is the area of the

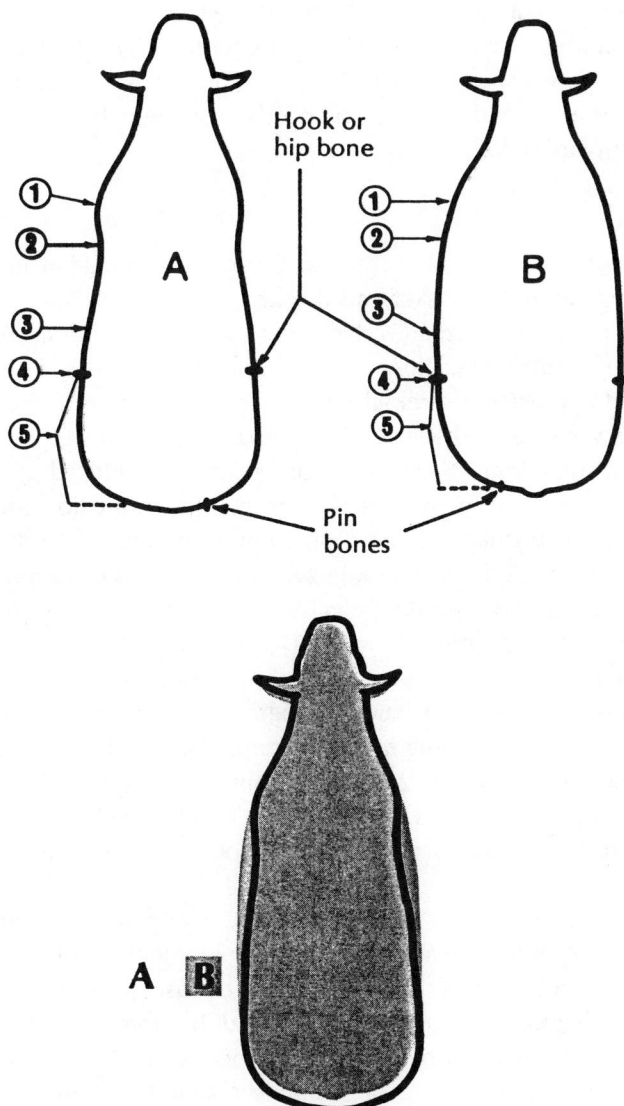

Fig. 1-5. Top view of the modern (A) and the old-fashioned (B) market beef steers, with superimposed comparison. (Courtesy, Purdue University)

animal where the least red meat or muscle is located. Animal B is short and tapered from the hooks to the pins. The short, tapering appearance from the hooks to the pins is quite often associated with animals that are short bodied, light muscled and overfinished. Note the overall general appearance of Animal B in the top view. Animal B in comparison to Animal A could best be described in the following manner: B is a shorter, shoved-together, closer-coupled, lighter-muscled animal than A with excessive finish over the shoulder and fore-rib. B is heavier middled and shorter rumped than A and, in general, needs more length, stretch, trimness and muscling to compare to A.

Side View

Modern Animal A (Fig. 1-6) is clean and free of extra leather in the throat area (1), is clean and neat in the brisket and dewlap area (2) and shows extreme muscle development in the forearm area (3). Note the substance of bone (4) that Animal A possesses in the cannon bone area; as mentioned previously, large bone size is positively related to muscling. Animal A is very clean and neat in the middle (5) and at the same time shows extreme length of middle. This could well be referred to as a very tight middle and a long, clean underline. Animal A is tucked up a bit in the rear flank (6), which is quite acceptable under present-day evaluation standards. In past years it was practically unknown for a champion animal not to have a deep, full flank. Research has yielded considerable evidence that this extra depth and fullness in the flank was due to nothing but fat-waste.

Note the correctly structured hind leg (7) of Animal A. Although not essential, properly structured hind legs help an animal express and display its muscling and correctness. Animal A does not have the extremely deep bulging rear quarter (8) desired by livestock (cattle) evaluators in the past. Once again, the skeletal structure and muscle development do not permit an animal to have the extreme depth and bulge to the rear quarter that was once desired. This is a highly unnatural condition and is mainly brought about by excessive fat deposits in the area of the cushion of the round.

Note that Animal A is not as neatly laid in at the tailhead (9) as might be desired; but again, the skeletal structure and muscular development simply do not lend themselves to natural smoothness in this area. A meaty, muscular, desirably finished steer might tend to be slightly plain around the tailhead, but remember that in most cases extra fat makes an animal smooth, and this is no exception. Animal A has a very long, straight topline (10), which is a desirable characteristic in modern-day cattle. Usually, strength of top is related to above-average muscling throughout the body, especially when it is found in an animal that possesses the body length that Animal A exhibits.

Old-fashioned Animal B (Fig. 1-6) shows excessive leather and finish in the throat area (1) and is extremely full and heavy in the brisket area (2). This extreme fullness in the brisket area is practically all waste and is highly undesirable from the

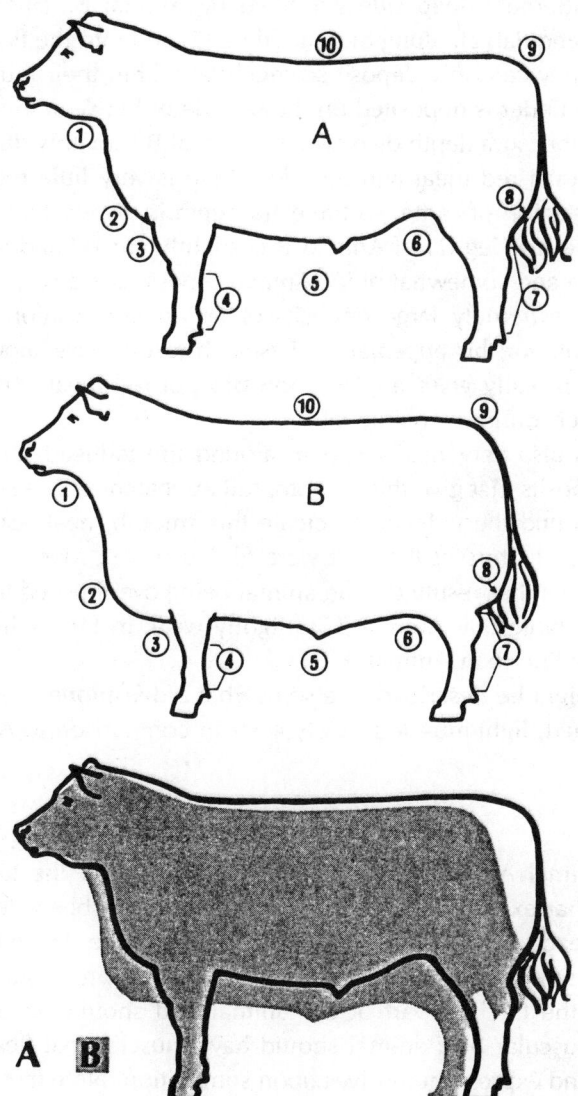

Fig. 1-6. Side view of the modern (A) and the old-fashioned (B) market beef steers, with superimposed comparison. (Courtesy, Purdue University)

retailers' standpoint. Animal B does not show extreme muscle development in the forearm area (3). Animal B exhibits average or less-than-average bone size in the cannon bone area (4), which might tend to indicate the lack of muscling mentioned previously for this animal.

Note the extremely deep side exhibited by Animal B. This is mainly due to excessive waste and flab all along the underline (5) and into the flank area (6). When animals fatten, they can only deposit so much fat within their muscles (intramuscular), and the remainder is deposited on the outside or between muscles (intermuscular). This is why the extra depth displayed by Animal B is mainly due to extra flab and fat-waste rather than red meat and muscle. There is very little muscle in the lower half of the mid-section of cattle, so the extra depth is of very little value.

Note that the rear leg (7) of Animal B is slightly curved and set under its body. The hock is large and somewhat puffy. Animal B possesses a very deep, bulging rear quarter (8), and extremely large deposits of fat on the cushion of the round are mainly responsible for this appearance. Research results have shown that this depth and bulge, which really gives a false impression of red meat and muscle development in the lower round, is mainly fat.

Animal B is also very neatly laid in around the tailhead (9), which is due to excessive fat deposits that give this smooth, full appearance. The skeletal muscles do not come up around the tailhead to create this smooth, neat appearance, so once again livestock evaluators of the past were fooled by fat. Many times the desire to have a pretty animal has resulted in an animal being overfinished to give it a smooth, full appearance. Note that Animal B is slightly weak in the topline (10) and has a much shorter topline than Animal A.

Animal B might be described as a somewhat old-fashioned, short-bodied, deep-sided, overfinished, light-muscled, wasty steer in comparison to Animal A.

Rear View

Modern Animal A (Fig. 1-7) is slightly rough around the tailhead (1), which simply means that excessive amounts of finish have not been deposited on either side of the tailhead to give it the smooth, neat appearance desired in the past. Note the extreme muscle development and expression in the rear quarter (2). The stifle muscle area is the thickest part of the animal and should be deep, bulging and expressive. A muscular beef animal should have muscles that flex upon movement and that bulge and express themselves upon stimulation. Note that the muscle seams can still be seen in Animal A both in the inner and outer rounds. Animal A exhibits the heavy bone and straight legs (3), which are characteristic of most muscular modern-day cattle. Note that Animal A is "cut up" in the twist, so to speak. The deep, full twist commonly referred to in the past was a result of fat-waste. During the feeding period, fat deposits early in the twist and cod area. Note that Animal A is clean and neat in the twist and cod area (4) and exhibits an excessive amount of muscle.

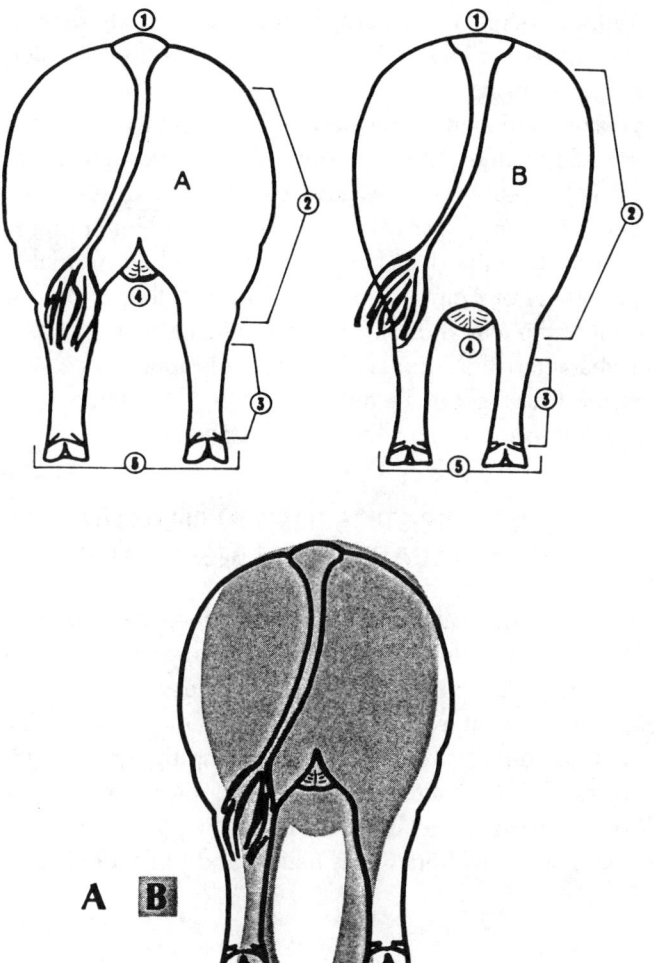

Fig. 1-7. Rear view of the modern (A) and the old-fashioned (B) market beef steers, with superimposed comparison. (Courtesy, Purdue University)

Observe the set to the feet and legs of Animal A. The feet (5) are set wide and out on the corners of the body. Animals that are heavy muscled must have wide set legs and thickness between them to carry the vast amounts of red meat and muscle tissue. This is portrayed by Animal A.

Old-fashioned Animal B (Fig. 1-7) has a counter-sunken tailhead (1) that is surrounded by excessive fat deposits. The fat deposits give the tailhead a neat, smooth-looking effect. Animal B lacks muscle development and expression in the stifle muscle area (2). Note the "pear-shaped" round of Animal B. This type of muscle structure is usually found in light-muscled, overfat animals. The presence of muscle seams cannot be found in the round of Animal B, which would indicate that the seams have been filled with excessive amounts of fat, giving the rear quarters a round, smooth appearance. When compared to the round, shapely, symmetrical appearance of Animal A, Animal B has a rather square look, which is usually an indication of excessive finish.

Note the average bone size (3) displayed by Animal B, which is usually associated with light-muscled cattle. Animal B possesses a deep, full twist and a very full cod (4). This area is loaded with fat, which is highly responsible for the deep, full-looking appearance. The muscle structure in the twist and inner thigh area in no way forms and develops to give the deep, full look displayed by Animal B.

The feet and legs (5) of Animal B are set close together under the center of the body. The close set of the feet and legs usually indicates a lack of muscle in animals that possess this characteristic. Animals that have this appearance and lack muscling tend to fatten faster, thus becoming the overfinished, light-muscled cattle that are undesirable according to present-day evaluation standards.

CHARACTERISTICS USED IN SELECTING
AND EVALUATING LIVE BEEF CATTLE

Frame can definitely be evaluated visually, and the repeatability and heritability for this parameter appear significantly high. Larger-framed cattle grow faster, are leaner and produce more pounds of edible beef per day of age than smaller-framed cattle. More important, recent evidence notes that frame (skeletal dimensions) associated with weight can be most effective in changing the growth curve of cattle. In addition, current research indicates a positive relationship of frame with pelvic size in females. Table 1-1 shows the relationship between visual frame scores and height measurements taken at the withers when heifers and bulls are 205 and 365 days of age.

TABLE 1-1. Relationship Between Visual Frame Scores and Height Measurements of Heifers and Bulls at 205 and 365 Days of Age

Visual Frame Score	Height Measurements, 205 Days of Age		Height Measurements, 365 Days of Age	
	Heifers	Bulls	Heifers	Bulls
1	35.1	36.0	39.0	41.0
2	37.1	38.0	41.0	43.0
3	39.2	40.0	43.0	45.0
4	41.2	42.1	45.0	47.0
5	43.3	44.1	47.0	49.0
6	45.3	46.1	49.0	51.0
7	47.4	48.1	51.0	53.0
8	49.4	50.1	53.0	55.0
9	51.5	52.2	55.0	57.0

Feet and Legs and Substance of Bone

Research results indicate a positive relationship between circumference of bone and percentage of muscling in the beef carcass.

Some livestock judges say size of bone is not important in the selection of market animals (steers) and should be given only minor consideration when breeding cattle are selected. If this is true, you might as well pick the finest-boned animals available, since extra energy and nutrients are necessary to make extra bone. People often say, "You can't eat bone." True, but increased production of lean beef in heavy-boned animals far outweighs increases in percentage of bone. Feed efficiency and growth rate are usually higher.

The best place to observe bone substance of size is the cannon bone of the front leg (Fig. 1-8). This is the bone from the knee down to the pastern joint. Usually,

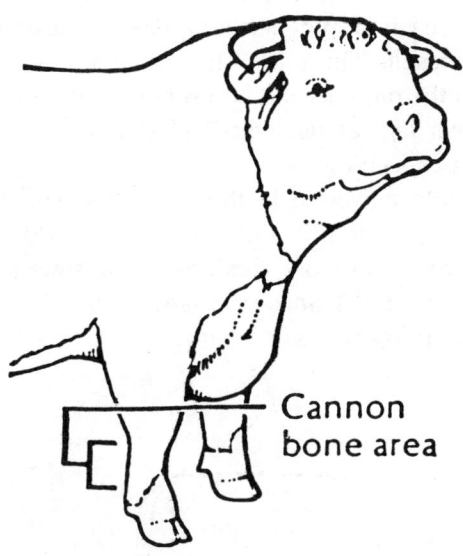

Cannon bone area

Fig. 1-8. The cannon bone of the steer's front leg is one of the best indicators of bone size. (Courtesy, Purdue University)

cattle with large-sized cannon bones will be large framed and heavily muscled. However, there is as much individual animal variation within a breed as there is between breeds for this characteristic. Bone size also indicates durability and ruggedness. In areas with extremely high- and low-temperature variations, or where cattle under range conditions frequently have to walk 3 or 4 miles to salt and water, ruggedness and durability are still very desirable characteristics.

Longevity is almost completely dependent on structural soundness. Longevity is an important trait in all phases of beef production as it affects the number of replacements necessary to maintain numbers in both commercial and purebred herds. Unsound feet, legs and joints affect the ability of bulls to follow and breed cows in the breeding pastures, thus reducing the productivity of the herd.

The feet should be large and show good depth of heel. They should be symmetrically shaped, with toes equal in size and shape. A bull with short, shallow outside toes behind will wear off his feet unevenly and will require frequent trimming to keep him traveling correctly. The joints should be well-defined, with no indication of puffiness or swelling.

Side View

The livestock judge must be especially aware of substance of bone and set to the feet and legs of livestock. Cattle that set improperly will have more problems, especially under conditions of stress, such as breeding or climbing hills. For a proper set to the feet and legs, a plumb line straight down from the pin bone at the point of the animal's buttocks should intersect the hock and dewclaw of the rear leg (Fig. 1-9).

If the animal is too heavy for his/her bone size, a "sickle-hocked" condition in the hind legs frequently results (Fig. 1-10). The weight forces the feet under the body, putting greater strain on the muscles and bones of the stifle joint and the hip and leg region. Strain on the rear legs of sickle-hocked animals can interfere with activity while they are breeding or feeding.

"Post-leg," or too little curvature in the hind legs (Fig. 1-11), seems to be a heritable trait, especially prevalent in fine-boned cattle. This condition is undesirable, as post-legged animals put more pressure on the stifle joint, with less chance for "play," or absorption of the strain over a larger area. The result of this is more muscle stretching and torn ligaments and tendons.

Hook or hip bone

Stifle joint

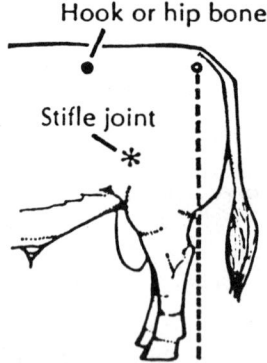
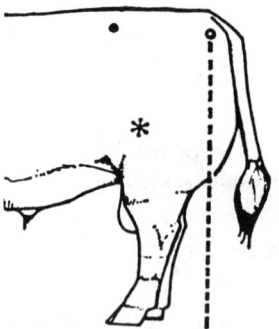
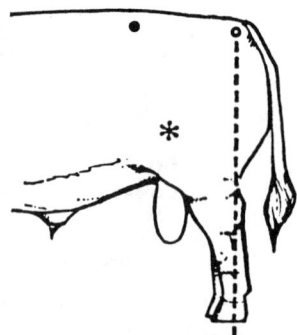

Fig. 1-9. Good example of a proper leg set. Note how a line down from the pin bone almost intersects the hock and the dewclaw. (Courtesy, Purdue University)

Fig. 1-10. This animal is sickle-hocked. The hind legs are too far under. (Courtesy, Purdue University)

Fig. 1-11. Post-legged animal. The hind legs are too far back, thus making them straight. (Courtesy, Purdue University)

Front View

Fig. 1-12 shows the correct placement of the front feet and legs. The legs come out of the center of the shoulder and straight down to the ground. The front legs are set out on the corners of the body, allowing for maximum support and stability during locomotion. When the body weight is equally distributed on the limbs, this puts the stress and strain on the strongest and most durable parts of the feet, legs and joints, resulting in very few structural soundness problems. The size and shape of the toes on each foot are usually very uniform on a structurally sound animal.

Two common structural unsoundnesses occur in the front limbs. In Fig. 1-13, the animal is toed-out, or splay-footed, in front. The knees are closer together than normal and from the knee down the legs twist to the outside of the body. This condition is usually magnified by large, deeper outside toes in front. Fig. 1-14 shows the opposite structural problem in front. The front legs are wide at the knees, giving them a somewhat bowed appearance. This condition is called toed-in, or pigeon-toed. In this condition the inside front toes are quite large, causing the animal to roll

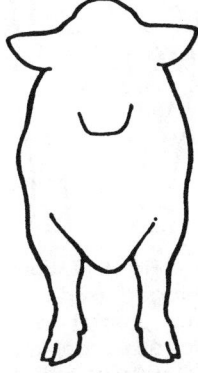  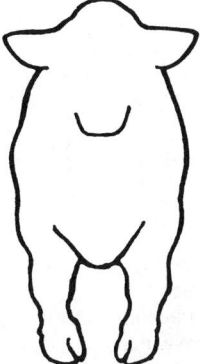

Fig. 1-12. Correct front leg structure. (Courtesy, Purdue University)

Fig. 1-13. Toed-out (splay-footed) in front. (Courtesy, Purdue University)

Fig. 1-14. Toed-in (pigeoned-toed) in front. (Courtesy, Purdue University)

to the outside or in the direction of the smaller outside toes. This is usually accompanied by a coarse, protruding shoulder, which is called wing-shoulder. As the front feet roll to the outside, they throw the shoulders out and cause them to become quite coarse and protruding. This trait seems to be heritable, as it is transmitted from parent to offspring. Quite often bulls that have coarse, heavy shoulders will also sire calves that are similar in structure, thus causing increased difficulty at calf birth.

Rear View

Fig. 1-15 shows the correct placement of the rear feet and legs. The legs come out of the center of the rear quarter and straight down to the ground. The rear legs are set out on the corners of the body, allowing for optimum support and stability during locomotion. Note the uniformity in size and shape of the toes on each foot.

The most common unsoundness that occurs in the rear limb structure is the cow-hocked condition, which is illustrated in Fig. 1-16. The hocks are closer together than normal, causing the lower portion of the limb to turn outward, resulting in the toed-out condition. This condition is usually magnified by larger, longer, deeper outside toes.

Fig. 1-17 shows the bow-legged condition from the rear view, which causes the animal to be wide at the hocks and close, or toed-in, at the heel. This condition is usually magnified by larger, longer, deeper inside toes behind.

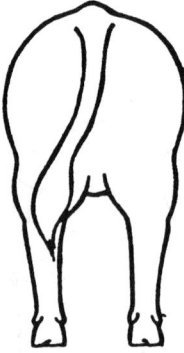

Fig. 1-15. Correct rear leg structure. (Courtesy, Purdue University)

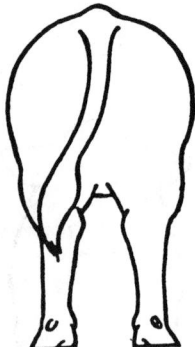

Fig. 1-16. Cow-hocked (close at the hocks) and toed-out behind. (Courtesy, Purdue University)

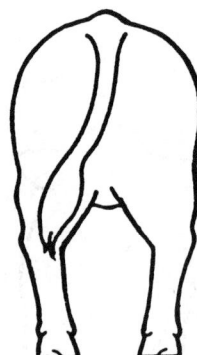

Fig. 1-17. Bow-legged (wide at the hocks) and toed-in behind. (Courtesy, Purdue University)

Breed Character

Breed character is nothing more than a trademark or an identity of the breed the animal represents. Breed character is expressed by color and color patterns; typical heads — referred to by size and shape; by shape, size and set to the ear; and by the characteristics established by the respective breed associations to represent the "ideal" of the breed. Information concerning breed type and character requirements for each breed can be obtained from the respective breed associations. Addresses of the various breed associations are provided at the end of Part 1.

Sex Character

What is sex character? A male should look like a male, and a female should look like a female. Bulls should appear masculine, and cows and heifers should appear feminine.

Masculinity

Male sex hormones are responsible for masculinity. Thus, if an animal reaches sexual maturity at an early age, it is usually referred to as an early-maturing animal. Animals that show signs of early sexual maturity are usually not good prospects for continued rapid growth or large, mature size. This seems appropriate, since it is an established fact that the production of male sex hormones stops long bone growth due to ossification of the break joint or epiphyseal groove, where increased length of the long bones takes place. After bulls pass the yearling stage, it is desirable that masculine traits be present as an indicator of fertility. The head should be masculine, while the neck should display a cresty, thick appearance. The testicles should be large and uniform in size and should descend properly. It is recommended that a 12-month-old bull should have a minimum of 30 centimeters of testicular tissue to insure desired fertility and reproductive ability. The sheath should be tight and free of obstructions. A pendulous sheath is undesirable, and in many cases, bulls possessing this trait have been problem breeders. Reproductive problems, as well as disease and injury to the penis, have been more frequent in bulls possessing a pendulous sheath, or penis.

Femininity

Female sex hormones are responsible for femininity. Females of breeding age (15 months or older) should be sharp, clean fronted and feminine headed and should look lady-like. Thin, lean necks and long, smooth muscling and promise of udder development indicate fertility and milking ability in young females. However, referring to breeding, calving and performance records to evaluate reproductive efficiency is more accurate than looking at heads or determining femininity. Females should have the reproductive capacity to carry a large, thrifty calf, and they should also possess a high, wide placement of the pin bones to facilitate ease at calving time. It is recommended that a heifer have a minimum of 150 square centimeters of pelvic opening area at breeding time (usually 15 months of age) to insure minimum difficulties at calf birth. Pelvic opening areas of 170 square centimeters or larger at breeding time are the most desirable from the standpoint of ease of calving in first-calf heifers.

SELECTING YOUNG CALVES

The decisions you make when you are judging, selecting and evaluating cattle should be based on practical observations and accurate information — in other words, they should be economically sound. Whenever any point concerning any type of livestock ceases to be of practical value, it no longer has a major place in livestock selection. Refer to Table 1-2 for heritability estimates for economic traits in beef cattle.

Quite often young calves are selected according to the same standards that are used in the judging and evaluation of finished steers or replacement stock for the

TABLE 1-2. Heritability Estimates for Economic Traits in Beef Cattle[1]

Trait	Rating	Approximate Average Heritability
		(%)
Fertility	Low	10
Calving interval	Low	0–15
Birthweight	Medium	35–40
Cow maternal ability	Medium	20–40
Weaning weight	Medium	25–30
Weaning conformation score	Medium	25–30
Yearling weight (365 days)	High	60
Feedlot and carcass traits		
Feedlot gain	High	45–60
Efficiency of feedlot gain	High	40–50
Final weight off feed (13–15 months)	High	50–60
Slaughter conformation score	Medium	35–40
Carcass grade	Medium to high	35–50
Dressing percentage	High	45
Rib-eye area	High	70
Rib-eye area per 100 pounds of carcass	Medium to high	30–50
Fat thickness	High	45
Fat thickness over rib per 100 pounds of carcass	Medium to high	25–40
Tenderness of lean	High	40–70
Retail product, percent	Medium	30
Retail product, pounds	High	60
Summer pasture gain of yearling cattle	Medium	25–30
18-month weight of pastured cattle	High	45–55
Cancer eye susceptibility	Medium	20–40
Mature cow weight	High	50–70

[1]Summarized from many published sources. Wider ranges indicate traits for which fewer estimates have been made and for which probable average heritability is less precisely known.

breeding herd. Remember, the young calf that is 6 to 9 months old has to have the potential for growth of bone, muscle and some fat tissue. If the calf is early-maturing (Fig. 1-18), it will complete the bone and muscle growth early in the finishing period and then will deposit fat, which will create a smooth, deep-sided, wasty-looking appearance. How can you tell the early-maturing cattle from cattle that will continue to grow? Look at the head, the frame, the bone length and the muscle length. Then, relate this appearance to the age and weight (performance record) of the animal in question. If the animal in question has a short, square, mature-looking head, this indicates an early-maturing calf. Why? The bones in the skull are some of the first to reach maturity in beef animals. Quite often early-maturing calves have reached their rapid growth phase earlier in their life cycle and have deposited more fat prior to weaning time.

Fig. 1-18. Early-maturing calf at three months of age.

Other criteria to be used in the selection of a calf should be age and weight. Select calves that will weigh from 1,000 to 1,200 pounds as finished animals at 15 to 20 months of age. Cattle that are lighter or heavier than the 1,000- to 1,200-pound weight range are usually either underfinished or overfinished. If cattle are younger than 15 months of age at the completion of the feeding period, they will not usually grade USDA Choice or produce a carcass that is in the greatest demand today. If cattle are older than 20 months at the completion of the feeding period, then they were too old when originally selected. For example, a calf born in January 1987 and shown in August 1988 would be 20 months old. If the calf is older than this, it was born in 1986, and there is no way that an animal of this kind is practical and profitable for anyone.

The young calf should have size, scale, frame and growth potential (Fig. 1-19). Substance of bone, as well as structural correctness, is very important. Large bone size is related to muscling and growability of cattle. If a young calf is not structurally correct on his/her feet and legs, don't expect improvement with age, because very few will. When selecting young calves, look for length of rib and length of rump. Length of body and width or thickness of body also should be given more consideration than depth of body. However, don't forget any of the three body dimensions — length, width and depth — when you are selecting young calves. The thickness of body will be a good indicator of muscling in the young calf. Be sure the young calf has a wide chest floor and a good arch of rib to give body capacity to utilize feed and to convert the feed to red meat and muscle in an efficient manner. The young calf should walk wide on both ends. This width of walk is a good indicator of body thickness, capacity and dimension of the skeletal make-up. Select a young calf that has a strong topline, is not sway-backed and is free of waste and indicators of waste. Young calves that have large amounts of skin (leather) in the throat and neck region and big, heavy briskets, along with deep, heavy, wasty middles, have tremendous predisposition for waste. In other words, they are destined to become wasty-appearing cattle as they reach weights of 1,000 to 1,200 pounds. The young calf should have a long, level rump and a high, wide placement of the pin bones — located on either side of the tailhead. The tailhead should be somewhat prominent and should set up on top of the rump, not down into the rump and back. This allows for greater muscle volume and dimension in the rear quarter of the steers and a larger pelvic cavity in heifers, which will usually permit greater ease of calving.

The two factors most highly related to faster-gaining, more efficient, more profitable cattle are length of body and length of leg. When you are selecting or judging cattle, length of leg and body, along with the three body dimensions of length, width and depth, should be the prime factor for consideration.

EXAMINING LIVE BEEF CATTLE

In evaluating a class of beef cattle, you should have in mind a definite system that fits your needs and permits you to analyze the class most effectively. The system given here is only one suggested form. However, it is not as important to use this system as it is to master whatever system you may develop.

The first impression is usually a lasting impression and should be made about 20 to 30 feet from the cattle. Stay away from the class for at least 3 minutes and get a good impression of the animals as a whole. Students are often prone to crowd classes of livestock, and this is the surest way to get a false impression. View the cattle from the side and note points such as size; scale, length of body; strength of top; length of rump (length from the hooks to the pins); trimness in the throat, in the brisket and along the underline; muscle development in the arm, in the forearm, over the back and loin and through the round; depth of rib; substance of bone; set to the feet and legs; and balance (Fig. 1-19). Then, view the cattle directly from behind, noting the spread and thickness over the crops, back, loin and rump; the spring of the fore-ribs and rear ribs; the uniformity of width from front to rear; the trimness of middle; the length of rump; muscling down the top and through the rear quarter; freedom from excess finish in the twist area and at the base of the round; thickness through the stifle; overall shape; and set to the rear feet and legs. You should look at the cattle from the front, noting breed type and sex character about the head and neck; width and depth of chest (constitution); substance of bone; set to the front feet and legs; muscling through the shoulder, arm and forearm areas; trimness in the throat and through the brisket; and neatness through the shoulder. Concentrate and train yourself to remember the outstanding good and bad points about each animal. At this point you should have the animals tentatively placed according to general appearance.

Fig. 1-19. This modern Salers calf has size, scale, frame and growth potential. (Courtesy, American Salers Association)

Now, a closer observation should be made — feeling (handling) the cattle carefully for firmness, uniformity, smoothness and correctness of finish. Start at the shoulders. Feel the covering (finish) over the top of the shoulders, crops, fore-ribs and rear ribs, back, loin, loin edge, rump and round. On steers, feel the ribs, elbow pocket, rear flank and cushion of the round to determine the degree to which the animals have been fattened. Keep in mind that there are several places on the skeleton of a beef animal where there are no muscles covering the bones of the skeleton, only the tendon to which the muscles are attached and the hide. These are excellent places to determine the degree of finish that an animal is carrying. These areas are (1) the point of the shoulder, (2) the ribs, (3) the hip or hook bones, (4) the vertebral column and (5) the pin bones. When handling the animal, observe the muscling in the shoulder, arm and forearm region; over the back and through the loin; and through the rear quarter, especially the stifle area. All this time you should concentrate on the amount and smoothness or roughness of the finish in each part. Do not poke the cattle or pull the hide, as this is irritating to the cattle and of little benefit to you. Use one hand, with all fingers extended together, to handle cattle. The only time both hands should be used is when you are measuring the length of rump (length from the hooks to the pins).

When convenient, it is helpful to see the cattle at the walk, for this brings out defects and good points that never show up when the cattle are at rest. When observing cattle at the walk, be sure to note freedom of movement, style, tightness of frame, strength of top, areas that indicate muscle development, correct set to feet and legs, substance of bone, general balance and harmony of all parts. Observation at the walk is used extensively by professional judges and is an excellent way, especially with larger classes, to segregate animals according to general size, muscling, type and overall correctness.

After you have a true mental image of every animal in the class, it is well to set up a tentative placing by picking out the top animals and the bottom animals or perhaps a top and bottom with a close middle pair. The remainder of the time should be spent in analyzing close pairs and checking the entire class to see if there is sound reason for your placing. When reasons are to be given, you should record sufficient notes to recall to mind each animal, especially if there is a considerable lapse of time before your reasons are to be given. You should remember two things: (1) the first impression is usually the best and (2) the reasons that you give to support your placing must be accurate.

MODERN BEEF BREEDS

Angus

The Angus breed has the distinction of being very stylish and alert and unusually smooth, with excellent balance. The general contour of the lines of this breed is round and smooth. The quality of bone and body and even distribution of hard, firm finish are outstanding. Any roughness or patchiness is a serious fault and meets with considerable disfavor among judges and breeders. The balance, moderate bone, trim middles and outstanding carcass quality (marbling) of Angus cattle have made them excellent killers. The bone is medium in size with neat, clean joints. They are noted for early sexual maturity and high fertility, which greatly contribute to their excellent maternal instincts.

Angus cattle are polled and black in color (Figs. 1-20, 1-21, 1-22, 1-23 and

Fig. 1-20. This typical Angus steer is what packers want to buy. They know Angus steers will grade and fit into boxed beef. (Courtesy, American Angus Association; photo by American Angus Association)

Fig. 1-21. This Angus cow is watching over her young bull calf. Neither the cow nor the calf ever strays too far from the other. The calf could grow up to become a sire for a herd or be steered and put in a feedlot to feed out. (Courtesy, American Angus Association; photo by American Angus Association)

Fig. 1-22. Angus and Black baldie steers are typical sights in feedlots. Packers will pay a premium for these steers, since they know Angus and Black baldie steers will grade. (Courtesy, American Angus Association; photo by American Angus Association)

Fig. 1-23. Angus bulls like this one will produce top offspring. They can service cows naturally, or their sperm can be collected for artificially inseminating cows. (Courtesy, American Angus Association; photo by American Angus Association)

Fig. 1-24. Angus heifer. She has not had her first calf yet. After her first calf, she will be called a first-calf heifer. If mated to the right bull, she will produce good offspring. (Courtesy, American Angus Association; photo by American Angus Association)

1-24). The American Angus Association disqualifies from registry any animal with a white switch, white ankles, white legs or white spots on the body. White markings are objectionable except on the underline behind the navel and are accepted there only to a moderate extent. A white scrotum in bulls is undesirable.

Mature bulls usually weigh from 1,600 to 2,000 pounds, (Fig. 1-23), and cows (Fig. 1-21) in average condition from 1,000 to 1,400 pounds.

Barzona

The Barzona is a medium-sized beef animal, with actual adult size varying somewhat with the environment. It is a smooth, long-bodied, well-balanced animal with a longish head. Normal color is solid, medium red, but it may vary from a deep to a light red. Occasionally white may appear on the animal's underline or switch. The dark-pigmented, loose hide and straight, short hair help make the Barzona very heat tolerant and insect resistant. It may be either horned or polled. It has relatively long legs of strong, medium-sized bone and tight, very hard feet. As a freemover, it travels long distances easily. It efficiently grazes large areas and is naturally able to utilize a very high percentage of browse in its diet.

The Barzona cow is a naturally fertile animal that matures early (Fig. 1-25). She has a high-quality udder with a strong, high attachment and small teats, and she milks well. The Barzona is a thrifty, feminine, good-doing cow with a strong, protective mothering instinct and a long, productive life. In any environment she weans a

Fig. 1-25. Present-day Barzona cow and her calf. (Courtesy, Barzona Breeders Association of America)

large, growthy, heavier-than-average calf. She is alert, curious but quiet and easily handled.

The Barzona bull is a hardy, vigorous, sexually aggressive animal that can settle a large number of cows over a widespread area. He reaches puberty early and has unusually high-quality semen. Mature bulls weigh between 1,600 and 2,000 pounds, and mature cows 1,000 to 1,400 pounds.

Blonde d'Aquitaine

The Blonde d'Aquitaine breed of cattle has been known in Europe since the eighteenth century. The principal breeding area is in Southeast France from the Pyrenees Mountains through the Garonne Valley. This beef breed is one of the largest in Europe.

Purebreds vary in color from cream to buff. The color is usually recessive in the first cross with other breeds.

Long body, moderate bone, smooth shoulders and extra length from hooks to pins are physical characteristics of the breed. Blonde cows are fertile in milk production. Blonde cows adapt to different types of terrain and all types of management, and they have one of the longest productive life spans of any of the European breeds (Fig. 1-26). The qualities of ease of calving, high performance and feed efficiency have accounted for the popularity of the breed in France and in the crossbreeding programs of the importing countries.

The average weight of mature bulls is 2,530 pounds (Fig. 1-27). The largest bulls of the breed will weigh just over 3,000 pounds. Mature cows weigh from 1,600 to 1,800 pounds.

Fig. 1-26. Modern Blonde d'Aquitaine cow and her calf. (Courtesy, American Blonde d'Aquitaine Association)

Fig. 1-27. Modern Blonde d'Aquitaine bull. (Courtesy, American Blonde d'Aquitaine Association)

Brahman

The best examples of the modern Brahman breed are considerably lower set than Brahmans first brought into the United States. They are extremely deep bodied and are very muscular throughout (Figs. 1-28 and 1-29). The Brahman head is long when compared to that of the other beef breeds, and the horns turn up rather than down

Fig. 1-28. Registered Brahman cow. (Courtesy, American Brahman Breeders Association)

Fig. 1-29. Registered Brahman bull. (Courtesy, American Brahman Breeders Association)

and out as in the horned European breeds. The hump, of course, is very pronounced on both the bull and the cow, but breeders insist on the hump being carried gracefully rather than falling off to one side. Brahman cattle stand on very good feet and legs and walk with ease. They are comparatively trim in their middles, are rather thin in their hides and have a characteristically high dressing percentage when condition is considered.

The most preferred color pattern for Brahman cattle is that of silver grey, and usually there is a tendency toward darker color on both the front and rear quarters. Some breeders have selected for a sandy or reddish color (Fig. 1-30).

Fig. 1-30. Painting of a typical Red Brahman bull. (Courtesy, American Brahman Breeders Association)

Very few cattle producers raise Brahmans as market animals to be sold as such with relatively pure breeding; but rather, Brahmans are crossed with other cattle and the produce is marketed. Few purebred Brahman cattle are slaughtered at present, as they are exceedingly valuable as breeding stock. The improvement that can be worked in a small-boned, nondescript herd of cows by a well-bred Brahman bull is quite phenomenal.

Brahman cattle grow larger than the other breeds used in the coastal areas. They develop rapidly and continue to grow to five or six years of age. In ordinary pasture condition, bulls will weigh approximately 1,800 pounds, and cows, 1,200 pounds. In fitted condition, mature animals exceed these weights by 500 or 600 pounds.

Brangus

The Brangus breed was developed from a blend of Brahman and Angus breeds. The cattle are based on foundation stock of $3/8$ Brahman and $5/8$ Angus. Animals recognized as Brangus can be produced by (1) breeding an animal having $1/4$ Brahman ancestry and $3/4$ Angus ancestry with an animal that is $1/2$ Brahman and $1/2$ Angus; (2) breeding an animal having $3/4$ Brahman ancestry and $1/4$ Angus ancestry to a purebred Angus; or (3) intermating Brangus individuals.

Brangus is a registered trade name and can be applied only to cattle registered with the International Brangus Breeders' association. Fifty-four breeders founded the association in Vinita, Oklahoma, in 1949. Only Brangus cattle are registered in the permanent registry of the association.

All cattle to be registered as Brangus or enrolled as foundation stock are inspected by an association appraisal committee.

The Brangus breed is black and polled — both inherited dominant qualities (Figs. 1-31 and 1-32).

Mature bulls weigh between 1,800 and 2,000 pounds, while mature cows weigh between 1,300 and 1,400 pounds.

Fig. 1-31. Registered Brangus cow. (Courtesy, International Brangus Breeders Association, Inc.)

Fig. 1-32. Registered Brangus bull. (Courtesy, International Brangus Breeders Association)

Chianina

The American Chianina Association does not attempt to establish weight or color specification for the Chianinas being bred in the United States, but the following breed characteristics of the Italian Chianinas are noteworthy and are obviously present in the American Chianinas.

Structurally, when compared to other types of cattle, they are a taller, longer and cleaner type (Figs. 1-33 and 1-34). They exhibit extreme uniformity of depth, trimness of the middle and a cleanness of dewlap and brisket. Strong legs and hard hooves are the result of centuries of selection for work. Chianinas tend to be ectomorphic in shape (long boned and long muscled) in their conformation, thus taking advantage of two economically important traits: ease of calving and, apparently, a freedom from double muscling. There is black pigmentation of the tongue, palate, nose, switch and anal orifice and around the eyes. This pigmentation greatly reduces the threat of pink eye, cancer eye and other skin diseases. The black pigmentation is dominant on the first cross with other breeds.

Italian Chianinas are the largest of all cattle breeds. Mature bulls stand 72 inches at the withers and weigh up to 4,000 pounds (Fig. 1-34), while cows stand 60 to 68 inches at the withers and weigh up to 2,400 pounds (Fig. 1-33).

Gelbvieh

The German Gelbvieh is bred in the Franconian region of Bavaria. The main

Fig. 1-33. Painting of a Chianina cow and her calf. (Courtesy, American Chianina Association)

Fig. 1-34. Painting of a Chianina bull. (Courtesy, American Chianina Association)

areas are around the cities of Würzburg, Bamberg and Nuremburg. The origin of the Gelbvieh breed goes back to the reddish-brown Keltic-German Landrace. This breed was then crossed with Simmental and Shorthorn in the early 1800's. Since then, the breed has been selected for unicolor and growth and later for carcass quality.

The Gelbvieh is a solid-colored golden red breed. The hair coat is fine and dense. All skin areas, including the udders and the area around the eyes, are well pigmented. The hooves are very dark and very hard. It is very long bodied and well-balanced and shows excellent beef characteristics (Figs. 1-35 and 1-36).

The German Gelbvieh is a well-muscled, fast-growing breed with good size and length. Classified as dual purpose, it has a reputation for carcass quality and milk production.

In 1952, a planned program was started for the selection on milk production. This program has been successful, and the German Gelbvieh breeders claim their breed to be the best unicolor, dual-purpose breed.

The average breeding bull weighs 2,220 pounds, reaching 2,900 pounds, and reaches a height of over 57 inches at the withers. The average mature cow (Fig. 1-35) weighs right at 1,400 pounds and is just over 52 inches in height at the withers.

Fig. 1-35. Gelbvieh cow and her calf. This Gelbvieh calf benefits from his mother's high milk production. (Courtesy, American Gelbvieh Association)

Fig. 1-36. Gelbvieh feeder steer. Feedlot gain, feed conversion ratio and carcass merit are typically superior in Gelbvieh calves. These characteristics stem from their progeny-tested German background. (Courtesy, American Gelbvieh Association)

Hereford

Hereford cattle are commonly called "Whitefaces" because of the uniform presence of white markings. There is white hair on the face and jaws, under the neck, on the brisket and chest floor, under the belly and on the flanks and tail. White markings occur on the legs, extending in many individual cattle to the knees and hocks. The white stripe (feathering) on top of the neck extending to the crops is characteristic of the breed. White "line-backed" or "red-necked" individuals or those having too much white on the legs and flanks are objectionable. Black in the tail and a "smutty" nose are considered undesirable features. An ideal Hereford has well-developed, wax-colored horns, which come from the head at right angles and point slightly forward, with a uniform drooping curvature from the base to the tip. Horns that turn upward or backward and horns with black tips detract from the attractiveness of the head.

Due to their color markings, long, curly hair and thick beef conformation, Herefords are flashy in appearance, with excellent substance of bone, muscling and thickness of body (Figs. 1-37 to 1-43). Hereford cattle are distinguished for their large size, tremendous scale, uniform type and color, smooth shoulders, excellent heads, great chest and body capacity, strong constitution and vigor. The lines of this breed are intermediate between the rounding lines of the Angus and the more rectangular lines of the Shorthorn. Long bodies, well-sprung ribs, long rumps and thick, full, muscular hindquarters, coupled with outstanding quality and balance, identify the Hereford type. Accepted types of Herefords possess distinctive heads with strong character.

Fig. 1-37. Model Hereford bull. (Courtesy, American Hereford Association)

Fig. 1-38. Model Hereford cow. (Courtesy, American Hereford Association)

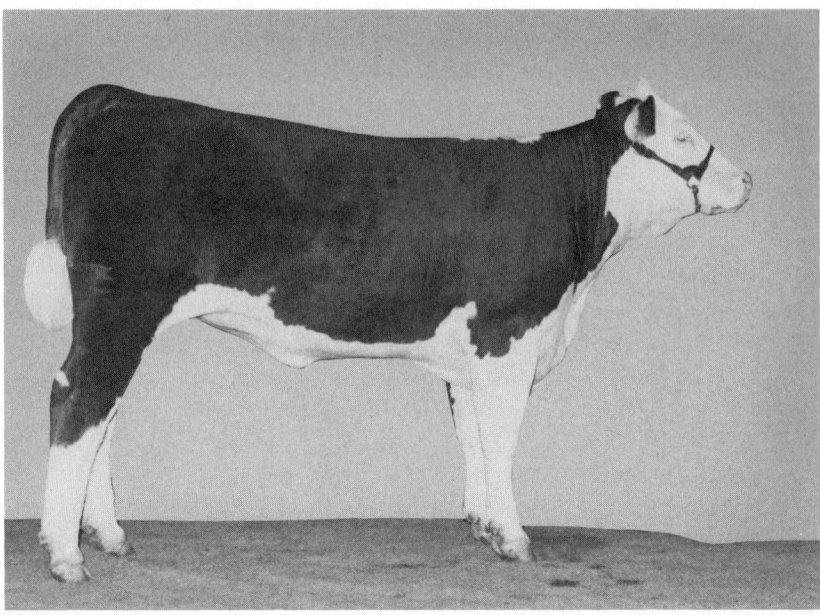

Fig. 1-39. Grand Champion Female, "L3 Genetic Lady 4," at the 1985 Illinois State Fair Register of Merit Hereford Show in Springfield. (Courtesy, American Hereford Association)

Fig. 1-40. Grand Champion Bull, "K&B L1 Momentum 1ET," at the 1985 Illinois State Fair Register of Merit Hereford Show in Springfield. (Courtesy, American Hereford Association)

Fig. 1-41. Grand Champion Steer of the 1985 Hereford - Polled Hereford Steer Show at the American Royal Livestock Show in Kansas City, Missouri. (Courtesy, American Hereford Association)

Fig. 1-42. Grand Champion Female, "Bright Miss Mark K200," in the Hereford Show at the 1985 National Western Stock Show in Denver. (Courtesy, American Hereford Association)

Fig. 1-43. Grand Champion Bull, "C L1 Express 3223 ET," in the Register of Merit Hereford Show at the 1985 American Royal Livestock Show in Kansas City, Missouri. (Courtesy, American Hereford Association)

In the selection of Herefords, considerable emphasis is placed on heads, constitution, size, scale and ruggedness. A moderate-length, average-width head with a strong muzzle is preferred. Mature Hereford bulls in good condition should weigh from 1,800 to 2,100 pounds or more, and mature cows in breeding condition, from 1,100 to 1,500 pounds.

Charolais

The coat color of Charolais cattle is white or a very light straw color. As a result, Charolais cattle are often referred to as the "big, white cattle" or the "silver cattle." Most Charolais have white, slender, tapered horns. However, some polled animals, usually dehorned as calves, are also being registered in the United States.

The purebred Charolais are big, long-bodied, heavy-muscled animals (Figs. 1-44, 1-45, 1-46 and 1-47). As vigorous, hardy, thrifty animals, they demonstrate a high degree of feed efficiency and an exceptionally high weight per day of age.

In conformation, the Charolais present a distinctive appearance because of the heavy muscling in the loin and the round, a trait that is derived from generations of selection for this quality. The insistence on a high degree of muscling in the hindquarters and loin area stems from the French method of cutting, marketing and preparing beef. Extreme double muscling is not a desirable trait.

Mature purebred bulls will usually weigh from 2,000 to more than 2,500 pounds (Figs. 1-44 and 1-46), depending on condition, while mature cows will run from 1,250 to as much as 2,000 pounds or more (Figs. 1-45 and 1-47).

Marchigiana

The Marchigiana (Figs. 1-48 and 1-49) was developed in Italy as a separate breed

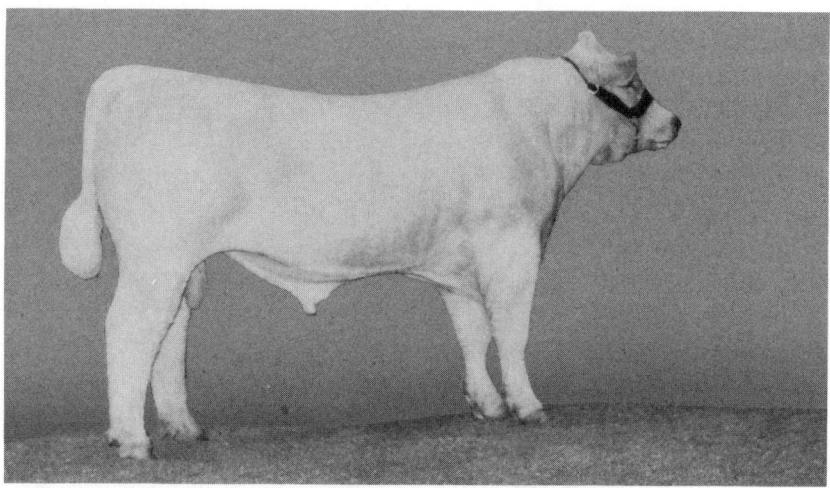

Fig. 1-44. National Grand Champion Charolais Bull, "Hickory Lane Professor," at a North American International Livestock Exposition. (Courtesy, Hickory Lane Farm, Blair, Nebraska)

Fig. 1-45. Outstanding Charolais cow that has proven herself not only in the show ring but also in the cow herd. (Courtesy, Hickory Lane Farm, Blair, Nebraska)

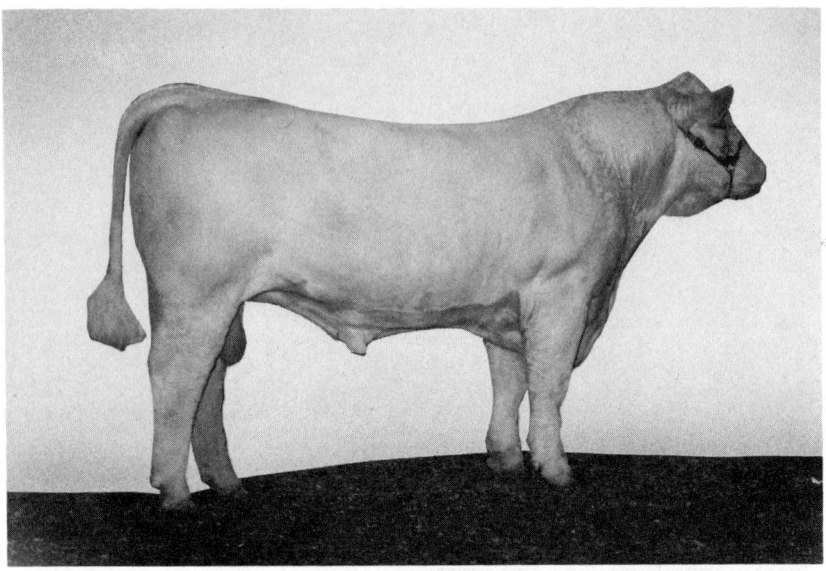

Fig. 1-46. "Immortal," undefeated champion Charolais bull. (Courtesy, American-International Charolais Association)

Fig. 1-47. Typical modern-day Charolais cow. (Courtesy, American-International Charolais Association)

Fig. 1-48. "Noccio," a long Marchigiana bull with a fair degree of muscling and an excellent disposition. He has adequate bone and excellent length. (Courtesy, Marky Cattle Association)

Fig. 1-49. "Nando," a masculine, docile Marchigiana bull with excellent length and muscling, correct feet and legs and adequate bone. (Courtesy, Marky Cattle Association)

during relatively recent times. The breed has been successively intermixed with the Chianina and Podalic types. The conditions under which the cattle are kept vary widely, from the lower plains to the mountain slopes. Cereals and legumes provide fodder in some provinces, while natural grazings and hay are furnished in other provinces.

These tan-colored cattle are producers of quality carcasses with a high percentage of lean meat. In crossbreeding programs, the progeny appear to have good growth rates.

The average mature Marchigiana bull weighs from 2,500 to 3,000 pounds. The average mature cow weighs between 1,400 and 1,500 pounds.

Milking Shorthorn

The Milking Shorthorn is similar to the beef-type Shorthorn in color and horns, but it is more angular and less thickly fleshed (Figs. 1-50 and 1-51). The primary objective of breeders of this Shorthorn type is to develop cattle that will produce large quantities of milk and steer calves that will gain rapidly and yield acceptable beef. In selection, greater emphasis had been placed on the milking qualities until about 1970 when beef qualities received more emphasis in selection programs. Since that time, Milking Shorthorns have been used to cross with beef-type Shorthorns, as breeders felt a need to select for trimmer, growthier, higher-cutability cattle.

Fig. 1-50. "Prairie Monster 39818," three-time All-American Milking Shorthorn bull and the 1984 National Grand Champion Bull. His sire, "Clayside Prince Becenten 375581," was a four-time All-American and a two-time National Grand Champion Bull. (Courtesy, American Milking Shorthorn Society)

Fig. 1-51. The cow, "Innisfail Wild Rose 158th 380786," classified as excellent, is a two-time All-American Mature Cow. She was Grand Champion Female at the 1983 and 1984 National Shows at Madison and Grand Champion Female at the 1982 Western National Show, held in conjunction with the Los Angeles County Fair, Ponoma, California. (Courtesy, American Milking Shorthorn Society)

Increased milk production comes from the cattle with increased scale. This same increase in milk and scale produces more beef. Such a combination meets the needs of the commercial beef producer. While this breed was developed as a dual-purpose breed, individual breeders have excelled in both milk and beef production. Horned Milking Shorthorns are more numerous than polled.

Mature bulls weigh from 2,000 to 2,500 pounds, while mature cows weigh from 1,400 to 1,800 pounds.

Maine-Anjou

The Maine-Anjou breed evolved in the western part of France about the middle of the nineteenth century. The climatic conditions of the area in which the cattle were developed are quite typical of the temperate regions of the North American continent.

Of the French breeds, the Maine-Anjou is probably the largest, and although used for both meat and milk, it is best developed as a beef producer (Fig. 1-52).

French breeders favor a red coat. However, there are some animals that have white spots or that are dark roan. They are very long, rather upstanding, with a particularly long rump. They are moderately deep with average bone. The skin is

Fig. 1-52. "Cunia," the premier sire of the Maine-Anjou breed, is best known for siring top females, show-winning steers and easy-born calves. He is registered with the American Shorthorn Association. (Courtesy, American Maine-Anjou Association)

loose, medium thick and lightly pigmented. The bone structure is about average for beef breeds.

Their growth rate and their production of carcasses of prime quality are the reported strong points of the breed.

Mature weights of bulls are 2,500 pounds or above, and for cows, slightly less than 2,000 pounds.

Murray-Grey

The Murray-Grey originated in Australia, probably from a cross of Angus and Shorthorn. The breed is slightly bigger than but is similar in conformation to the Angus.

In color, the Murray-Grey ranges from silver grey to dark grey. The Murray-Grey is polled and is widely known as a docile breed.

The Murray-Grey Beef Cattle Society was formed in the United States in 1962. Registration is available to all cows having not less than $7/8$ Murray-Grey breeding and to all bulls having not less than $15/16$ Murray-Grey breeding. Recording is also available to both sexes having not less than one-half Murray-Grey breeding.

At maturity, bulls average 2,000 or more pounds, while cows weigh from 1,300 to 1,400 pounds.

Normande

Normande is the spotted or speckled breed from France (Figs. 1-53 and 1-54). The coloring is so unique that it is not difficult to pick out Normande cattle in a pasture mixed with other breeds of cattle. Although the cattle are primarily dark red

Fig. 1-53. Normande cows like the one above are known for their exceptional mothering ability and docility. The average cow produces 9,240 pounds of milk per year. (Courtesy, American Normande Association)

Fig. 1-54. Outstanding Normande sires like the one above are being used in the United States to "breed up" herds to purebred status. Extremely fertile and prepotent, Normande bulls should satisfy crossbreeders who are interested in putting more important maternal characteristics into their herds. (Courtesy, American Normande Association)

and white in color, there is tremendous variation between animals — some have quite dark coats, while others have light ones. Colored patches around the eyes make the animals appear bespectacled, as well as providing resistance to pink eye and cancer eye.

The Normande breed, noted for its docility and hardiness, is a natural for cross-breeding. Normande cows have strong maternal characteristics of calving ease, milking ability and mothering ability, plus pigmentation around the eyes and teats.

At maturity, purebred Normande bulls average between 2,200 and 2,500 pounds, while cows usually weigh between 1,400 and 1,500 pounds.

Pinzgauer

The Pinzgauer is one of the oldest dual-purpose breeds of cattle in the world today. The herdbook for the breed is over 1,400 years old.

Pinzgauers have proven to be highly adaptable. Although they evolved from centuries of breeding in the Austrian Alps, they have performed very satisfactorily in the hot, humid climates of Africa and South America.

The Pinzgauer is a stretchy animal with a deep chestnut coat marked predominantly with a white tail and some white on the underside of the barrel. All areas of the mucous membranes and skin are pigmented. The legs are very correct, with hard, strong, dark hooves being a very prominent feature. The Pinzgauer has one of the highest muscle-bone ratios of any purebred breed in the world.

Purebred Pinzgauer bulls average about 2,600 pounds at maturity, while purebred Pinzgauer cows weigh approximately 1,500 pounds.

Polled Hereford

The Polled Hereford is similar to the Hereford in body and breed characteristics, except for the horns (Figs. 1-55, 1-56 and 1-57).

Fig. 1-55. Productive Polled Hereford cow and her young bull calf. (Courtesy, American Polled Hereford Association)

Fig. 1-56. Modern Polled Hereford bull. (Courtesy, American Polled Hereford Association)

Fig. 1-57. Modern Polled Hereford steer. (Courtesy, American Polled Hereford Association)

Romagnola

The Romagnola is an Italian breed resembling the Chianina and the Marchigiana (Figs. 1-58 and 1-59). This ancient breed was developed from crossing the Podalic cattle and local animals and then stabilizing these descendants in the local environment. For about 100 years, Romagnolas have been subjected to careful selection for both meat and milk. The cattle are found on rich lowland soils, on lower hills and on mountain slopes.

In comparison to some larger exotic breeds, Romagnola cattle are of medium height and bone, but they are well-muscled and early-maturing.

Mature Romagnola bulls average 2,400 pounds or more, and mature Romagnola cows average 1,400 pounds or more.

Fig. 1-58. "Monello," a tremendously thick, heavy-muscled, adequate-boned Romagnola bull with a correct set of feet and legs. (Courtesy, American Romagnola Association)

Fig. 1-59. "Marco," a smooth-shouldered, excellent-pigmented Romagnola bull, is one of the longest ever produced. (Courtesy, American Romagnola Association)

Salers

The Salers breed was developed on the mountains of the southwestern center of France where the rough climate, the rather poor soil and an elevation of from 2,500 to 7,000 feet above sea level make it necessary for a breed to be extremely hardy and self-supporting.

The cattle are generally mahogany in color, with white spots appearing sometimes between the umbilical cord and the rear udder of the cows. The hair is thick and often curly. A short, triangular face, broad forehead and narrow, lyre-shaped horns are characteristic. The neck is generally long and slender. The cattle have a straight, level topline; wide, deep chest; well-sprung ribs; wide rump; and a long, thin tail with nicely shaped tailhead (Figs. 1-60, 1-61 and 1-62). The thighs are long and well-muscled without excess. The dark hooves are especially hard and resistant.

In the history of the breed, no double-muscling or any other genetical defect has ever been reported.

Fig. 1-60. Modern Salers cow and her calf. (Courtesy, American Salers Association)

Fig. 1-61. Modern Salers bull. (Courtesy, American Salers Association)

The Salers breed, a dual-purpose one, is renowned for high milk yield, good growth rate and remarkable breeding qualities.

The average weight of mature bulls is 2,530 pounds (Fig. 1-61), while cows average 1,540 pounds (Fig. 1-60).

Fig. 1-62. Outstanding modern Salers heifer. (Courtesy, American Salers Association)

Scotch Highland

The Scotch Highland, developed in the Hebrides Islands near the west coast of Scotland, is one of the oldest European breeds. Developed under rigorous climatic conditions and with scant feed supplies, Scotch Highland cattle are very hardy and exceptionally good at foraging. Some ranchers in the northern plains of the United States cross Scotch Highland cattle with other breeds to infuse a greater winter hardiness in the offspring, although only a small number of these animals have been imported into the United States.

Small to average in size, the Scotch Highland has long, coarse outer yellow hair and a soft, thick undercoat (Figs. 1-63 and 1-64).

At maturity, bulls weigh 1,200 pounds, and cows weigh 900 pounds.

Fig. 1-63. Modern Scotch Highland bull that possesses those character-istics highly desired by the breed associations. Scotch Highland bulls sire smaller calves than do most other beef breeds. The calves are "easy born" and have a lot of hardiness and vigor. (Courtesy, American Scotch Highland Breeders' Association)

Fig. 1-64. Scotch Highland cows like the one above are known for their mothering instincts. They have earned the reputation of being providers for and protectors of their young. The unassisted calving record and the high percentage weaned calf crop speak highly of the breed. (Courtesy, Amer-ican Scotch Highland Breeders' Association)

Simmental

The Simmental breed is one of the oldest and most widely distributed of all breeds of cattle. Originating in the Simme Valley in Western Switzerland, the breed, under a variety of local names, evolved as a triple-purpose animal with recent emphasis on milk and meat production.

Simmental cattle vary from yellowish-brown or straw-colored to dark red, combined with white markings. The head, as well as the underside of the brisket and belly, is generally all white. The eyes often have red spots around them. The legs and tail are generally white. There may be white patches on the body, particularly behind the shoulders and on the flanks. The horns are fine and white, curving outward from the side toward the front with the tips turned slightly upward. In the United States, the breed registry has no color restrictions.

Simmental cattle are large, well-muscled and rugged in appearance (Figs. 1-65 and 1-66), but they are exceptionally docile. The cows in particular are easy to handle under pasture and range conditions.

Milk production is one of the most important genetic traits, with the average Simmental cow producing 9,000 pounds of milk annually.

Mature bulls weigh between 2,300 and 2,400 pounds, while mature cows weigh between 1,600 and 1,700 pounds.

Fig. 1-65. Typical Simmental cow. (Courtesy, American Simmental Association)

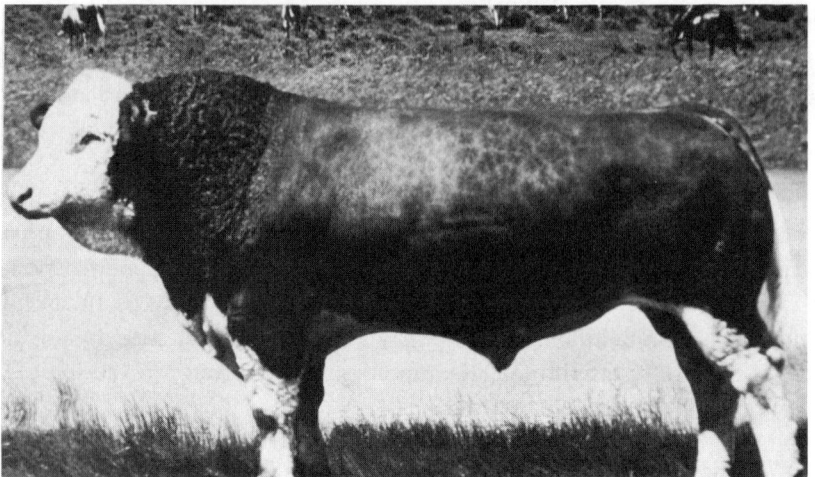

Fig. 1-66. Outstanding Simmental bull. (Courtesy, American Simmental Association)

Shorthorn / Polled Shorthorn

Shorthorns are accepted in various colors — namely, red, white, roan and red and white combined in different patterns. There are more reds and roans than other colors, with an almost solid red being the preferred color. However, as long as the color is accepted by the breed association, it should not be a factor in placing a class of Shorthorns. On the other hand, color often enters into the selection of stock for a breeding herd. Breeders discriminate against "loudly" spotted animals and against white legs on red and roan Shorthorns. A desirable animal has a buff-colored muzzle without any smoky or blackish tinge. The horns, provided they are left intact, should be shaped so as to extend straight out from the side of the head, curving gracefully forward and drooping slightly downward. The horns are short and yellowish, with occasional dark tips. Breeders dislike jet black tips and chalky white horns.

Shorthorns are large, rugged-framed, heavy-boned, growthy cattle with more skeletal size than other British breeds (Figs. 1-67, 1-68, 1-69, 1-70, 1-71, 1-72, 1-73 and 1-74). Modern-type Shorthorns are stretchy, massive and trim throughout, with clean, high-quality bone. Naturally maternal, Shorthorn cows possess excellent high-quality udders that are strongly attached. Judges discriminate against patchy, over-finished Shorthorns that are wasty through their briskets and along their underlines. Shorthorn breeders and livestock judges are selecting the longer, stretchier, trimmer, heavier-muscled, minimum-fleshed kind in hopes of improving the overall muscling and cutability of Shorthorn cattle.

Mature bulls in good flesh weigh from 1,800 to 2,000 pounds or more, and cows in breeding condition weigh from 1,100 to 1,500 pounds.

The ideal Polled Shorthorn should have the same breed and body type as the Shorthorn. The primary difference is that the animal is polled.

Fig. 1-67. Ideal Shorthorn bull. (Courtesy, American Shorthorn Association)

Fig. 1-68. Ideal Shorthorn cow. (Courtesy, American Shorthorn Association)

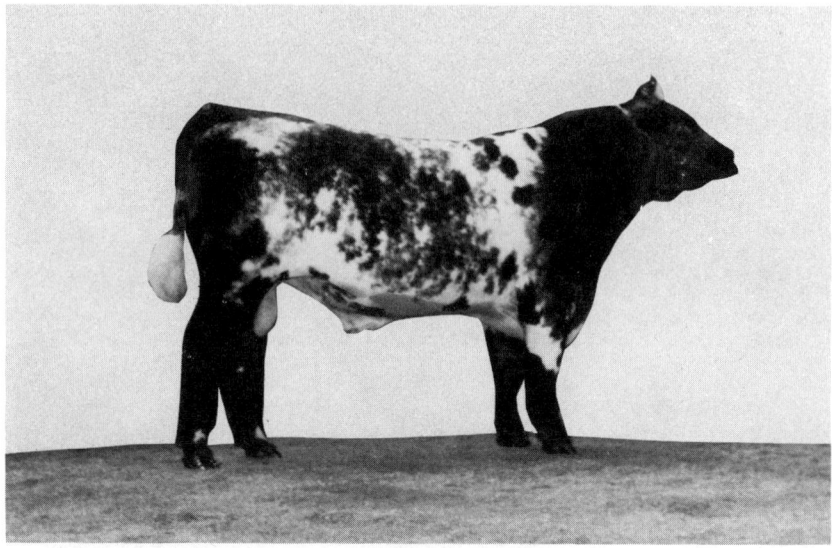

Fig. 1-69. Grand Champion Shorthorn bull of the P.A.C.E. Shorthorn Show at the 1985 North American International Livestock Exposition. (Courtesy, American Shorthorn Association)

Fig. 1-70. Grand Champion Shorthorn Female of the P.A.C.E. Shorthorn Show at the 1985 North American International Livestock Exposition. (Courtesy, American Shorthorn Association)

Fig. 1-71. Modern Shorthorn cow and her calf. (Courtesy, American Shorthorn Association)

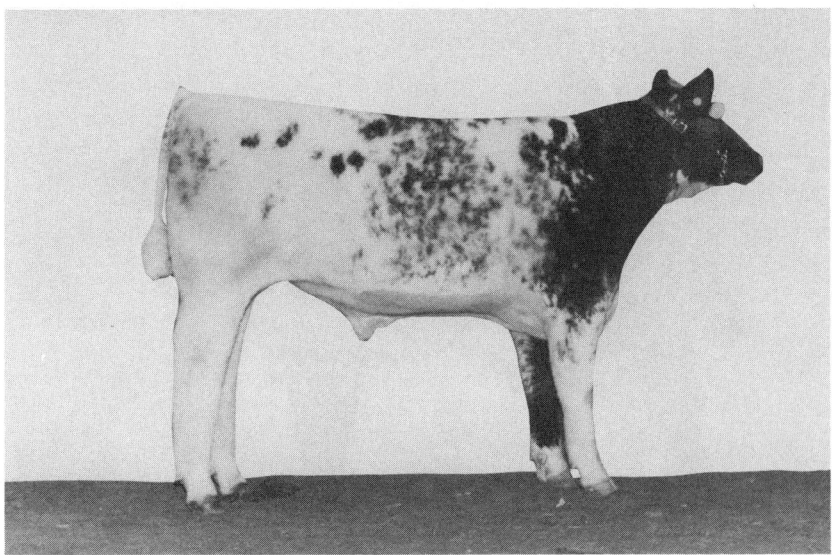

Fig. 1-72. This extremely modern Shorthorn steer was selected as the Grand Champion Steer over all breeds at the 1985 Oklahoma 4-H and FFA Junior Livestock Show. (Courtesy, American Shorthorn Association)

Fig. 1-73. "Lefty 83rd," an outstanding Polled Shorthorn bull and a half-brother to "Curt X" pictured in Fig. 1-74, was Reserve Junior Champion Bull at the 1985 Fort Worth Stock Show. (Courtesy, American Shorthorn Association)

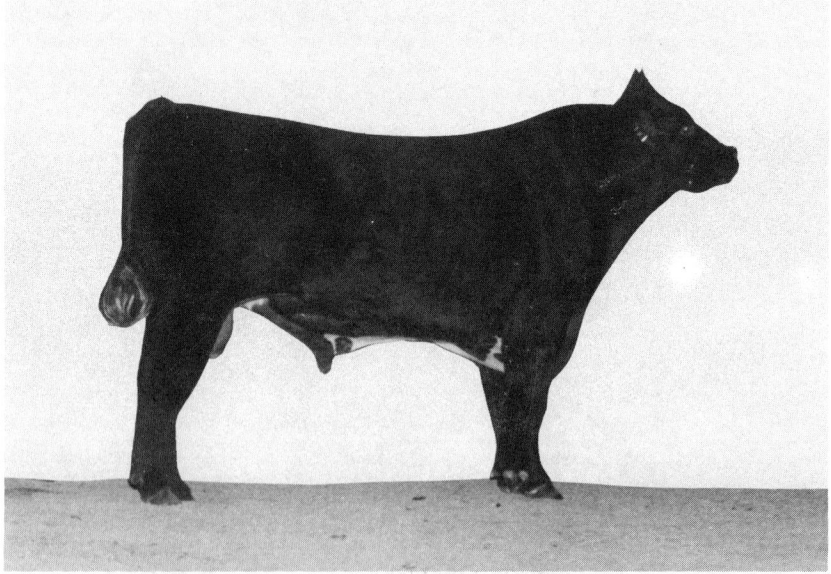

Fig. 1-74. "Curt X," a tremendous Polled Shorthorn bull and a half-brother to "Lefty 83rd" pictured in Fig. 1-73, was selected as the Grand Champion Shorthorn Bull at the 1985 Fort Worth Stock Show. (Courtesy, American Shorthorn Association)

Tarentaise

The Tarentaise breed was derived from an ancient Alpine strain in France. Now found in other areas of France in addition to the southeastern Alpine regions, the cattle are generally crossed with other stocks to improve milk production.

The Tarentaise cattle have a solid wheat-colored hair coat, ranging from a light cherry to a dark blonde. Bulls tend to darken around the neck and shoulders as they mature (Fig. 1-75). They frequently have a darker dorsal stripe, which is valued by breeders in France. All body orifices, including muzzle, eyes, vulva, etc., exhibit black pigmentation. Lack of such pigmentation disqualifies an animal for registration in France.

Fig. 1-75. "Luke," a modern Tarentaise bull, is a proven calving ease sire who should produce top-quality replacement cows. (Courtesy, American Tarentaise Association)

In conformation the breed is unique because it has a certain refinement and trimness of bone but with a surprising degree of muscling, particularly in the hip and round.

The Tarentaise strengths are (1) muscling; (2) milk production comparable to larger, mature weight exotic breeds known for milk production; (3) ease of calving; and (4) a combination of traits entirely unlike those offered by any other continental breed imported to date.

At maturity, bulls weigh from 1,700 to 2,000 pounds, while cows weigh between 1,150 and 1,250 pounds.

Beef Friesian

The Beef Friesian breed is one of the oldest established dual-purpose breeds in Europe, stemming back over 200 years to The Netherlands.

Beef Friesians are extremely fertile, cycling early and settling quickly. Both bulls and cows have the size and scale to be bred at 12 months of age. Cows are known for their calving ease, which is in direct correlation to days in gestation period. Calving records on 22,083 cows in the British Isles and three years of records in the United States show an average gestation period of only 276 to 279 days.

Free from double muscling, Beef Friesians transmit an exceptionally rapid growth factor, as they have been bred for years to convert grass and hay into quality red meat economically.

Mature bulls weigh from 2,400 to 2,750 pounds, while cows weigh approximately 1,500 pounds.

Brown Swiss

The Brown Swiss is one of the oldest, if not *the oldest,* dairy breeds, being descended from cattle in the valleys and on the mountain slopes of Switzerland (see Part 3).

Brown Swiss cattle are solid brown in color, ranging from light to dark. Those in the United States are larger, stretchier cattle with higher milk production, more desirable udders and quality bone than their European counterparts having the same origin. Many of their original characteristics have been maintained or improved; these include size; ruggedness; strong, sound feet and legs that wear; quality udders that last; docile temperament; heat tolerance; thriftiness and gainability.

Research stations and ranchers have found that using Brown Swiss dairy bulls on all types of commercial beef cows gives these cows heavier weaning weights, better feedlot performance and more valuable carcasses. The real advantage comes from using F_1 heifers as brood cows in the cow herd. They have size; they calve with ease; and they have more milk for their calves. In addition, they are better rustlers; they are healthier, earlier maturing and more fertile; and they live longer. Their well-shaped quality udders with uniform teat placement give few spoilage problems.

Mature bulls weigh over 2,000 pounds, while mature cows weigh around 1,500 pounds.

Beefmaster

The Beefmaster breed began as a three-way cross in 1931 when Tom Lasater began the crossbreeding program by using purebred Brahman and Hereford herds. Lasater later brought in purebred Shorthorn cattle to achieve a three-way cross estimated at 50 percent Brahman, 25 percent Hereford and 25 percent Shorthorn; this resulted when he combined first-cross Brahman × Hereford animals with first-cross Brahman × Shorthorn animals.

Developed in a rugged environment, these persistent rustlers are self-reliant and highly resistant to many diseases, insect pests and predators. Because Beefmasters thrive in varied environments, they are favored by ranchers in extremely harsh environmental conditions.

Beefmasters vary in color; however, a solid color, usually reddish-brown, is preferred.

Beefmaster are bred for their disposition, fertility, weight, conformation, hardiness and milk production. These animals are gentle, intelligent and responsive. They have been intensively selected for milk production from the beginning. Beefmaster calves often weigh 700 pounds or more at weaning without creep feeding. This, to a large degree, reflects the milking ability of the dams.

At maturity, bulls weigh from 2,000 pounds to 2,600 pounds (Fig. 1-76), and cows weigh from 1,400 to 1,800 pounds.

Fig. 1-76. This modern, well-muscled Beefmaster bull was selected as one of the top polled Beefmaster bulls in the breed. He has a strong topline and a clean front. (Courtesy, Foundation Beefmaster Association)

Galloway

The Galloway breed was developed in Southwest Scotland where the climate is moist and chilly. This polled breed has long, black hair that is soft and wavy with a thick, mossy undercoat.

In general, great emphasis is placed on the Galloway's hardiness, top carcass quality, significant foraging ability, quick response to good feeding conditions, good coats, wide muzzles and sound feet and legs. The newborn calves are able to survive more severe weather conditions than those of most other breeds.

The breed was first imported into the United States in 1870. The number raised here has never been large but has been increasing in recent years.

A mature Galloway bull weighs from 1,800 to 2,000 pounds; a mature Galloway cow weighs from 1,200 to 1,400 pounds.

Limousin

The Limousin breed originated in West Central France. Prehistoric cave drawings in the hill country of France depicting cattle of the same conformation of the modern Limousin are estimated to be 7,000 years old. While Limousin cattle were originally selected as both meat and work animals, recent emphasis has been as meat animals.

The purebred French Limousins are trim, long bodied and well-muscled (Figs. 1-77, 1-78 and 1-79). They are a rich red-gold over the back, which shades to a light buckskin or tan under the belly and around the eyes and muzzle.

Limousins are noted for their calving ease, high fertility and carcass leanness. Practical research reports show Limousin-cross carcasses have outstanding quality, with greatly improved cutability and excellent red-meat yield per day of age.

U.S. and Canadian tests show efficient, low-cost feedlot gains on Limousin cross-breds, as much as 25 percent better than conventional feeder cattle.

At maturity, Limousin bulls weigh about 2,000 pounds (Fig. 1-78), and cows about 1,300 pounds (Fig. 1-79).

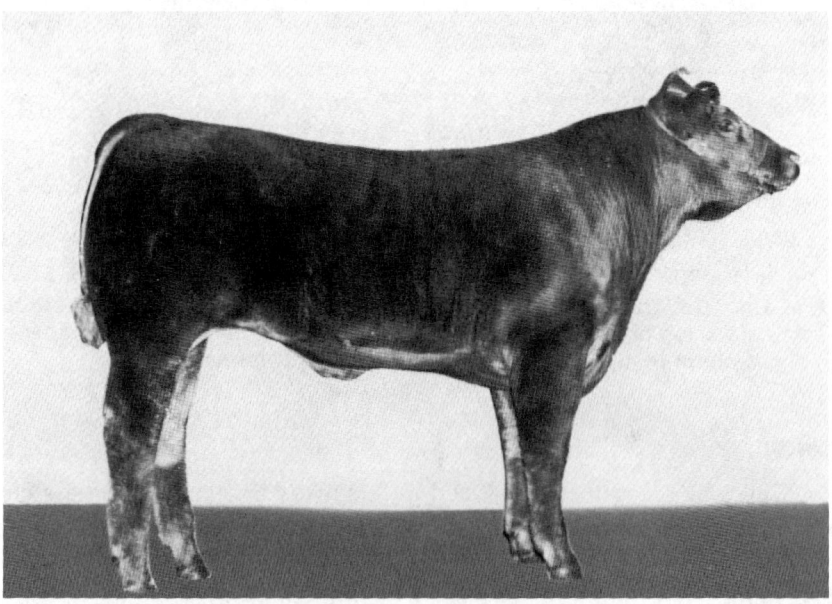

Fig. 1-77. This 1983 Fort Worth Grand Champion Steer was a ¹⁵/₁₆ Limousin with performance plus. Show steers with eye appeal and profitability like this one are a Limousin extra. (Courtesy, North American Limousin Foundation)

Fig. 1-78. This modern Limousin bull has the frame size and muscling to increase the size and performance of his offspring. (Courtesy, North American Limousin Foundation)

Fig. 1-79. Limousin cows like the one above are noted for their calving ease, milking ability and reproductive efficiency. (Courtesy, North American Limousin Foundation)

Lincoln Red

The Lincoln Red breed originated in Lincolnshire, a county located in the north-eastern part of England. Because of the location, over the years the breed developed ruggedness and hardiness to the harsh winds of the North Sea.

From the use of both red and black Angus cattle, the Lincoln Red Shorthorn was the first horned British breed to introduce the polling factor in 1939. This gave the breed a significant lead in the trend towards polled cattle. As the breed continued to add to its distinctive characteristics of rapid growth and large live weights, it eventually became known as the Lincoln Red.

Its solid cherry red color, particularly around the eyes and on the udders, lessens the danger of pink eye and sunburned udders.

A feed-efficient animal, the Lincoln dam is easily adaptable to regional grazing and climatic conditions. She has a strong, broad muzzle and well-placed legs. Her feet are sturdy and virtually self-maintaining.

Mature weights of bulls are in excess of 2,000 pounds; mature weights of cows are between 1,300 and 1,400 pounds.

Norwegian Red

The Norwegian Red cattle originated in the middle of the nineteenth century from imported Ayrshire cattle used on local cattle. There was little interest in this breed until about the turn of the century when efforts were concentrated on improvement of the mature Norwegian breeds. In 1939, the Hedmark and the Horned Low-

Fig. 1-80. "Nypan" is a modern Norwegian Red bull. Norwegian breeders would say he is harmonious because all the parts of his body are in harmony. U.S. breeders would say he is smooth and well-balanced. (Courtesy, North American Norwegian Red Association)

land breeds were fused to form the Norwegian Red breed. Since then, the breed has steadily gained ground; in 1972, it represented 60.1 percent of the recorded cows in Norway.

The Norwegian Red breed possesses great body and rump length, small calving size, good mothering ability, outstanding fertility, excellent weight gain, exceptional feed conversion of high roughage and modern-day carcass characteristics.

It is the first of the exotic breeds to be imported directly to the United States from its native country.

The mature bull weighs from 2,200 to 2,650 pounds (Fig. 1-80), while the mature cow weighs from 1,200 to 1,430 pounds.

Devon

The Devon breed is popularly known as "Red Rubies" because of its dark cherry red color. The skin is orange-yellow, with pigment being especially noticeable around the eyes and muzzle. The medium-length horns are creamy white, usually with dark tips. The hair is medium thick and often long and curly during the winter; however, coats are short and sleek in the summer.

The head is broad and medium length. In typical representatives of the breed, the topline is level, the ribs are well-sprung and the chest is deep. The body is long and moderately deep, and the loin and hindquarters usually carry thick natural muscling. Devon cows have udders of good texture and moderate size. Beef characteristics predominate in most present-day Devons, although a few still approximate the dual-purpose type.

Devons have long been known for their docility. They are excellent rustlers and active grazers. They are noted for being growthy, having high cutability and producing high-quality, well-marbled meat with fine texture and good flavor.

Mature bulls in good working condition weigh from 1,800 to about 2,400 pounds, and mature cows usually range in weight from 1,000 to 1,500 pounds.

South Devon

The South Devon breed originated in the western part of England. The date of origin is not definitely known, but herds have been established since the sixteenth century.

Red in color, the breed is a very gentle one and has outstanding mothering ability. Having had a fine performance record for more than 100 years, with outstanding qualities of beef, milk and butterfat, the South Devon is an ideal breed for crossing with existing beef breeds to increase milking ability and calf weaning weights. The largest of all the British breeds, it is the only breed in the United Kingdom that qualifies for a premium on milk as well as for the official beef calf subsidy on both its heifer and steer calves.

Today's cow-calf producer is looking for a breed that has ease of calving and that will stand up to the rigors of the range. The South Devon has been outstanding in

North America. Crossbred calves from South Devon sires have had exceptional performance in all respects.

When mature, bulls weigh from 2,000 to 2,800 pounds (Fig. 1-81), and cows weigh from 1,500 to 1,600 pounds.

Fig. 1-81. Typical South Devon Bull. (Courtesy, North American South Devon Association)

Red Angus

The Red Angus (Fig. 1-82) is derived from the Black Angus breed. Large English longhorns, predominantly red in color, were brought into Scotland and crossed with the black native polled breed. The resultant offspring were all black polled, but all carried the red gene. Subsequently, interbreeding produced an average of one red calf in four, in accordance with Mendel's law of heredity. Crossbreeding increased the number of animals carrying the red gene in the breed.

Red Angus cattle are extremely sound. Without a doubt, high-weaning weights make strong argument for the breed. These calves mature early and put on enough finish to "grade" at an early age. Red Angus cows make excellent mothers because of their gentle disposition and ability to produce enough milk for lusty calves.

Udders of Red Angus cows do not sunburn. The red color tends to reflect much of the sun's heat, thus making Red Angus cattle adaptable to sub-tropical climates.

The Red Angus Association of America is now moving toward the execution of a recently adopted computer program that will project the performance philosophy of the Red Angus breed and incorporate the research information and the policies recommended by the Beef Improvement Federation.

At maturity, Red Angus bulls weigh 2,000 pounds or more, while cows weigh from 1,200 to 1,400 pounds.

Fig. 1-82. Modern Red Angus bull. (Courtesy, Red Angus Association of America)

Red Poll

In the early 1800's, the Red Poll breed originated from crossings in the eastern coastal area of England. By 1846, nearly complete merging of the ancient inbred stocks of the shires of Norfolk and Suffolk was substantially completed, when the breed was recognized by the Royal Agricultural Society.

The Red Poll breed varies from light red to very dark red in color. Any shade of red is acceptable as long as it does not approach fawn or yellow. The skin is usually buff or flesh-colored, and natural white is present in the tail switch. Natural white is permissible in limited amounts on the underline; however, cattle with solid black, bluish or cloudy noses are not eligible for registration.

The breed is a dual-purpose one and has been so documented for at least 165 years. The carcasses are high in proportion of lean, are low in outer fat covering when finished and have acceptable marbling in the meat.

Some Red Poll herds have long been managed as beef cattle in typical cow-calf herds. The mothering ability of the cows is such that creep feeding is seldom necessary. With increasing specialization in the dairy business, small-farm dairy herds have been greatly reduced in recent years. This has resulted in increased emphasis on cow-calf operations in the Red Poll breed. In 1964, about 64 percent of registrations came from such cow-calf herds, maintained primarily for beef production. A strain with moderately high milk production has also been preserved.

When mature, bulls weigh from 2,000 to 2,200 pounds, and cows average 1,430 pounds.

Santa Gertrudis

The Santa Gertrudis breed was developed at the King Ranch in Texas by Robert J. Kleberg, Jr., who crossed beef-type Shorthorn cows and Brahman bulls. In 1920, a bull calf of $3/8$ Brahman and $5/8$ Shorthorn ancestry, with many of the desired qualities, was born. Named "Monkey," he proved to be an outstanding prepotent sire, who, before his death, produced more than 150 useful sons. He became the foundation sire, and all Santa Gertrudis cattle are descended from him.

To be a certified purebred Santa Gertrudis, an animal must represent at least four top crosses of Santa Gertrudis, and it must be inspected by a classifier from the breed association. Individual pedigrees are not required for certification because multiple-sire breeding units are used in many herds.

Cherry red in color, Santa Gertrudis cattle have short, straight hair in hot climates and long, straight hair in cold climates. The majority of Santa Gertrudis cattle are horned, but polled individuals occur and are acceptable. Loose hides, with surface area increased by neck folds and sheath or navel flaps, are characteristic of the breed (Figs. 1-83 and 1-84).

Mature Santa Gertrudis bulls weigh between 2,200 and 2,600 pounds; mature Santa Gertrudis cows weigh between 1,400 and 1,800 pounds.

Fig. 1-83. Modern Santa Gertrudis cow. (Courtesy, Santa Gertrudis Breeders International)

Fig. 1-84. Modern Santa Gertrudis bull. (Courtesy, Santa Gertrudis Breeders International)

Welsh Black

The Welsh Black breed originated in Wales in Great Britain as a "hill" breed amidst exposure to the Atlantic Ocean on the Welsh sea coast. Distinguished as being strong and robust, in the early days the Welsh Black was a triple-purpose breed — bred for meat, milk and work.

These solid black cattle are rather long-bodied, heavy-boned and somewhat upstanding.

Mature bulls weigh from 1,800 to 2,200 pounds, and cows weigh from 1,200 to 1,500 pounds.

FUNDAMENTAL BASES FOR SELECTION OF BEEF CATTLE

Livestock selection and evaluation constitute an art, and yet there are certain scientific findings that have had considerable influence on the selection of good stock. Too often you are instructed in the art of selecting animals without being given the fundamental reason for such teaching. An appreciation of the reasons why a thick, muscular round, a wide chest, a long body, etc., are desired will give you a broader and more permanent knowledge of livestock selection. The cattle feeder, the purebred breeder and the packer have learned by years of actual experience the correlation between body form and type and the right kinds of animals for slaughtering, efficient fattening, reproducing and growing. You may not have the opportunity to learn these things in the "school of experience." Therefore, it is helpful for you to obtain a basic knowledge of the anatomy, muscle structure and wholesale meat cuts of cattle, as well as other fundamental facts associated with reproductive efficiency upon which selection is based.

Anatomy

Anatomy is of major importance in the judging of beef cattle. The anatomy and skeletal structure of a beef animal exert a considerable amount of influence on the muscle structure of the animal. Cattle feeders and producers are interested in the kind of cattle that will produce the greatest amount of lean, tender muscle in the shortest period of time. Cattle with the correct anatomy and skeletal structure will produce carcasses that yield the desired kind of beef, and they will do it more economically. Looking at the organs of a cow, bull or steer will explain in part the reasons for a broad chest floor, a long body, a wide spring of ribs, heavy bone, correct feet and legs and many other desirable features. Fig. 1-85 is an illustration of the vital organs and skeletal structure of a cow. The shoulder, chest floor, fore-ribs

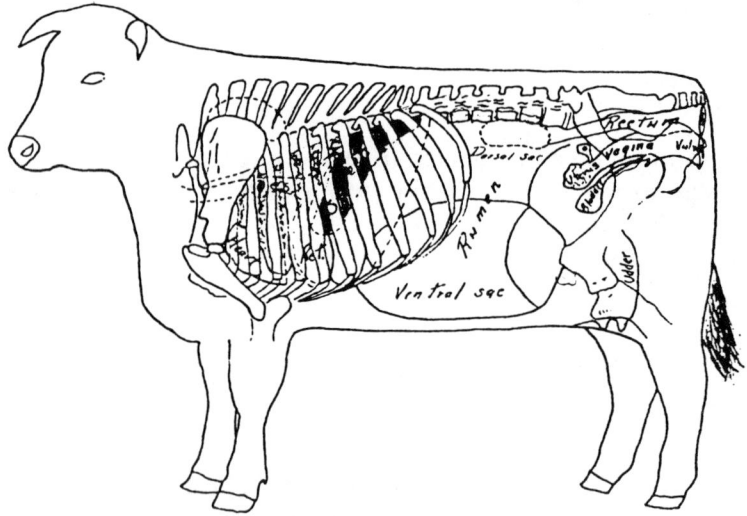

Fig. 1-85. Location of internal organs of the cow (left side).

and top of the shoulders encase the heart and lungs. The vitality and vigor of an animal are in a large part determined by the size and normal functioning of these organs. If an animal has a wide, full heart girth, adequate spring of the fore-ribs and a wide chest floor and chest, then the growth and function of these vital organs will not be handicapped by the skeletal and muscular tissue. The region extending from the mid-ribs to the hindquarters encloses the digestive and reproductive systems. **Digestive capacity and efficiency** are indispensable factors in the growth and fattening of animals. Feeders know that a "fish-backed," flat-ribbed steer is a poor feeder.

Cattle are ruminants; thus, the capacity to handle large amounts of roughage, as indicated by wide-sprung ribs and long bodies, is a prime factor in making them profitable. In the past, it was thought that the greater the depth of body of an animal, the greater its capacity. However, research has pointed out that the additional depth was primarily due to increased fat deposits all along the underline of the animal. The modern beef animal gets its increased capacity from additional length of body rather than from depth of body. As a result, the modern beef animal is longer bodied, trimmer and cleaner all along its underline and has a greater percentage of its carcass weight in the most valuable wholesale cuts.

It is needless to amplify the importance of *reproductive ability.* Livestock breeders, exhibitors and judges are becoming more aware that a beautiful prize-winning animal that is sterile or infertile is a detriment to his/her breed. Reproductive ability is dependent on many factors too numerous to discuss here. However, when a bull or a cow is selected with digestive, lung and heart capacity and *sex character,* it usually follows that prolificacy is present. Reproductive ability is indicated indirectly by breed character, constitution, roomy middle, wide-sprung ribs, long rump with pin bones that are set wide and high and masculine or feminine character. Breeding cattle that are too highly conditioned (overfinished) at young ages or, for that matter, any time during their showing career may never fully recover from the effects of overconditioning; and as a result, numerous fertility, breeding, calving and management problems often occur.

Although you may not always be conscious of the fact, the judges are trying to select animals with ample heart, lung, digestive and reproductive capacity and the kind of carcass that is pleasing to everyone concerned. It is an understanding of the gross fundamentals of anatomy, muscle structure and wholesale carcass cuts that teaches you, the student of livestock judging, both how and why certain animals are selected and others rejected.

Wholesale Cuts

The justification for the existence of beef cattle is edible beef and beef by-products. A thorough understanding of the location and relative importance of each wholesale cut aids greatly in the evaluation of the points on a live animal. The locations of the various wholesale cuts are illustrated in Figs. 1-2 and 1-3. The highest-priced wholesale cuts are represented by the ribs, short loin, loin end, rump and round. On the live animal these involve, respectively, the regions known as the ribs, back, loin and hindquarters. The lowest-priced cuts are the chuck, including the neck, and the brisket, foreshank, plate and flank. Corresponding parts on the live animal are the neck, shoulder region, brisket, forelegs and lower middle.

The lower middle or belly is relatively unimportant on the carcass but very important in the judging of live steers. A trim middle is indicative of a high dressing percentage and a good killing steer. Steers that have too much "fill" resulting in paunchy middles are discriminated against by judges and by packer-buyers.

From the viewpoint of the packer, the hindquarter is more important than the forequarter; whereas, from the viewpoint of the breeder, the forequarter has relatively higher value than the hindquarter. The kind of feeder cattle that produce choice carcasses are those that have excellent muscle development in both the forequarters and the hindquarters. A good breeding animal as well as a good steer must possess a strong constitution and a well-proportioned head.

EVALUATING BEEF CARCASSES

In order to make a prediction of the slaughter value of a particular animal, you should be familiar with:

1. Various USDA grades of slaughter animals (Table 1-3).
2. Location of the various wholesale cuts in the carcass and in the live animal (Figs. 1-2 and 1-3).
3. Relative importance of the various cuts with respect to value (Fig. 1-3).
4. Factors that influence dressing percentages (Table 1-4).
5. Yield of retail beef (Table 1-5).

Slaughter Grades

The various grades for slaughter animals are based on the prediction of the grade of the carcass that will result. Table 1-3 shows the USDA classes and grades of slaughter cattle.

The two major factors influencing the grade of a carcass are conformation and quality.

TABLE 1-3. USDA Market Classes and Grades of Slaughter Cattle[1, 2]

Market Class[2]	Grade[3]	
	Quality	Yield
Steer	Prime, Choice, Select, Standard, Utility	1–5
Heifer	Prime, Choice, Select, Standard, Utility	1–5
Cow	Choice, Select, Commercial, Utility, Cutter, Canner	1–5
Bullock	Prime, Choice, Select, Standard, Utility	1–5
Bull	Not eligible for quality grade	1–5

[1]Adapted from the *USDA Official Quality and Yield Grade Standards.*

[2]The difference between bullock and bull carcasses is based solely on evidences of skeletal maturity. Bullock carcasses must have A, or skeletal maturity; bull carcasses have evidence of B, or older skeletal maturity. When officially graded, bull or bullock carcasses are identified for their class: Example, Bull or Bullock.

[3]Quality grades are shown for each market class. Yield grades 1 through 5 are applicable to all slaughter cattle. Differences between Commercial and Standard are based on maturity of animals: Commercial—more mature animals; Standard—younger animals.

Conformation

This is appraised both in the live animal and in the carcass by the thickness and the uniformity of thickness, as the carcass is viewed from the side and over the top. Fig. 1-86 illustrates extremes in conformation.

The desirable carcass shape represented by the solid lines in Fig. 1-86 indicates the thickness and uniformity of thickness desired to provide a maximum quantity of thick beefy cuts. These characteristics can be seen in the live animal and thus can be an indication of the shape of the carcass that will result. It is important that the thickness and shape of the carcass be due in large part to muscling rather than fat, since excessive fat cover detracts from yield or retail cuts (see Table 1-5).

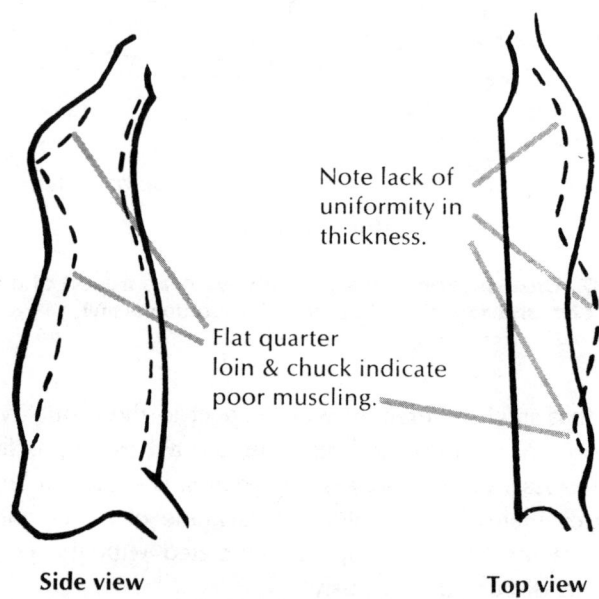

Note lack of uniformity in thickness.

Flat quarter loin & chuck indicate poor muscling.

Side view   Top view

Fig. 1-86. Desirable (solid lines) and undesirable (broken lines) confor-mation in beef carcasses. (Courtesy, Purdue University)

Quality

To appraise the quality of a beef carcass, you should consider the following: (1) marbling, (2) texture, (3) color, (4) firmness and (5) maturity.

Viewing the rib-eye cross-section of the carcass between the 12th and 13th ribs, which is the point where carcasses are broken into forequarters and hindquarters (Fig. 1-87), is the best way to evaluate the marbling, texture, color and firmness of the beef. To determine the maturity of the carcass, view the hardness of the bone at various points on the split carcass along the spinal processes. As animals mature, the bone becomes white and flinty, as compared to the red color observed in the split bone of young animals.

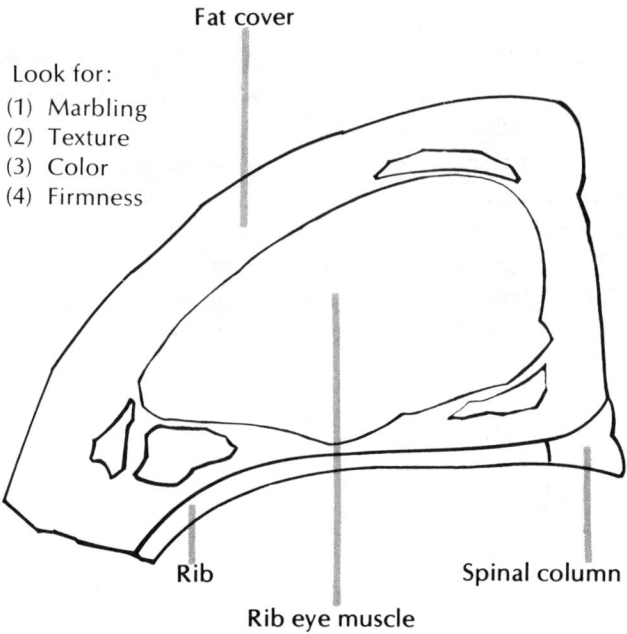

Fat cover

Look for:
(1) Marbling
(2) Texture
(3) Color
(4) Firmness

Rib

Spinal column

Rib eye muscle

Fig. 1-87. Cross-section of the beef rib-eye area traced at a point between the 12th and 13th ribs. (Courtesy, Purdue University)

It is obvious that small differences with respect to these quality characteristics (marbling, texture, color, firmness and maturity) are almost impossible to detect in the live animal. A cross-section of a beef rib-eye area is shown in Fig. 1-87.

Therefore, to estimate the ultimate carcass grade of a live animal, you must evaluate those characteristics that may be associated with the conformation and quality requirements discussed previously.

In determining carcass value, note the relative importance of the different parts of an animal (Fig. 1-3). Thus, in the selection of the live slaughter animal, a great deal of emphasis is placed on the thickness and beefiness of the animal in the region of the higher-priced cuts located over the top and through the rear quarter.

Yield Grade vs. Dressing Percentage

The two commonly used beef carcass yield factors, yield grade and dressing percentage, are often confused with each other. They are actually measures of very different traits and require different interpretations in carcass evaluation.

Yield grade is a cutability measurement that gives an indication of the amount of salable meat that can be obtained from a carcass. It is defined as the *percent of carcass weight in boneless, closely trimmed retail cuts from the round, rib, loin and chuck*, which accounts for more than 80 percent of the value of a beef carcass. Fortunately, it is not necessary to prepare all these cuts in order to determine carcass cutability. The Livestock Division of the Consumer and Marketing Service, U.S. Department of Agriculture, has developed a prediction equation using four easily obtained carcass measurements.

Percent cutability is easily converted to yield grade. The yield grades range from 1 to 5, with 1 representing the highest cutability group and 5 the lowest. Carcasses with Yield Grade No. 1 will have over 52 percent of their carcass weight in boneless, closely trimmed retail cuts from the round, loin, rib and chuck, while No. 5 will yield less than 45.4 percent of the high-valued cuts (Table 1-5).

Dressing percentage, on the other hand, is the *yield of dressed carcass.* An animal with a high dressing percentage will not necessarily have a high percent cutability or good yield grade. This is because dressing percentage improves when amounts of finish or fat are increased, while excess fat causes cutability to go down. As cutability goes down, the value of the carcass to the retailer decreases because the cutting losses are greater.

Thus, dressing percentage is not a good indication of carcass value and should not be given much emphasis in carcass evaluation and selection programs. Yield grade (cutability), on the other hand, is one of the best tools available for estimating carcass value and is used extensively in beef carcass shows. Both yield grade and dressing percentage can be important measurements when they are clearly understood and used in their proper perspective. (Table 1-4).

TABLE 1-4. Influence of Carcass Grade on Dressing Percentage of Cattle[1]

Grade	Dressing Percentage	
	Average	Typical Range
Prime	63	60–67
Choice	59	57–64
Select	57	55–61
Standard	55	53–58
Commercial	54	52–59
Utility	49	45–53
Cutter	45	41–47
Canner	42	37–44

[1]Adapted from *USDA Official Beef Grading and Slaughter Standards.*

Dressing Percentage, or Yield of Dressed Carcass

In addition to quality and conformation, the yield of carcass weight relative to live weight is important in determining the ultimate value of the animal to the packer. Thus, dressing percentage is important. The factors influencing dressing percentage to the greatest degree are: (1) amount of fill, (2) finish or degree of fatness, (3) muscling and (4) general refinement.

An example, a steer weighing 1,000 pounds and yielding a 600-pound carcass would have a dressing percentage of 60 percent. *Fill is the amount of feed and water in the digestive tract at slaughter.* Therefore, it is fairly obvious that animals heavily filled with feed and water at the time of slaughter will have a low dressing percentage. The influence of finish on dressing percentage can be shown by the typical ranges in dressing percentage for the various grades shown in Table 1-4.

General refinement refers to the coarseness of bone, head, hide, etc., and can influence dressing percentage to some degree.

Yield Grade, or Cutability

The yield of retail beef is extremely important in determining the ultimate value of a beef carcass. This yield value, also referred to as cutability, is greatly influenced by the relative amounts of lean and waste fat in the carcass. The factors that are used to ascertain the retail yield grade (see Table 1-5) of a carcass are: (1) muscling (area of loin eye), (2) fat covering, (3) kidney knob weight and (4) carcass weight. These factors may be used in assigning a carcass to a category according to the retail yield that might be expected (see Table 1-5). The yield may be superimposed on quality grade to more clearly indicate the actual value of a carcass with respect to yield of quality beef.

TABLE 1-5. Percent of Carcass Weight in Boneless, Closely Trimmed, Retail Cuts from Round, Loin, Rib and Chuck for Corresponding Yield Grades[1]

Yield Grade	Yield of Cuts
1.0–1.9	54.6–52.6
2.0–2.9	52.3–50.3
3.0–3.9	50.0–48.0
4.0–4.9	47.7–45.7
5.0–5.9	45.4–43.3

[1]Adapted from the *USDA Official Yield Grade Standards for Beef.*

Fig. 1-88 illustrates the difference that might be observed in the relative amounts of fat and lean in a carcass.

The objectives with respect to rib-eye cross-section measurements currently being sought in the beef industry are: (1) 2 square inches of loin-eye area per 100 pounds of carcass weight and (2) not more than 0.10 inch of fat cover depth per 100 pounds of carcass weight over the rib. Both measurements are made at the twelfth rib break.

In evaluating the live animal for retail yield or cutability, then, you must make an effort to predict the relative amount of muscling and fat cover in the live animal as well as the live weight.

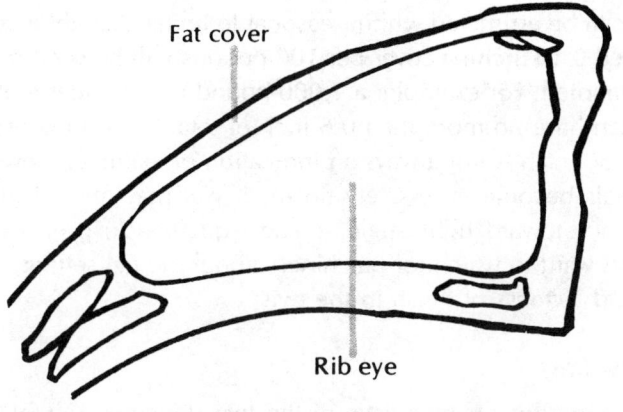

Fat cover

Rib eye

	U.S. Choice (Quality Grade) No. 1 (Yield Grade) Choice No. 1	Choice No. 4
Carcass Weight (lbs.)	600	600
Rib Eye Area (Sq. inches)	15.0	11.5
Fat Cover (Inches)	0.5	1.0
Kidney nob (%)	2.0	4.0

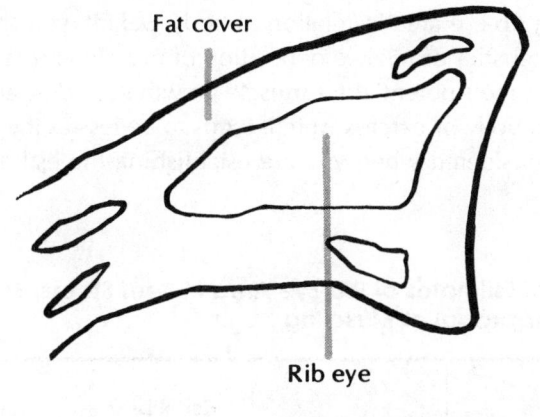

Fat cover

Rib eye

	U.S. Choice (Quality Grade) No. 4 (Yield Grade)

Fig. 1-88. Examples of quality and yield grade in beef carcasses. (Courtesy, Purdue University)

Estimating the Fat Cover

This ability is gained only through experience. However, with a certain amount of practice, this can be estimated within reasonable limits. Slaughter steers or heifers that do not exceed 0.10 inch fat cover per 100 pounds will be very firm to the touch when they are handled. For example: a 1,000-pound choice steer that yields a 600-pound carcass can have no more than 0.6 inch of total fat cover over the rib.

Smoothness of finish is not always an indication of leanness; however, as a rule, when beef animals become excessively finished, you may find roughness in cover and some tendency toward right angle spread (squareback) over the loin area. In addition, you will want to watch for patchiness about the tail setting, a wasty brisket and underline and softness of finish in the twist.

Estimating Muscling

Again, estimating this characteristic in the live slaughter animal requires some experience. However, the area of rib eye in a beef animal, which is used as an estimate of carcass meatiness, is much more closely related to the weight of the animal than fat cover. It is possible then to establish a benchmark for estimating size of loin eye. Slaughter steers within the typical finished weights of near 1,000 pounds will average approximately 1.0 square inch of loin-eye area per 100 pounds of live weight. Thus, a 1,000 pound slaughter steer indicating an average amount of muscling will have around 10 square inches of total loin-eye area.

From this benchmark, you can work both ways to estimate the size of the rib-eye area. Table 1-6 illustrates this point.

In estimating rib-eye area in relation to body weight, you should understand the following characteristics of relative deposition of muscle and fat: as animals become older and heavier, their potential for muscle growth subsides, and as a result, rib-eye area in relation to body or carcass weight tends to be less as the animals mature. This fact should be considered when you are establishing the estimated rib-eye area per

TABLE 1-6. Typical Estimates of Rib-Eye Area in Beef Steers, Based on Live Weight and Appraisal of Muscling

No.	Live Weight	Relative Muscling Appraisal	Est. Rib-Eye Area /100 Lbs. Live Wt.	Est. Total Rib-Eye Area	Rib-Eye Area/100 Lbs. of 600-Lb. Carcass
1	1,000	Heavily	1.2	12.0	2.00
2	1,000	Average	1.0	10.0	1.66
3	1,000	Light	0.8	8.0	1.33

100 pounds live weight. If the animal tends to be lighter than 1,000 pounds, adjust your estimate before you calculate the total rib-eye area in the animal.

The foregoing discussion indicates the importance of assigning descriptive terms such as "heavily muscled," etc., to derive an estimate of loin-eye area (see Table 1-6). It now becomes important to learn those characteristics that may be viewed in the live animal and serve as indications of muscling. In appraising a beef animal for muscling, view the animal from the rear and look for extreme natural thickness through the rump as well as bulge and thickness through the rear quarter. As you stand behind a heavily muscled steer, all you see is "rear end." Look for natural thickness over the back and loin (do not be confused by heavy fat cover) and finally look for boldness of shoulder and evidence of muscling in the forearm.

The preceding discussion has emphasized the importance of the relative amount of fat cover and muscling in the carcass in carcass value estimation. Essentially, then, everyone in the beef industry is attempting to select, produce and slaughter real "meat-type" beef cattle. This is an important challenge to the entire beef industry.

BEEF CARCASS EVALUATION TECHNIQUES

Beef enjoys a position as a prestige meat item. This is because beef is and has been a quality product. The present-day consumer is looking for high-quality lean meat with a minimum of waste, either fat or bone, and it should be the industry's goal to produce this product. The ideal beef carcass should have a high yield of high-quality muscle (lean meat) with a minimum of waste.

Subjective Carcass Evaluation

Subjective evaluation of the beef carcass is based on estimations of conformation, finish and quality.

Conformation

A desirable beef carcass is thick, meaty and heavily muscled throughout with a high proportion of its weight in the high-priced cuts, namely the round, loin, rib and chuck. The carcass is well-balanced and has its weight evenly distributed between front and rear quarters. Heavy-fronted beef carcasses are to be discriminated against. The carcass need not be excessively deep in the side as a very deep-sided carcass will have a higher percentage of plate, flank and brisket (low-value cuts) and a lower yield of high-value cuts.

The round should be relatively short-shanked, bulging or plump and thick through the cushion. The loin and rib should be wide and full and deep in the chine, characteristics of a well-developed loin-eye muscle. A meaty carcass should have 2 square inches of loin eye per 100 pounds of carcass weight. A heavy-muscled, full-clodded chuck and relatively short neck are desired.

Finish

There are four types of finish: external, internal, intermuscular and intramuscular. External, internal and intermuscular finish contribute to waste, while intramuscular fat is a quality factor. The amount of fat deposited on a carcass is of major importance in determining retail yield from the carcass.

A carcass should be completely covered with a thin layer of external finish. This is advantageous in preventing excess cooler shrinkage during the aging period; however, any finish above this minimum amount can be regarded as waste. As a general rule, the properly finished carcass should have no more than 0.1 inch of external finish per 100 pounds carcass weight. Internal finish includes kidney fat (kidney knob), pelvic fat, overflow and heart fat; it is all waste. Large deposits of seam fat (intermuscular) also detract from retail yield.

The term "character of fat" refers to the color, firmness and texture of the fat. A white to creamy white hard fat is desired. Cattle that have been fed on green, lush forages will usually have a softer, yellow-colored, more oily finish. This finish is mainly discriminated against because of the soft, oily character rather than the yellow color. It does not detract very much from meat quality.

Quality

In beef, the term "quality" refers particularly to maturity and to marbling, color, texture and firmness of the lean. Quality attributes of a beef carcass, except for maturity, are observed in the eye muscle of the ribbed carcass. The rib eye of a high-quality beef carcass is bright cherry red, firm to the touch and fine-textured. It is extensively and finely marbled with a white, firm fat. Marbling is associated with palatability; most studies show that it accounts for about 12 percent of the variability in tenderness and somewhat more for juiciness. For determining USDA carcass grade, the 10 degrees of marbling are: (1) abundant, (2) moderately abundant, (3) slightly abundant, (4) moderate, (5) modest, (6) small, (7) slight, (8) traces, (9) practically devoid, and (10) devoid.

Since the age of the animal at slaughter is closely associated with eating qualities, especially tenderness, in beef judging, maturity is evaluated. The following are maturity characteristics of young animals: (1) soft, red porous chine bones; (2) large, white, soft buttons; and (3) relatively narrow, red ribs. The following are maturity characteristics of mature animals: (1) hard, white, flinty chine bones; (2) ossified buttons; and (3) wide, white ribs. The relationship between marbling and maturity in determining the quality of a beef carcass, and ultimately the final USDA quality grade, is present in Fig. 1-89. As the animal matures, a progressive color change takes place in the lean tissue, going from the pale pink of veal to the deep, dark red of beef from old cows.

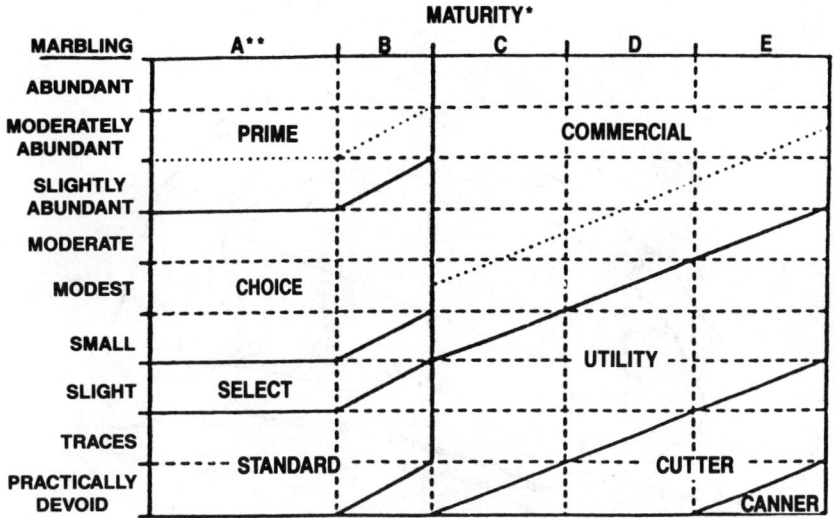

*Maturity increases from left to right (A thru E).
**'A' maturity portion is the only portion applicable to bullock carcasses.

Fig. 1-89. Relationship between marbling, maturity and carcass quality grade. (Assumes firmness of lean is comparably developed with degree of marbling and that the carcass is not a "dark cutter.") (Courtesy, USDA)

Objective Carcass Evaluation

Fat Thickness

Thickness of external fat should be measured to the nearest 0.1 inch at the twelfth rib, at a point three-fourths of the length of the rib-eye muscle from the chine bone and perpendicular to the outer surface of the fat (Fig. 1-90). If fat thickness at this location is not representative of that over the rest of the carcass, the fat thickness measurement should be adjusted so that it is more representative of overall carcass fat. On most carcasses, no adjustment is needed. An accurate assessment of fatness is important because degree of fatness is the major factor affecting carcass cutout.

One fat thickness measurement, taken perpendicular to the outer fat surface at a point three-fourths the length of the rib eye from the chine bone end, has been found to be highly correlated with the average for the three measurements for beef cattle.

Rib-Eye Area

Area of rib-eye muscle (longissimus dorsi muscle) by itself is not a good indicator of carcass leanness because it is closely related to carcass weight. However, in combination with fat thickness, carcass weight and percent kidney fat, rib-eye area is

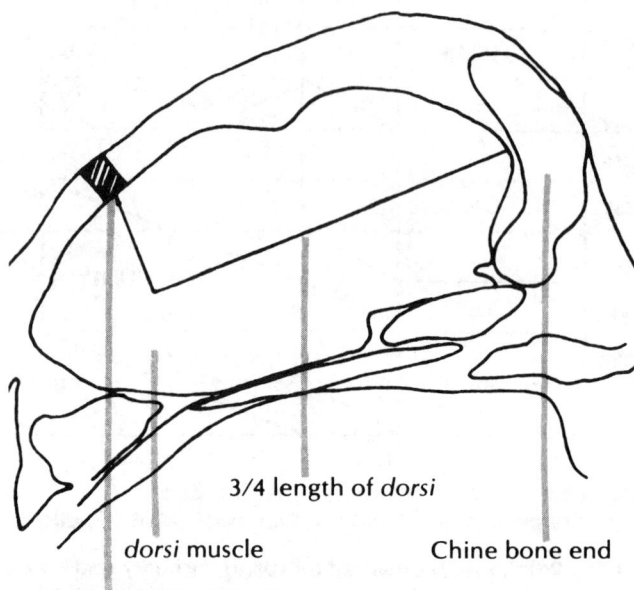

3/4 length of *dorsi*

dorsi muscle   Chine bone end

Fat thickness measurement

Fig. 1-90. Correct way to obtain fat thickness measurement. (Courtesy, Purdue University)

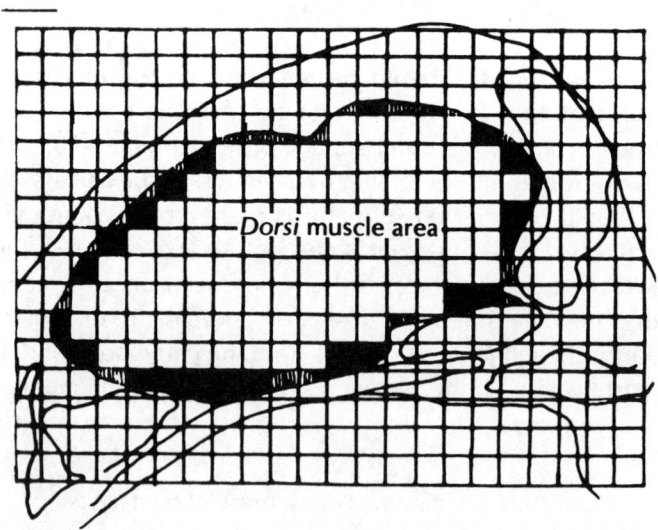

Dorsi muscle area

Fig. 1-91. Measurement of rib-eye muscle (longissimus dorsi) with plastic grid overlay. (Courtesy, Purdue University)

useful in the accurate prediction of cutability. Rib-eye area is measured on the cross-section exposed between the twelfth and thirteenth ribs when the carcass side is ribbed. Two methods of measuring area are: (1) tracing the muscle outline on acetate paper and determining the area with a compensating polar planimeter and (2) using a plastic grid overlay (Fig. 1-91).

To arrive at the dorsi area by using the grid method, count all squares half or more filled with dorsi muscle.

In Fig. 1-91, the darkened squares around the outside of the dorsi are half or more filled and are counted. The striped squares are less than half filled and are not counted.

This dorsi has 118 half or more filled squares. Convert this to square inches by dividing by 10: 118 ÷ 10 = 11.8 square inch dorsi area.

The method described above is the quickest method of obtaining dorsi area. However, a permanent record can not be obtained by this method.

Percent Kidney, Heart and Pelvic Fat

The percent of fat in the kidney, heart, and pelvic regions should be estimated to the nearest 0.5 percent by a qualified person.

You can estimate the total pounds of fat in both sides and determine the percent fat from Table 1-7. A lower percent is indicative of higher cutability.

TABLE 1-7. Pounds of Fat per Various Weight Carcasses and Fatness

Carcass Weight	Pounds of Fat for 2%	Pounds of Fat for 3%	Pounds of Fat for 4%
(lbs.)			
500	10	15	20
550	11	16.5	22
600	12	18	24
650	13	19.5	26
700	14	21	28

Carcass Weight

Hot carcass weight is usually obtained from the tag on the carcass in the cooler. Heavier carcasses tend to have lower cutability. The most widely used equation for prediction of retail yield was developed by the USDA from data on steers, heifers and cows of all grades. The USDA prediction equation is: percent closely trimmed, *boneless* retail cuts from the round, loin, rib and chuck = 51.34 − 5.78 (fat thickness over rib eye, inches) − 0.462 (kidney, heart and pelvic fat, percent) + 0.740 (rib-eye area, square inches) − 0.0093 (hot carcass weight, pounds).

The same information may be used to predict cutability group (yield grade). The

TABLE 1-8. Beef Cattle—Live Animal

Animal Identification	Live Weight		Dressing Percentage or Hot Carcass Weight		Loin-Eye Area	
	Estimate	Actual	Estimate	Actual	Estimate	Actual

Cutability formula: % cutability = 51.34 − (inches of fat thickness × 5.78) − (% kidney, heart and pelvic fat × 0.462) + (square inches loin-eye area × 0.740) − (warm carcass weight × 0.0093).

equation is: cutability group = 2.5 + 2.5 (fat thickness, inches) + 0.2 (percent kidney, heart and pelvic fat) + 0.0038 (hot carcass weight, pounds) − 0.32 (rib-eye area, square inches). There are five cutability groups (1 – 5), each representing a range of 2.3 percent in retail cuts. The lowest numbered group, 1, represents the highest cutability (see Table 1-5).

Table 1-8 is an example of a form used in the evaluation of live beef animals and later their carcasses.

Information contained in Table 1-9 involves a quick method that you can use to determine the USDA yield grade for live cattle and beef carcasses.

EXAMPLE FOR CALCULATING LIVE PRICE
FOR BEEF CATTLE

The following is an example for calculating the live price for a group of beef cattle.

and Carcass Evaluation Sheet

Backfat Thickness		Estimated Percent Kidney, Heart and Pelvic Fat		USDA Carcass Quality Grade		USDA Yield Grade or Cutability Percentage	
Estimate	Actual	Estimate	Actual	Estimate	Actual	Estimate	Actual

1. Estimate dressing percentage, carcass quality grade and cutability score (yield grade).
2. Provide current carcass price per hundredweight, as based on carcass weight, sex and grade.
3. A standard dressing percentage (60 percent) should be used to calculate final price.

Base steer carcass prices, dollar per hundredweight (chilled carcass weight)

Grade	Carcasses Weighing Under 750 Lbs. (Hot)	Carcasses Weighing over 750 Lbs. (Hot)
Prime	$110	$109
Choice	108	107
Select	101	100
Standard	96	95
Utility	84	84

For heifer carcasses, subtract $1 per hundredweight from the preceding quoted price. The preceding prices are based on a cutability score of 3.2. For every 0.1 change in cutability score, add or subtract $0.35 from each of the preceding base prices.

EXAMPLE: If the average live weight was 1,100 pounds and the dressing percentage is estimated to be 63, then the average carcass weight = 693 pounds. If the cattle are estimated to grade Low Choice in quality and a cutability score of 5.0, the estimated value (based on previous chart) = $108.00 − $6.30 = $101.70 per hundredweight carcass. Estimated live value per hundredweight = ($101.70) (0.60 standard dressing percentage) = $61.00, (round to the nearest $0.10). If the actual value per hundredweight = $61.50 per hundredweight, then the contestant's score = 60 − [($61.50 − $61.00/0.05)] (1 point off) = 60 − 10 = 50.

4. Assume that the cost of slaughter offsets the value of by-products.
5. Express the final estimated average price for the lot of live cattle on a dollar per hundredweight *live weight basis.*

This price determination is based on both the USDA quality grade and the USDA yield grade (cutability) and is expressed on a live weight basis. It is particularly valuable as a teaching tool to illustrate the difference in the value of high- and low-cutability cattle and/or high- and low-quality cattle on a live weight basis. The final values can be determined after slaughter and, of course, should receive preference over the estimates made on the live animal.

Example for Beef Carcass Pricing

This example is intended to provide a practical problem in the pricing of carcasses at a wholesale level and from a retail point of view. In order to simplify the problem, a number of assumptions should be made — some of which may not be exactly correct, but ones that will help minimize misunderstandings concerning desired responses.

The following information should be given:

1. Hot carcass weight.
2. A formula to convert cutability score to percent total salable retail product.
3. A pricing sheet (of retail and wholesale prices) for different quality levels of beef that would be cut for retail sale. The price would be a weighted average for all retail cuts rather than one for each cut.

It would be assumed that a 20 percent markup has been included in this price and that it would represent the value that a retailer would anticipate receiving at a point in time. The question to be answered is this: What wholesale price can the retailer pay for each of the carcasses (dollar per hundredweight) to remain competitive and stay in business?

Grade each carcass and then calculate, from the pricing information, a separate wholesale price (dollar per hundred) that the retailer can pay for each side in order to realize the marked-up retail price.

Express the final estimated price for every two carcasses on a dollar per hundred-weight *wholesale carcass basis.*

TABLE 1-9. Method of Determining USDA Yield Grades for Live Cattle and Beef Carcasses[1]

1. Determine a **preliminary yield grade** from the following schedule:

Fat Thickness over Rib Eye[2]	Preliminary Yield Grade
0.2 in.	2.5 in.
0.4	3.0
0.6	3.5
0.8	4.0
1.0	4.5
1.2	5.0

2. Determine the **final yield grade** by adjusting the preliminary yield grade for percent kidney, heart and pelvic fat and for rib-eye area.

   a. **Adjustment for percent kidney, heart and pelvic fat**—For each percent of kidney, heart and pelvic fat greater than 3.5 percent, add 0.2 to the preliminary yield grade. For each 1 percent less than 3.5 percent, subtract 0.2 from the preliminary yield grade.

   b. **Adjustment for area of rib eye**—For each 1 sq. in. more area than indicated on the following weight/rib-eye area schedule, subtract 0.3 from the preliminary yield grade. For each 1 sq. in. less area than indicated on the schedule, add 0.3 to the preliminary yield grade.

Warm Carcass Weight	Rib-Eye Area
500	9.8
600	11.0
700	12.2
800	13.4

[1]Adapted from the *USDA Official Yield Grade Standards for Beef.*

[2]Fat thickness is a single measurement taken over the rib eye between the twelfth and thirteenth ribs and three-fourths the length of the rib eye from its chine bone end. The measurement may be adjusted upward or downward to account for unusual distribution of finish on the carcass.

Example for One Side of Beef

1. Carcass information supplied to contestant: Hot carcass weight (pounds) is 660.
2. Carcass information estimated by contestant: USDA cutability score is 3.0; USDA carcass quality grade is Choice +.
3. Assume for this example that:
   a. A base cutability score of 3.2 equals 70 percent of chilled carcass weight that represents final salable product consisting of a standard trim and an almost entirely boneless (loin steaks bone in) product.

    b. For every increase or decrease of 0.1 cutability score, subtract or add 0.3 percent.

    c. Average retail price per hundredweight for various levels of quality. These prices include a 20 percent markup to account for operational costs and "normal" profit.

Grade	"Trimmed" Retail Value	"Trimmed" Wholesale Value
Prime	$117.60/cwt. ÷ 1.2 =	$98.00/cwt.
Choice+	$115.68/cwt. ÷ 1.2 =	$96.40/cwt.
Choice Average	$115.68/cwt. ÷ 1.2 =	$96.40/cwt.
Choice–	$114.00/cwt. ÷ 1.2 =	$95.00/cwt.
Select	$108.00/cwt. ÷ 1.2 =	$90.00/cwt.
Standard	$105.00/cwt. ÷ 1.2 =	$87.50/cwt.
Commercial	$108.00/cwt. ÷ 1.2 =	$90.00/cwt.
Utility	$103.20/cwt. ÷ 1.2 =	$86.00/cwt.

4. Estimate price.

    a. Establish percent salable product for the carcass: $70\% + [(32 - 30) (0.3)] = 70.6\%$.

    b. Establish retail value for the carcass: Choice+ = $115.68.

    c. Convert retail price to wholesale value of salable portion of carcass by subtracting the percent marked-up portion. This is simply done by dividing the retail price by 1.2 (because retail price is 120 percent of wholesale price) – $115.68 ÷ 1.2 = $96.40.

    d. Multiply salable product wholesale value by average retail yield percent: $96.40 × 70.6% = $68.05.

5. Make these assumptions:

    a. Retail prices provided are realistic for the market price.

    b. Markup to include costs of operation and "normal" profit is realistic. It is calculated as a percent increase from wholesale price of salable product.

    c. Weight of carcass is not a factor in pricing (i.e., heavyweight carcasses are not penalized).

6. Determine your score. A perfect score for each carcass would be 40 points and for each $0.10 per hundredweight deviation (or portion thereof) from the official price, 1 point would be subtracted. In the above example, if the actual price was $68.50 per hundredweight, then the contestant's score for this carcass would be $40 - [(\$68.50 - \$68.05) ÷ .10] = 40 - 5 = 35$. The maximum score for the class would be 80 points because there should be two carcasses to be priced.

BEEF CATTLE JUDGING TERMINOLOGY

The following beef judging terms or expressions can be used to present your reasons for placing beef classes.

General Expressions

1. Modern type.
2. A bull with excellent lines.
3. Excellent, well-balanced type.
4. A big, growthy, muscular, burly bull with a tremendous spring of rib and a natural thickness.
5. A big, stretchy heifer but carrying excessive amounts of condition.
6. Carrying uniform thickness from front to rear.
7. A very desirable beef type (meaty type).
8. Deep-bodied, overfinished.
9. Adequate depth with a minimum amount of outside cover.
10. A bull of excellent size with a tremendous amount of substance to carry him.
11. A breedy-headed, strong-jawed, massive, rugged individual.
12. A very promising calf.
13. A big, stout, rugged bull standing on excellent substance of bone.
14. A large bull, a bit off-type, but possesses a good head and lots of substance.
15. Very typey and smooth but definitely lacking some size and scale.
16. The right kind.
17. Very correct, stretchy, muscular and desirably finished.
18. Very heavy boned and rugged for his age; extremely smooth and uniform in finish.
19. Excellent balance.
20. Thick and muscular with adequate depth of body.
21. A bull of similar depth and thickness but lacking uniformity of lines from front to rear.
22. Big, husky fellow.
23. An outstanding individual.
24. Tremendous depth.
25. Desirably finished.
26. Fitted to perfection.
27. The low-set, thick, smooth kind, lacking the size, scale and muscling of the other animals in the class.
28. A big, stout, muscular youngster.
29. A very neat, trim, well-balanced calf with a desirable amount of outside cover.
30. A bull of desirable type.
31. A very impressive type, modern-day bull.
32. Square and blocky, undesirable in type.
33. A female of extreme balance and symmetry.
34. More bloom.
35. Very attractive, straight-lined calf, indicating much promise for the future.
36. A true champion by modern standards.
37. Calves that are thick, meaty, evenly finished and correct in their lines.

38. Modern type and loaded with red meat.
39. A big, long, trim, correctly finished steer.
40. Handicapped seriously by age.
41. The look of a real breeding bull.
42. Carries plenty of size and scale.
43. A world of substance.
44. A more stylish, free-moving heifer.
45. Growthier and longer-bodied with more thickness and muscling through the rear quarter than No. 1; also, stronger headed but lacking in depth of rib and straightness of topline.
46. A little longer-legged, longer-framed steer that is more evenly covered.
47. A low-set, short-bodied, tight-wound, old-fashioned steer carrying too much bark over the ribs, back and loin.
48. A heifer of great scale, smoothness and uniformity of fleshing.
49. A stretchier steer that has more length from the hooks to the pins and is heavier muscled through the stifle and lower quarter.
50. A cow possessing adequate size and scale.
51. A well-balanced female that holds her head high and carries herself with ease.
52. A smart-headed individual.
53. Evenly balanced and carries herself with ease.
54. A bull that takes a longer stride and moves more true on his feet.
55. Strong and masculine about the head and neck, denoting breed type and sex character.
56. A bull of great size and scale, combined with excellent muscling over the ribs, back and loin and down through the quarter.
57. A big, stout, attractive heifer.
58. A two-pair class, each pair being a different type.
59. The most correctly balanced bull in the class.
60. The most correctly moving bull in the class.
61. A neat, trim, well-balanced, stylish heifer.
62. More Shorthorn (or any other breed) breed character throughout the head and neck region.
63. A bull that breaks in the loin when he walks and is extremely tight in the heart.
64. Most desirable type.
65. A steer with more spread and thickness over the top, who shows greater evidence of muscling through the stifle region.
66. A big, massive bull with a lot of width and constitution through the chest floor.
67. A neat-shouldered, long-ribbed, long-quartered, heavy-muscled bull.
68. An exceptionally long-bodied, trim-fronted steer with a lot of class and is harder, firmer and more correct in the finish over his ribs, back and loin and longer from the hooks to the pins than any other steer in the class.

69. A tight-framed, short-coupled, deep-bodied steer that is carrying an extreme amount of waste through the brisket and along the underline.
70. Structurally the most correct animal in the class.
71. The heaviest-boned bull in the class that stands squarely on his feet and legs.
72. A bull lacking the size, scale, substance, balance and muscling of the other bulls in the class.
73. A trimmer-middled steer that is stronger topped and more nicely balanced throughout.
74. A long, narrow-bodied, upstanding individual lacking the thickness and muscling of the other steers in the class.
75. Early-maturing kind.
76. Rangy steer.
77. More symmetry and balance throughout.
78. Outstanding because of his size, thickness and trueness of lines.
79. A short-legged, small-framed, fine-boned heifer lacking the size, scale, substance and muscling of the other heifers in the class.
80. Extreme variations in type in the class.
81. A heifer of excellent type and well-bloomed.
82. A bull that is short and steep in the rump and set under on his rear legs.
83. A small, fine-boned, post-legged heifer that shows extreme puffiness about the hocks.
84. A wing-shouldered bull that rolls over on his front feet when he walks.
85. A female that would appear to have more genetic potential to produce those larger, meatier offspring at weaning time.
86. Lacking the production potential of my top three individuals in terms of broodiness and capacity.
87. A cow that shows more growth potential to develop into a bigger, more rugged, more profitable herd matron.
88. A bull that appears to possess more genetic potential to produce the right kind.
89. A bull that would add more size, scale and profit potential to the commercial herd.
90. A bull that appears to have more genetic potential to stamp his offspring with more growth, size and scale.
91. A bull that shows that extra growth potential in terms of developing into a more rugged, stouter, more modern individual that would be more apt to push forward that conventional herd.
92. Lacking size, scale and growth potential; would be the least useful when mature in terms of siring those larger, faster-growing offspring for that profit-minded producer.
93. More promise and future outcome.
94. More genetic potential to produce the right kind.

95. A female that lacks the production potential of my top females.

96. Taller, longer skeletoned, more structurally correct.

97. A taller, stretchier, framier individual that is jacked up taller on both ends on more length of leg.

98. Larger, growthier individual that has more skeletal size and framework.

99. A more progressive female in terms of her longer bone and longer, smoother muscle structure.

100. A longer-boned, longer, smoother-muscled female and thus more progressive in her type or kind.

101. A bull that appears to have a greater weight per day of age, thus giving him more growability and profitability to all segments of the industry.

102. More apt to stretch out, clean up and modernize the offspring of that conventional cow herd.

103. Greater weight per day of age and thus more ability to sire calves with that extreme size and muscle.

104. More ability in siring a calf crop with more growability and doability.

105. Should stamp that extra muscle, thickness and size into his calf crop faster than any other individual in the class.

106. A higher, wider set to those pins, which may facilitate greater ease at calving time.

107. More useful heifer to that practical profit-minded cattle producer.

108. More size, scale and dimension to his skeletal make-up.

109. A female that appears to have the genetic potential to drop those faster, more efficient-gaining offspring.

110. Fundamentally more sound; should be more useful in her future productive life.

111. More growth potential and future outcome for that profit-minded producer.

112. Rangy, gaunt-appearing heifer.

113. Fresher, more youthful-appearing steer, heifer, bull.

114. More desirable composition of gain.

115. More strength and broodiness about the head.

116. More length and elevation to his structural make-up.

117. A larger-framed individual.

118. Smaller-framed, tighter-wound, earlier-maturing individual.

119. The least potential to develop into a productive herd matron.

120. (Least or most) genetic promise.

121. Longer and more feminine in her muscle structure from front to rear.

122. More muscle volume stretched over a larger skeletal framework.

123. Able to convert that high-priced cereal grain to a salable, usable product with a greater degree of efficiency.

124. Greater weight per day of age.

125. Taller from the ground up.

126. A more meat animal shape all down his top.

127. A bull that stands up on a more rugged substance of bone.
128. Lacking growability, doability and profitability to all segments of the industry to merit a higher placing in this class today.
129. Wider at the pins, which would facilitate greater ease at the time of parturition.
130. Greater capacity to carry that large, growthy Angus offspring.
131. More genetic potential to jack up, clean up and modernize his offspring when mated to that conventional cow herd.
132. More flexible in his kind as far as going on to higher and/or heavier weights.
133. Size, scale, frame and all growth factors.
134. In a more desirable degree of condition for the breeding pasture.
135. More fit and true in breeding condition.
136. More desirable composition of gain.
137. Faster-growing.
138. Larger frame and skeleton.
139. Bigger or larger when he reaches maturity.
140. Conventional, light-muscled steer.
141. Higher volume.
142. Thicker in the lower one-third of the body.
143. A bull that handles with more muscling today (need to say where).
144. Overpowering.
145. More profit potential.
146. More profitable to the progressive breeder.

Head and Breed Character

1. A short, thick, heavy neck.
2. Breedy-headed.
3. Bold-headed.
4. Strong breed character.
5. A masculine head and neck.
6. An abundance of breed type.
7. A bull whose head denotes real masculine sex character.
8. A burly, masculine-headed bull.
9. More Angus (Charolais, Hereford, Shorthorn, etc.) breed type and sex character about the head, ears and neck.
10. A more alert-headed heifer, bull, steer.
11. A wide head with a moderately dished face.
12. Wide between the eyes and a broad muzzle.
13. A clean-cut, moderately dished face with ample width.
14. A plain-headed heifer, bull, steer.
15. A narrow-faced bull.
16. A long, plain-headed heifer that lacks the breed type and sex character of the other animals in the class.

17. Lacking Hereford (Angus, Charolais, Shorthorn, etc.) breed type.
18. A well-proportioned head with a sharp, clean-cut poll and a soft set to the eye.
19. A bull that lacks prominence about the poll.
20. A strong-headed bull or heifer.
21. A growthier-headed steer that is longer and cleaner through the neck.
22. A short, broad, undesirable head with an extreme dish to the face.
23. A pug-nosed, flat-polled, thick-necked heifer.
24. A sweet-headed heifer.
25. A stout-headed, strong-jawed, burly-necked bull.
26. Lacking sex character and breed type about the head and neck.
27. A more up-headed heifer or bull.
28. A low-headed bull or heifer.
29. A modern-day, impressive, burly-type bull.
30. Cleaner, more slender-necked heifer.
31. Mature-headed.

Forequarters

1. Neat and smooth about the shoulders.
2. An extremely good-fronted individual.
3. A powerful front-ended bull or steer.
4. Not as clean in the throat, as neat in the brisket or as smooth at the shoulders as No. 1.
5. Smoothly laid in at the shoulders.
6. A bull whose neck blends in more smoothly with the shoulders.
7. A broad-chested, clean-fronted bull with extreme muscle development through the arm and forearm region.
8. A steer that is out in his shoulders and sloppy through the brisket.
9. A wide chest floor indicating excellent constitution.
10. Extremely coarse through the shoulders.
11. A trim-fronted, smooth-shouldered bull with adequate depth and spring of the fore-ribs and heart girth.
12. Tight in the heart (or tied in back of the shoulder).
13. Pinched in the heart girth.
14. A narrow-chested individual.
15. Open-shouldered.
16. Too narrow-fronted.
17. Full in the heart with excellent spring of ribs.
18. Neatly laid in back (or behind) of the shoulder.
19. Sharp over the top of the shoulder.
20. A neck and shoulder that blend smoothly.
21. Open and coarse over the top of the shoulder.
22. A bold-fronted bull, very muscular through the arm and forearm area.

23. A bolder-chested, wider-sprung, deeper-ribbed individual that has more total reproductive capacity.
24. A thicker-topped, broodier, roomier, fuller-chested individual.
25. A large, roomy-middled, high-capacity individual.
26. More smoothly laid in at the shoulders, thus the producer may have fewer calving problems.
27. Bold and powerful-fronted.
28. Narrow-made, sissy-fronted bull.
29. Tighter-shouldered.
30. More width, length and depth to her chest chamber and reproductive tract.
31. Low-headed, coarse-fronted, unbalanced steer.
32. Heavy-fronted.
33. Short, thick neck.
34. Splay-fronted.

Ribs, Back and Loin

1. Straight-lined, strong-topped individual.
2. More spread and thickness over the top.
3. More spring in the rear rib, with more spread over the back and a thicker, meatier, heavier-muscled loin.
4. Strong top.
5. Weak top.
6. Long-ribbed, strong-topped, long-loined steer.
7. A little easy in her topline.
8. A high-loined bull.
9. A weak-loined heifer that breaks in the loin when she moves.
10. An extremely strong-loined individual.
11. Tremendous spring of rib.
12. Topline straighter and more nearly level at the move than No. 2.
13. Lacking spring of ribs.
14. A bull that holds his top up well.
15. Even topline.
16. Down a bit in her top.
17. Lacking strength of top.
18. A narrow loin.
19. Wider over his ribs, back and loin and more uniform in his spread and thickness from front to rear.
20. A high rear rib.
21. Extreme muscle movement in the back and loin region at the walk.
22. More correct in his finish over the ribs, back and loin.
23. Rough and nonuniform in his finish over the ribs and is starting to roll at the loin edge.

24. Handles soft over the fore-rib and is bare over the lower rear rib.
25. A hard, firm, correctly finished, good-handling steer over the ribs, back and loin.
26. A flat-ribbed bull that lacks the spread and thickness of the other bulls in the class.
27. A wide-ribbed, harsh-handling steer that lacks uniformity of thickness and strength of top.
28. A symmetrical, well-balanced steer that is nicely turned over the top and correct in his finish.
29. A square-topped steer that shows evidence of excess condition over the top of the ribs, back and loin edge.
30. A narrow, fish-backed, light-muscled steer.
31. A leveler-topped, leveler-rumped individual.
32. Strength of top.
33. Squarer-ribbed.

Hindquarters

1. Extremely heavily muscled through the rear quarters.
2. Deep and full in the twist, indicating excessive fat deposits in the area.
3. Bulging rear quarters.
4. Neat at the tail.
5. Neatly laid in at the tailhead.
6. A bit prominent at the tailhead.
7. Rough and patchy at the tailhead.
8. A trifle high at the tailhead.
9. A wide, meaty steer that is extremely thick and muscular through the center part of the quarter.
10. A long-rumped, heavily muscled, correctly finished steer.
11. Especially long from the hooks to the pins and carries down into a deep, thick, bulging, muscular rear quarter.
12. Long and level at the rump.
13. A narrow-rumped, light-quartered steer that lacks the muscle development exhibited by the other steers in the class.
14. A steep, short-rumped bull.
15. Sloping from the hooks to the pins.
16. A short-rumped steer that is patchy around the tailhead and carries an excessive amount of finish in the twist area and at the base (cushion) of the quarter.
17. Extreme muscle development through both the inner and outer quarter areas and free of any excess finish in the twist.
18. A deep, full flank that shakes and wiggles when he walks, which would indicate that the flank region is filled with large amounts of waste fat.
19. A steer that is tucked up in the flank, but that is especially clean and trim in this area.

20. Neatly laid in at the hips (or hooks).
21. Rather prominent and out at the hips or hooks (hooky or hippy).
22. Narrow at the pins and light in the quarter.
23. A thick, bulging, muscular stifle that flexes and expresses itself more promi- nently when the animal moves.
24. A bull that is thicker and deeper in his stifle than any other animal in the class.
25. A heifer that tapers from the hooks to the pins (pear shaped or pear hipped) and lacks the muscling and thickness through the lower quarter.
26. A heifer that is longer hipped, higher at the pins and meatier through the quarter than any other animal in the class.
27. More evidence of muscling through the quarter.
28. A longer muscle structure from hip to hock.
29. Longer from hip to hock, giving him more total muscle dimension through the quarter.
30. Thickness in inner and outer lower quarter.
31. Longer, higher-volume muscle structure through the stifle, quarter, etc.
32. Higher set to his pins, thus giving him more depth of quarter.
33. Width and muscle definition.
34. Longer through his stifle and longer from hip to hock, higher tail setting, thus more total dimension through the quarter.
35. Wider at the pins.
36. Muscling through the biceps and over the top of the rump.

Legs, Bone and Movement

1. A low-set, short-legged, light-boned heifer.
2. A big-framed steer that is up off the ground, straight legged and stands on more bone than any other animal in the class.
3. A more upstanding steer.
4. A bull that is too short legged and built too close to the ground.
5. Ample bone.
6. A bull that stands on adequate bone for his size.
7. Plenty of bone.
8. A straighter, stronger-legged bull that stands on more substance of bone than any other animal in the class.
9. Lacking substance of bone.
10. Fine-boned, light-muscled, overfinished heifer.
11. Squarely and correctly set on his feet and legs (Fig. 1-9).
12. Too close at the hocks. Too wide at the hocks.
13. Sickle-hocked (Fig. 1-10).
14. Cow-hocked (Fig. 1-16).
15. Too much "set" to the leg, set under the body too far (Fig. 1-10).
16. A bull that toed out in front (or behind) (Figs. 1-13 and 1-16).
17. A bull that needs a straighter "set" to his rear legs.

18. A bull that stands squarely on all four legs and moves very straight and true at the walk.
19. Legs set out on the corners where they belong.
20. Close behind (referring to placement of rear legs).
21. Toes in front (or behind) (Figs. 1-14 and 1-17).
22. A bull that rolls over on his front feet, which is especially noticeable at the walk.
23. Toes turned out in front (splay-footed) (Fig. 1-13).
24. Coarse-boned bull that lacks the quality and refinement of bone displayed by the other animals in the class.
25. Clean bone.
26. Neat joints.
27. Moves too wide at the hocks.
28. Puffy and swollen at the hocks.
29. Post-legged (Fig. 1-11).
30. Spraddles when she walks.
31. Very stifle legged, which usually accompanies the post-legged condition.
32. No movement or flex of the hocks at the walk.
33. A female that cocks her ankle when she moves.
34. A bull that is weak in his knees (buck-kneed).
35. A big, stout, strong-boned bull.
36. A bull that stands squarely on his legs (Fig. 1-9).
37. Crooked on his hind legs.
38. Bowed on her left rear leg (one, two or all legs) (Fig. 1-17).
39. Weak pasterns.
40. A little soft or ouchy (tender or peggy) on the front feet.
41. A bull that comes at you and goes away from you with more total muscle mass on both ends.
42. A longer, more powerful stride.
43. Lacking coordination of movement.
44. Sloppy moving, uncoordinated female that drops her pins at the move.
45. A bull that moves out wider off both ends, indicating more intermuscular dimension.
46. Freer wheeling.
47. A longer, freer-strided individual.
48. A bull that stands up straighter and moves out squarer off both ends.
49. More flex and give to his rear hock, enabling him to take a longer, further-reaching stride.
50. A bull that moves out on a more powerful, further-reaching stride.
51. A bull that stands wider on both ends, indicating more total dimension to the quarter.
52. Incorrect in his or her underpinning.

53. Gimpy moving.
54. Wider at his/her hocks.
55. Longer, freer, more determined stride.
56. Snappy moving.
57. Powerful-fronted (strided).
58. Splay-footed (Fig. 1-13).

Finish and Fleshing

 1. Correct in finish.
 2. Smoothly covered.
 3. More uniform in his condition.
 4. A harder, firmer, more uniform fat cover than any other steer in the class.
 5. A soft, flabby, overfinished steer.
 6. A steer that is rough and nonuniform in his condition.
 7. Uneven in covering or finish.
 8. Overfinished (over the hill, past 12 o'clock).
 9. A good-handling steer.
10. Nicer-handling, thinner-hided heifer.
11. Carrying the maximum acceptable amount of finish or covering.
12. More correct degree of finish.
13. Overconditioned.
14. Lacking condition.
15. The fattest steer.
16. The thinnest steer.
17. A thin-rind steer.
18. Too much outside bark (covering or finish).
19. A bull that is more correct and uniform in his natural fleshing than any other animal in the class.
20. A steer that is harder, firmer and more correct in the finish over his ribs, back, loin and through the quarter than any other steer in the class.
21. A hard, firm, good-handling, correctly finished steer.
22. A hard, firmly finished steer that is carrying too much finish over the fore-rib and is getting rough along the edge of the loin.
23. A hard, firmly fleshed steer that is bare over the rear rib (lacks finish over the rear ribs).
24. Neater and more smoothly laid in at the tailhead.
25. A more highly fitted animal than the other three.
26. Not as highly conditioned as the other animals in the class.
27. Patchy at the tailhead.
28. A very noticeable tie in the back.
29. More even distribution of fat cover from front to rear.
30. Freer from fat deposits in the udder.

31. More nearly market-ready in his degree of finish.
32. Has a more desirable composition of gain.
33. Harder, firmer handling with more even distribution of fat cover (condition).
34. Rough over the loin edge.
35. Harsh over the lower ribs.
36. More uniform over the ribs and loin edge (referring to fat cover).
37. More market ready in his conditioning.
38. Freer of waste from end to end (head to tail).

Quality, Style, Femininity and Masculinity

1. Quality and femininity to spare.
2. A very stylish heifer, bull, steer.
3. Smooth, clean-cut and breedy throughout.
4. Smooth and nicely balanced.
5. A thin, pliable hide.
6. Clean-cut head, clean bone and refined joints, indicating quality.
7. Lacking quality and refinement about the head, ear, hide, hair coat and bone.
8. A tight-framed, stylish heifer, bull, steer.
9. Extremely rugged, but a bit coarse in his features.
10. A great deal of smoothness and quality.
11. A combination of high quality, extreme balance and correct finish.
12. A very stylish, alert, smart-looking individual.
13. A big, raw-boned steer that lacks the quality of bone possessed by the other animals in the class.
14. A heavy-hided, harsh-handling heifer, bull, steer.
15. A coarse-headed, heavy-shouldered steer that lacks the refinement and style of the other animals in the class.
16. An extremely heavy-hided, harsh-handling steer or bull.
17. A more stylish, cleaner-conditioned, more well-balanced heifer that puts more good things together than any other female in the class.
18. An alert, up-headed heifer.
19. A high-headed, straight-lined, flashy heifer.
20. An alert female that is straighter lined and carries herself with more ease.
21. More reproductively sound in his testicular development and descent.
22. Balance and tightness of frame.
23. A bull that will settle more cows on first service.
24. Larger vulva, indicating more reproductive maturity.
25. A bull that will produce those higher selling offspring for the profit-minded producer.
26. More performance, growability.
27. More maternally mature in her udder development.
28. Higher quality.
29. Low-headed or heavy-headed.

30. Slack-framed.
31. More ruggedly made.
32. Small, underdeveloped testicles.
33. Uneven hang to his testicles.
34. Tight-sheathed bull or steer.
35. Heavy, pendulous sheath.
36. Heavy-naveled heifer.
37. Squarer, truer hang to his testicles.

Carcass

1. A higher-yielding, higher-grading steer.
2. A steer that will yield the most desirable carcass in the class.
3. A heavy-middled, low-yielding kind.
4. Not enough finish to put him into the Choice grade.
5. A steer that will not make the Choice grade.
6. A higher-killing steer.
7. A steer that will yield a more correctly finished, heavier-muscled carcass.
8. A steer that will yield a wasty, undesirable carcass.
9. A steer that will yield a carcass with excessive amounts of waste fat along the underline, over the top and at the base (cushion) of the quarter.
10. A steer whose carcass will have an abundant amount of kidney, heart and pelvic fat, as well as an excess amount of rind on the outside.
11. A steer that will hang a smoother, more correctly finished carcass that will be trimmer and heavier muscled throughout.
12. A steer that will yield an overfinished, wasty, light-muscled carcass.
13. An extremely thick, muscular, meaty steer, but he simply has too much outside cover to hang a desirable high-yielding carcass by today's standards.
14. A steer that will hang up the most desirable carcass in the class, one that will yield a minimum of waste and a maximum of red meat and muscle.
15. A steer that will hang a meatier, trimmer carcass that will yield a higher percentage of high-quality retail product than any other animal in the class.
16. A steer that will hang a longer, meatier, more muscular, more correctly finished carcass that will yield a high proportion of lean meat and muscle.
17. Most predisposed to fat (waste).
18. Least flexible in going to heavier weights.
19. More growability and doability for the efficient-minded feedlot producer.
20. A cleaner, conditioned individual that has a more desirable composition of gain.
21. More flexible for that profit-minded feedlot producer in terms of going to that heavier market weight and maintaining a desirable composition of gain.
22. Best chance of reaching that self-service meat counter with the highest-quality grade.

23. Lacking the size, scale and frame to go to heavier weights.
24. Trimmest, tightest-framed, most nicely balanced.
25. More potential to reach that marketing weight at a younger age.
26. A larger, growthier individual that appears to be more profitable in terms of feed efficiency and growability.
27. More growability; will have greater profit potential for the producer by more efficiently converting feed into a more salable product.
28. A steer that will hang a cleaner, trimmer, more nicely balanced carcass with more consumer appeal.
29. A steer that will hang a thicker, heavier-muscled carcass with more total pounds of the trimmed retail cuts and thus will have a higher merchandising value.
30. A steer that should yield a more massively muscled carcass with more total lean red meat and muscle mass that should be of greater value to both the packer and the producer.
31. More apt to slip into that Choice grade.
32. Higher percent trimmed retail cuts, thus a greater merchandising value.
33. Open-ribbed, harsh-handling.
34. A steer that should hang a higher-cutability carcass.
35. A steer that should hang a higher-grading carcass.
36. A steer that should hang a higher-quality carcass.
37. A steer that should pack more total pounds of those consumer-demanded cuts into that retail meat counter.
38. A steer that should have more pounds of higher-quality beef in the form of the higher-priced rib, loin and round cuts.
39. More likely to slip into that higher-quality grade.
40. More total red meat and muscle volume wrapped into one package.
41. When killed, carcassed and hung on the rail, a steer that should have a thicker, heavier-muscled, more shapely carcass.
42. Before being placed in the self-service meat counter, a steer that should have less fat trim.
43. A steer that should hang a carcass that could split a larger loin-eye area.
44. When killed and carcassed, a steer that should have more total muscle volume from rail to floor.
45. A steer that should yield a carcass with a higher ratio of lean to fat.
46. A steer that should yield a higher-cutability carcass, thus giving a greater return to both packer and producer.
47. A steer that should yield a greater amount of retail lean per day of age.
48. Before entering merchandising channels, a steer that should require less fat trim.
49. Higher retail lean per day of age.
50. If the market so demanded.

UTILIZING PERFORMANCE DATA
IN JUDGING CLASSES

Judging contests are an important educational tool. Someday, young cattle producers will be faced with selection decisions that affect the profitability of their operations. They should be prepared to use all the information available to them, including performance data. Therefore, so that future cattle producers may be better educated, performance data should be included in livestock judging classes and in all judging competitions. The combination of evaluation through performance records and visual appraisal better prepares students for realistic selection decisions.

Judging contests have long been used to exercise the decision-making abilities of young producers. When presented with a set of cattle, contestants make logical decisions as to the relative worth of each animal in the class. Often, the only knowledge about the class comes from visual appraisal. Contestants can visually appraise and estimate general size, weight and composition (lean-to-fat ratio), but they could increase their accuracy if they could consider performance information. Factors such as weight, rate of gain, frame score, backfat and scrotal circumference are easily measured, and contestants can make more accurate decisions by evaluating them. Estimated breeding values (EBV), ratios, accuracies, and expected progeny differences (EPD), which are easily obtained from most beef breed associations and progressive breeders, can help determine differences between individuals more accurately than visual appraisals alone.

A cattle producer who uses performance information to make selections is like any successful businessperson who uses the most accurate inputs possible to make economically sound decisions. The following information provides some examples of judging situations that include performance records. The possible types of classes are as numerous as the selection decisions producers must face every year. Before any judging exercise, contest or otherwise, you must define the class of cattle, obtain the appropriate performance data and make and justify decisions to meet the defined needs.

Defining the Class

Any selection exercise involves determining which animal comes closer to fulfilling a defined need. Before a sensible choice can be made, the judge should provide the answers to three questions:

1. How are the selected animals to be used (the selection purpose)?
2. Under what conditions are the selected animals expected to perform (the selection situation)?
3. From the selection situation, what are the most important functions the animals must serve (selection priorities or goals)?

The class *purpose, situation* and *priorities,* such as for this class of Shorthorn heifers, can be very simply stated:

> *Purpose* — Shorthorn replacement heifers.
>
> *Situation* — Small herd of registered purebreds producing commercial bulls.
>
> *Priorities* — Growth.

These selection criteria can be stated more elaborately, depending upon the advanced level of the student or contestant, or what concepts the instructor wishes to teach. Consider the following description for a class of Angus bulls.

> *Purpose* — Natural service sires in a two-breed rotational crossbreeding system.
>
> *Situation* — Midwest commercial operation (150 head, Hereford-cow base) integrated with corn production. Cows are medium mature weight and moderate in milk production. Cows forage on improved pasture and / or stalks. A percentage of the heifer crop is retained as herd replacement; the remaining heifers and all steers are finished for slaughter on the farm. Labor is a limited resource.
>
> *Priorities* — Maintain current mature weight. Increase longevity. Improve cutability. Decrease calving difficulty.

It is not necessary to use actual data; hypothetical situations and data can be applied to a class in a realistic manner. The educational value of the exercise is the critical issue.

Other examples of selection situations are:

Angus Heifers

> *Purpose* — Replacements for a seedstock herd that supplies bulls to commercial herds.
>
> *Situation* — Commercial herds operate with low feed and management resources (western range conditions). Feeders are sold at weaning; all replacement females are produced from calf crop.
>
> *Priorities* — Preweaning growth. Soundness. Fleshing ability.

Simmental Bulls

> *Purpose* — Herd bulls in commercial crossbreeding system.
>
> *Situation* — Small Angus X Hereford cross cows (900-pound average mature weight) make up the herd. High feed resources are available. Daughters will be retained in the herd. Calving difficulty has been a major problem.
>
> *Priorities* — Calving ease. Milking ability in replacements. Yearling growth in feeder cattle that are sold.

Polled Hereford Bulls

Purpose — To sire seedstock herd sire prospects and replacement females.

Situation — Progeny from these bulls are sold to or used primarily by other seedstock operations and by some commercial operations. Average feed resources and high management labor resources are available in both.

Priorities — A balanced performance program with progress in all areas, but no major setbacks in any one area. Marketing appeal necessary. Longevity.

A statement of selection priorities should not imply that they are the only criteria to consider. Certainly, if a bull has a severe structural problem or clearly lacks testicular development, he should not be chosen, even though soundness or fertility was not listed as a selection priority. The instructor may even choose to omit selection priorities from the class description. This would encourage the students or contestants to develop individual priorities based upon the selection situation and then to make corresponding selection decisions.

Considering Performance Data

Besides what students can see, what other factors should they consider? Students wishing to excel in beef cattle judging should be prepared to utilize the following performance information.

1. Birthdate.
2. Birthweight.
3. Birthweight EBV and accuracy.
4. Weaning weight, ratio (actual and adjusted) and number of contemporaries.
5. Weaning weight EBV and accuracy.
6. Yearling weight, ratio (actual and adjusted) and number of contemporaries.
7. Yearling weight EBV and accuracy.
8. Weaning or yearling hip height and frame score.
9. Maternal EBV for weaning weight or dam's MPPA (most probably producing ability).
10. Post-weaning average daily gain, ratio and number of contemporaries.
11. Fat thickness.
12. Yearling scrotal circumference.
13. Weight per day of age.
14. Birth, weaning and yearling weight EPD's.
15. Calving ease scores, EPD's and maternal EBV's for calving ease.

All of these records may not be available for every class. Some classes may have nothing more than birthdates and weaning weights.

Students should be trained to understand each factor and to use whatever is available in an optimum manner. It is impossible to describe every combination of class description and performance data set. However, many will be related to commercial production. Table 1-10 provides some guidelines students should consider when they are making bull selections.

When both EBV's and actual data are presented, emphasis is usually given to the EBV's, even if the accuracies provided are somewhat low. However, if EBV's are not provided, actual weights, within contemporary group ratios, should be given, and students should seriously consider them.

TABLE 1-10. Commercial Bull Selection Criteria[1]

Trait	Standards	
	Maternal Sire	Terminal Sire
Function		
Scrotal circumference (cm.) (min. at 1 yr.)	34 +	32 +
Calving ease score	Unassisted	Minor assist accepted
Birthweight (lb.)	65–85	80–110
Birthweight EBV[2]	102 +	95 +
Structural soundness	Excellent	Adequate
Milk Production		
Maternal EBV	102 +	Not important
Growth		
Weaning weight EBV	98–104	104 +
Yearling weight EBV	98–104	104 +
Market Acceptance		
Frame score (1 yr.)	4–6	5–7
Hip height (1 yr.) (in.)	47.0–51.0	49.0–53.0
Fat thickness (1,100 lb.)	0.2–0.4	<0.2

[1]Courtesy, Iowa State University, *Beef Improvement Federation Fact Sheet,* BIF:FS5, June 1985.

[2]In most cases, higher EBV's are associated with lower birthweights. However, for some associations the opposite is true. The instructor should clarify.

Simmental Bulls — Example

Decisions are not always simple, regardless of whether or not data are provided. To illustrate, let's look at a pair of Simmental bulls. The performance data are provided in Table 1-11. Two possible descriptions follow.

For Situation A, the judge may decide that Bull No. 1 is a more logical choice than Bull No. 2. This decision can be justified (perhaps through oral reasons) by stating that:

1 has greater weaning and yearling weight EBV's than 2. Thus, if 1 and 2 were bred to comparable cow groups, 1 should sire faster-growing calves that are heavier at weaning. Furthermore, 1 has a greater scrotal circumference than 2.

Although 2 has a smaller actual birthweight than 1, their birthweight EBV's are nearly equal, and so similar incidences of dystocia are expected from either bull on the average.

Situation B presents a different need to be met by the bull. A judge may prefer Bull No. 2 over Bull No. 1 in this case and would justify this because:

2 is expected to contribute greater maternal ability to the rotational system as he was estimated with a much greater maternal breeding value than 1. Also, 2 has a lower (near average) growth EBV than 1. Thus, with low feed resources available, replacement heifers sired by 2 should be of more moderate weight than those by 1. Finally, 2 has a slightly superior EBV for birthweight than 1. However, 2 is to be faulted for having only a minimum required yearling scrotal circumference.

Remember that there are no clear guidelines on placing classes, even though performance records are included. In fact, including weights and breeding values can create more ways to justify alternate decisions. The records may even be contradic-

TABLE 1-11. Simmental Bull Data[1]

Trait	Bull No. 1	Bull No. 2
Birthweight (lb.)	104	87
Birthweight EBV	99	101
Weaning weight EBV	109	100
Yearling weight EBV	105	100
Maternal weaning weight EBV	100	110
Scrotal circumference (cm.)	36	32

Description of Class	Situation A	Situation B
Purpose	Terminal cross sire	Sire in three-breed rotational system
Situation	Midwest cash corn crop and cow-calf operation (15% of herd—first-calf heifers)	Low feed resource, in western range conditions
	850–1,000 lb. average mature weight	Minimum labor available
	Feeder calves sold at weaning	1,000–1,100 + lb. average mature weight
Priorities	Growth	Limit mature cow size
	Fertility	Maternal ability
	Calving ease	

[1]Courtesy, Iowa State University, *Beef Improvement Federation Fact Sheet*, BIF:FS5, June 1985.

tory to the results of visual appraisal alone. However, these contradictions can provide a marvelous opportunity to discuss various producer goals, as well as how alternative selection practices can be used to reach them.

Combining Performance Data
with Visual Appraisal

At one time, cattle were selected by visual appraisal alone because nothing better was available. Fortunately, evaluation techniques have improved, and so visual appraisal can now be used as an aid to the more accurate performance selection. Composition, frame size, muscle expression, structure, and abnormalities can be determined visually, after the main decisions have been made on performance data.

It is important, especially in reason classes, for performance information to be made available so that visual traits can logically be combined with performance data. The following are a few examples of combined visual and performance traits.

1. Smooth, well laid-in shoulders and a long, narrow head, coupled with low or moderate birthweight, should indicate fewer calving problems.
2. Long-bodied, high-volumed heifer with high maternal EBV should indicate more future productivity as a cow.
3. Natural thickness over top and heavy weaning weight ratio and / or yearling weight EBV should indicate the bull will sire thick-made, heavy calves for the commercial market.
4. Structural correctness on feet and legs, in addition to moderate to large scrotal circumference in bulls, should indicate the bull will be a more successful natural breeder.

As in any judging situation, it is impossible to make a clear-cut choice that cannot be argued. Students should not be discouraged if someone else has a different opinion on how two individuals may best fulfill a given need. These decisions are always controversial. The ultimate goal is to make a sound, defendable decision based on fact and to learn from the judging exercise how to improve cattle production through better selection practices.

GLOSSARY

Adjusted weaning weight — Weight of a calf at weaning, adjusted to a standard 205-day weight.

Adjusted yearling weight — Weight of a calf at one year, adjusted to a standard 365-day weight.

Backcross — Second-generation crossbred resulting from mating an F_1 or first cross back to one of the parent breeds. Incorrectly called an F_2 cross.

Backgrounding — Conditioning process in the growing phase prior to fattening in the feedlot, either on grass or drylot growing ration.

Balance — Equal in width from point of shoulders to tailhead, even distribution of length in front, middle and rear.

Bangers — Cows that show reaction to brucellosis test.

Barren — Referring to a sterile female.

Bloat — Disorder characterized by gas distention in the rumen, seen on the animal's left side.

Bloom — Desirable condition of skin and hair.

Bos indicus — Species of cattle with hump over the shoulders.

Bos taurus — Species of cattle without hump over the shoulders.

Bottom side — Pertaining to dam's ancestry in a pedigree.

Brahman cross — Cross of the American Brahman with any other cattle breed.

Brahman hybrid — First cross of the Brahman with any of the British, European or dairy breeds.

Breed — Animals of like color, type and other characteristics, which are similar to those of their parents or past generations.

Breeder — The owner of a calf's dam at the time she was bred.

British breeds — Those breeds native to the British Isles, such as Hereford, Angus and Shorthorn.

Broody — Shows the appearance of being a good mother cow.

Buller — Cow that is in continuous heat due to cystic ovaries or other physical defects caused by hormonal imbalance; also seen in steers.

Bulling — Describing a cow in heat, apparent when a cow tries to ride other cows or stands while others try to ride her.

Bully — Animal that is well-crested, active, alert, has masculine characteristics.

By — Designates the sire.

Calves — Young cattle under one year of age and of either sex.

Castrate — To remove the testes of male cattle.

Clean — Negative in test for brucellosis; free of disease; describing animal believed to be free of congenital abnormalities.

Close breeding — Linebreeding or inbreeding, mating of related animals.

Combining ability — The level of hybrid vigor produced by a breed when crossed.

Complementability — The degree to which two or more breeds match so that the strengths of one breed overcome the weakness of the other.

Condition — Degree of fatness in animals.

Continental breeds — Those breeds native to Continental Europe, such as Charolais, Simmental and Limousin.

Cool out — Reducing energy level of ration, usually after show season; using oats and bran to lighten the feed.

Cows — Female cattle that evidence through age, weight, conformation and udder that they have produced one or more calves.

Crop — Depression behind the shoulder of a bovine.

Crossbreds — Animals that are any combination of two or more breeds.

Cryptorchids — Male cattle with undescended testes.

Cull — To eliminate an animal of low quality from a herd.

Dam — A female parent.

Dropped — Born; birth given to, calved.

Estrus — The recurrent period of sexual excitement in mature cows; period when cows will accept bull — heat period.

F_1 — First cross of two unrelated pure breeds.

F_2 — Crossbred resulting from the mating of two F_1's of the same type.

Family — Ancestry; line of breeding.

Fat cattle — Steers or heifers fattened on grain for slaughter.

Feed efficiency — Pounds of feed required to produce a pound of gain.

Feeders — Steers and heifers ready to enter feedlot finishing.

Feedlot — Group of pens or barn lot where steers and heifers are fattened for slaughter.

Fertility test — Evaluation of semen for live sperm count; evaluation to determine the ability to produce offspring.

Finish — Degree of external fatness.

Fitted — Describing an animal fattened, trained and groomed for show or sale.

Founder — Nutritional ailment due to overeating; foundered animals become lame with sore front feet and excessive hoof growth.

Freemartin — Heifer born twin to a bull, showing many male characteristics and usually incapable of reproducing because of the exchange of blood between the two fetuses and the presence of male hormone in the female.

Grass tetany — Magnesium deficiency of grazing cattle.

Gene — One of the biologic units of heredity contained in the chromosome, each of which controls the inheritance of one or more characteristics.

Get — Calves sired by the same bull.

Grade animals — Beef animals that have one or both parents that are not registered or recorded.

Heat — The 6 to 14 hours every 21 to 28 days when a cow will accept service of a bull.

Heavy (with calf) — Describing late stages of pregnancy in cows.

Heifers — Female cattle that have not born offspring.

Herd bull battery — The number of bulls in service in particular herds.

Herd sire — Principal breeding bull in a herd.

Hindquarter — The back part of the half of a carcass (beef), divided usually between the 12th and 13th rib, loin and round.

Hooks — Hip bones.

Hybrid or heterosis — Animal resulting from a cross between parents that are genetically unlike; first cross between two breeds; backcross, three-breed cross or four-breed cross.

Hybrid vigor — The degree to which the crossbred offspring outperforms its parent purebreds.

Inbreeding — Mating of related animals; close breeding or line breeding.

Linebreeding — Selective breeding; sire and dam related; mild form of inbreeding.

Loin-eye area — Area of the rib eye at the 12th rib; used in carcass evaluation to help determine meatiness of carcass; same as rib-eye area.

Motility — Activeness of bull's semen as seen through microscope.

Nick — The production of genetically outstanding calves.

Open — Not pregnant.

Out of — Designating dam.

Parturition — Calving.

Pasture bred — Referring to cows serviced by bull in pasture.

Pedigree — An ancestral record, a genealogical tree.

Performance test — Measure of performance, usually on bulls to determine their rate and efficiency of growth and carcass traits.

Pin bones — Posterior portion of pelvis, protrudes on each side of rectum.

Polled — Referring to cattle born without horns.

Post-legged — An animal that has extremely straight hindlegs.

Precondtioning — Preparing young cattle at or shortly after weaning prior to shipment to feedlot.

Prepotency — Ability to transmit individual's traits to offspring.

Progeny test — Evaluation of the offspring of sires to determine a sire's ability to transmit heritable traits such as gainability, conformation, meatiness, congenital abnormalities.

Purebred — Animal of recognized breed that is eligible for registry in the official herd book of that breed.

Ratio — Performance of an animal in relation to its contemporaries.

Reactor — Animal that shows a positive reaction to the test for Bang's disease or tuberculosis.

Registered — Describing a purebred animal for which a registration certificate and number have been issued by the breed association; referring to the recording on the records of the animal's name, along with the name of its sire and dam.

Rotational crossing — The systematic rotation of heifer replacements from one breeding unit to a succeeding unit for two or more rotations and utilization of purebred bulls of a different breed in each unit.

Safe-in-calf — Pregnant beyond doubt; usually reported after vet's examination.

Scour — Persistent diarrhea.

Second cross — Second-generation cross of two or more breeds.

Seedstock — Registered animals for establishing a breeding herd.

Service — The act of breeding.

Sickle-hocked — Having crooked hindlegs.

Spay — To remove ovaries.

Springer — Heifer or cow showing signs of advanced pregnancy, near to calving.

Stag — Male bovine castrated after sex characteristics have been developed.

Steer — Male bovine animal castrated before sexual maturity.

Stockers — Steer or heifer calves run on grass before going to feedlot.

Straightbred — Animal with breeding of only one breed.

Stringhalt — Tightening of tendons in rear legs.

Substance — Desirable combination of bone, frame and muscling.

Tattoo — Puncture in skin of numbers and/or letters usually used on registered animals, which are tattooed in the ear or ears to permanently identify the animals, in order to indicate year of birth. Not required or used by Brahman breeders.

Three-breed cross — Crossbred resulting from the crossing of three breeds, such as an F_1 or first cross mated to a third breed. Also called a three-way cross.

Trait — Distinguishing quality or feature.

Twist — Fleshing between hind legs where thighs come together.

Two-breed rotation — Systematic crossing of heifers produced in a two-breed cross to a bull of one of the parent breeds.

Type — Characteristics that identify or contribute to an animal's usefulness for a purpose — beef, dairy, dual-purpose.

Weight per day of age — Calculated weight of a calf for each day of its life.

BREED REGISTRY ASSOCIATIONS

Breed	Association	Officer and Address
Angus	American Angus Association	Richard Spader Executive Vice President 3201 Frederick Boulevard St. Joseph, Missouri 64501
Barzona	Barzona Breeders Association of America	Pete Jameson Secretary-Treasurer P.O. Box 631 Prescott, Arizona 86302
Beef Friesian	Beef Friesian Society	Maurice W. Boney Administrative Director 210 Livestock Exchange Bldg. Denver, Colorado 80216
Beefmaster	Foundation Beefmaster Association	Laurence M. Lasater President 200 Livestock Exchange Bldg. Denver, Colorado 80216
Blonde d'Aquitaine	American Blonde d'Aquitaine Association	Gerald Cunningham President Route B, Box 230 Grand View, Idaho 83624
Brahman	American Brahman Breeders' Association	Wendell Schronk Executive Vice President 1313 La Concha Lane Houston, Texas 77054

(Continued)

BREED REGISTRY ASSOCIATIONS (Continued)

Breed	Association	Officer and Address
Brangus	International Brangus Breeders' Association, Inc.	Jerry Morrow Executive Secretary 9500 Tioga Drive San Antonio, Texas 78230
Charolais	American-International Charolais Association	Joe Garrett Executive Vice President 11700 N.W. Plaza Circle Kansas City, Missouri 64195
Chianina	American Chianina Association	Robert Vantrease Executive Secretary P.O. Box 890 Platte City, Missouri 64079
Devon	Devon Cattle Association, Inc.	Martha J. Brooks Executive Secretary 5922 Jane Way Alexandria, Virginia 22310
Galloway	American Galloway Breeders' Association	Jim Carney Secretary-Treasurer Route #1, Box 106A Athol, Idaho 83801
Gelbvieh	American Gelbvieh Association	James A. Spawn Executive Officer 5001 National Western Drive Denver, Colorado 80216
Hereford	American Hereford Association	H. H. Dickenson Executive Vice President 715 Hereford Drive Kansas City, Missouri 64101
Limousin	North American Limousin Foundation	Greg L. Martin Executive Officer 100 Livestock Exchange Bldg. Denver, Colorado 80126
Lincoln Red	Canadian Lincoln Red Association	Box 447 Richmond Hill, Ontario Canada L4C 4Y8
Maine-Anjou	American Maine-Anjou Association	Steve Bernhard Executive Secretary 564 Livestock Exchange Blvd. Kansas City, Missouri 64102
Marchigiana	Marky Cattle Association	Don Lebsack President 111 Lustk Exchange Building 470 Marion Street Denver, Colorado 80216

(Continued)

BREED REGISTRY ASSOCIATIONS (Continued)

Breed	Association	Officer and Address
Milking Shorthorn	American Milking Shorthorn Society	Leslie Ann Stuff Office Manager P.O. Box 449 Beloit, Wisconsin 53511
Murray-Grey	American Murray-Grey Association, Inc.	Joan L. Turnquist Executive Secretary P.O. Box 30085 1222 N. 27th Street, Suite 208 Billings, Montana 59101
Normande	American Normande Association	Jack Barr Executive Vice President P.O. Box 350 Kearney, Missouri 64060
Norwegian Red	North American Norwegian Red Association	Box 5606 Kansas City, Missouri 64102
Pinzgauer	American Pinzgauer Association	Peg Meents Secretary Route 1, Box 104E Kelley, Iowa 50134
Polled Hereford	American Polled Hereford Association	T. D. Rich President 4700 East 63 Street Kansas City, Missouri 64130
Polled Shorthorn	American Polled Shorthorn Association	Roger E. Hunsley Executive Secretary 8288 Hascall Street Omaha, Nebraska 68124
Red Angus	Red Angus Association of America	Betty Grimshaw Administrative Director P.O. Box 776 4201 I-35 North Denton, Texas 76201
Red Poll	American Red Poll Association	Corrie H. Schueler Secretary-Treasurer P.O. Box 35519 Louisville, Kentucky 40232
Romagnola	American Romagnola Association	Route 1, Box 1058 Anderson, Texas 77830
Salers	American Salers Association	Tom Marcus 5600 South Quebec Suite 220A Englewood, Colorado 80111

(Continued)

BREED REGISTRY ASSOCIATIONS (Continued)

Breed	Association	Officer and Address
Santa Gertrudis	Santa Gertrudis Breeders' International	W. M. Warren Executive Director P.O. Box 1257 Kingsville, Texas 78363
Scotch Highland	American Scotch Highland Breeders' Association	Francine A. Hogate Secretary Box 81 Remer, Minnesota 56672
Shorthorn	American Shorthorn Association	Roger E. Hunsley Executive Secretary 8288 Hascall Street Omaha, Nebraska 68124
Simmental	American Simmental Association	Earl B. Peterson Executive Vice President P.O. Box 24 1 Simmental Way Bozeman, Montana 59715
South Devon	North American South Devon Association	T. E. Fitzpatrick Executive Secretary Box 68 Lynnville, Iowa 50153
Tarentaise	American Tarentaise Association	Box 446, Dept. C Reedpoint, Montana 59069
Welsh Black	Welsh Black Cattle Association of America	Sue Cash Secretary Route 1, Box 76B Shelburn, Indiana 47879

2

SHEEP

Sheep judging is difficult. There are so many different breeds and types of sheep that considerable study and application are necessary to become a competent judge of medium-wool, long-wool and fine-wool breeds. To master the art of sheep judging, you must know the breed characteristics, the body type, the type of wool peculiar to each breed and the relative importance of all these factors.

The major points to consider in judging sheep are type, fleece and skin, constitution, natural fleshing and finish, quality, muscling and muscle structure, sex character, balance, style, breed type and size. The relative importance of these factors will depend on the type, the breed and the specific purpose for which the sheep are produced. For example, fleece is of little or no importance in the placing of a class of wethers, but it is very important when sheep are being selected for wool production.

To be able to appraise sheep correctly calls for a study of the ideal mutton and wool types. Avail yourself of every opportunity to observe and handle sheep, making it a point to remember the characteristics peculiar to each breed. Acceptable judging ability comes only after arduous practice, plus intelligent and open-minded observation. Visit flocks at every opportunity.

In the beginning, you should become thoroughly familiar with the various names and locations of the parts of a sheep (Fig. 2-1). A working knowledge of these points will enable you to understand more completely the discussion that is to follow. This is basic knowledge that must be thoroughly mastered before you can expect to form mental images of ideal sheep types. Each sheep or class of sheep that you select or judge should be compared to the mental image that you have formed in your mind as being ideal for that particular type of sheep.

SCORING SHEEP

After you, the beginner, have the names of the different points of the sheep well

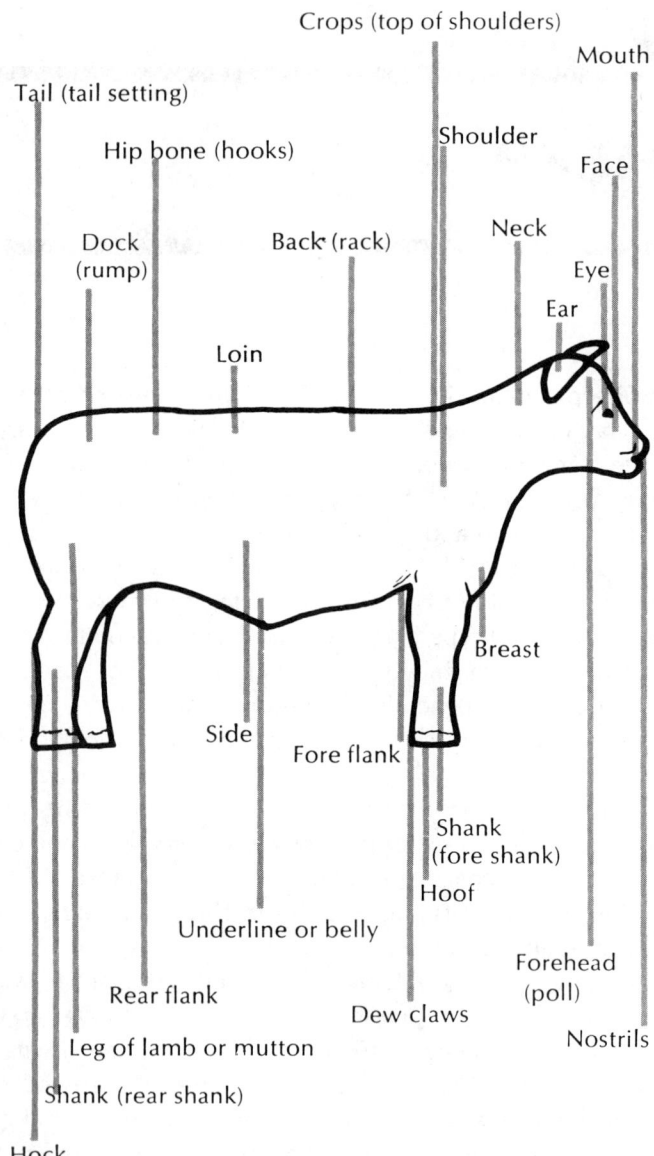

Fig. 2-1. Parts of the sheep.

in mind, you are ready to learn the ideal type. Often you start judging without a fair understanding of what the ideal sheep should look like. As mentioned previously, there are many breeds of sheep; but as far as the beginner is concerned, the medium-wool breeds are the most satisfactory breeds with which to work in learning to judge. Thoroughly study the sheep judging card, since it will help to fix in your mind the names of the various points of the animal together with the ideal type. The judging card is set up so that four animals may be scored on each of the described points. After you have filled out the judging card, the instructor should discuss the placings. When you have thoroughly mastered this method of selecting sheep, you are ready to leave the score card and to judge the class on the basis of a final placing.

Basically, the description of the body parts of the modern beef type (Part 1) also applies to the modern sheep type, and the selection and evaluation procedures are similar. Sheep simply come in a smaller package. However, sheep are covered with wool, so it is especially important that you handle the animal to be sure of its conformation, muscling and finish.

SHEEP JUDGING CARD

Points to Consider   Placing

	1st	2nd	3rd	4th
General Appearance Straight, strong topline; thick and muscular throughout; excellent spring of rib, long-bodied, long-rumped; symmetrical and stylish; stands squarely on adequate substance of bone; trim middle, neat in the breast area and along the underline				
Form 1. *Head* — Face short; mouth and nostrils large; eyes large and clear; forehead broad; ears alert, not coarse; wide between the ears				
2. *Neck* — Average thickness, medium length, blends smoothly into shoulders; throat clean				
3. *Shoulder* — Smooth; muscular (forearm also); minimum flesh covering; neat and smooth on top				
4. *Breast* — Wide, full; neat, clean, and trim with little waste in the breast area				
5. *Chest* — Wide, average depth; girth adequate; crops neat and reasonably full				
6. *Ribs* — Well-sprung, smoothly and uniformly covered with a minimum amount of firm flesh				
7. *Back* — Thick, muscular, straight, strong, uniformly covered with minimum fleshing				

(Continued)

SHEEP JUDGING CARD (Continued)

Points to Consider

		Placing		
	1st	**2nd**	**3rd**	**4th**
8. *Loin* — Thick, deep, long, muscular, straight, strong, uniformly covered with minimum fleshing				
9. *Hips* (or hook bones) — Neatly laid in; smoothly and uniformly covered with a minimum amount of firm finish				
10. *Rump* (or dock) — Long, thick and muscular from the hips (hook bones) the pins; level and free from patchiness and excess finish				
11. *Thighs* (or stifle muscle area) — Deep, thick, muscular, full . .				
12. *Twist* — Medium depth and free of excessive finish				
13. *Flanks* — Slightly tucked up, neat, trim and firm				
14. *Legs* — Straight, strong, placed out on the corners of the body; medium length; shanks should indicate average or above-average size or substance of bone				
Finish Smooth, uniform, firm covering, especially over ribs, back (rack) and loin; maximum of 0.3 inch of fat over the 12th rib on a 100-pound market animal; freedom from patchiness and rolls is desired; animal should be trim and free of waste in the breast, along the underline, in the rear flank, at the base of the leg, and in the twist or crotch area. (Degree of finish is of great importance in market classes but is not so important in breeding classes as long as the ability to take on finish is indicated. However, excessive finish in breeding classes is undesirable.)				
Quality Frame and finish smooth; hair silky; wool fine, soft and lustrous; pelt light and pliable; bone adequate size and clean-cut.				
Dressing Percentage Smooth, firm finish; neat, trim middle; medium-weight pelt. (Not considered in judging breeding sheep.)				
Breed and Sex Character (Applies only to breeding classes and will be discussed in the section dealing with the various breeds.)				
Wool Characteristic of the breed; uniform in length of staple and fineness; free from organic matter (dirt, feed, etc.), kemp and dark fibers; dense and showing good character and color.				

FIRST STEPS IN EVALUATING LIVE SHEEP

The general appearance of sheep is often deceiving due to the covering of wool. When sheep are observed from a distance, body defects are difficult to detect, especially when the fleece is long and cleverly trimmed. It is desirable to view the sheep prior to handling so as to get a general conception of size, scale, type, muscling, breed character, correct structure and balance. About three to five minutes should be spent in viewing the sheep at a reasonable distance (20 to 30 feet). A good rule of thumb to follow is to place the class without handling the sheep and then never move an individual up or down more than one placing, provided the sheep are in fleece less than 1 inch long. This will result in fewer "busts" and more correct placings with a minimum of pair switches.

The parts of the live lamb and their relative location, identification or association to the lamb carcass cuts, the carcass measurements and the relative value of the various cuts of lamb, are essential in judging breeding and market classes of sheep. Fig. 2-1 shows the parts of the live lamb and Fig. 2-2 shows the location of the wholesale and retail cuts.

The points mentioned in this discussion on market lambs, concerning conformation and meatiness or natural fleshing, are fully as important when sheep are being evaluated for breeding purposes. Additional items to consider or emphasize to a greater extent for breeding sheep are:

1. Breed and sex character about the head.[1]
2. Breed type for registered breeding stock.
3. Strength and straightness of feet and legs.
4. More emphasis on growth, straightness of lines, symmetry and style.
5. Fleece and wool value.

The following is the recommended procedure for examining sheep visually.

Front View

For the market lamb this view is of less significance than for breeding sheep (Fig. 2-3). The lamb should have a bold muscular shoulder, arm and forearm. It should be free of excess skin and finish in the throat and breast area. The chest should be wide and clean and set back into the body to give the animal constitution and capacity, as well as the clean, trim, muscular look desired in the modern-day, "meat-type" lamb. The head should be reasonably refined (not large) and should possess some quality and character. The front legs should set out wide on the corners of the body and

---

[1]For registered breeding stock.

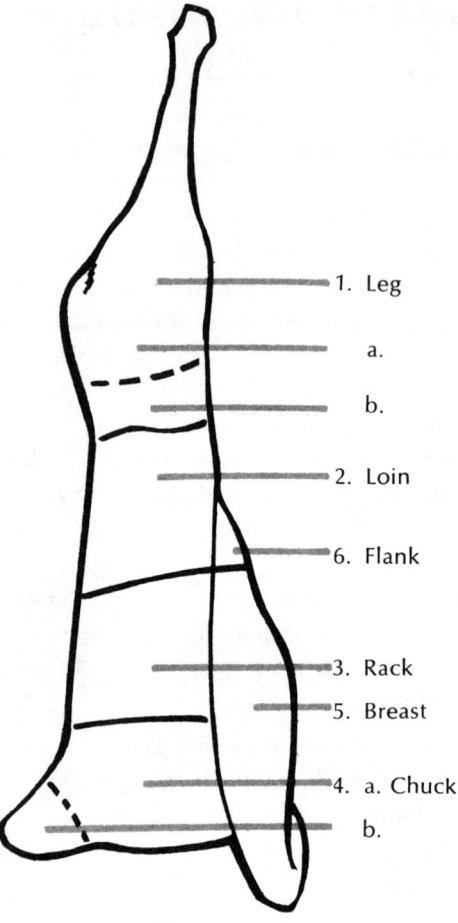

1. Leg

a.

b.

2. Loin

6. Flank

3. Rack

5. Breast

4. a. Chuck

b.

WHOLESALE CUTS	% OF CARCASS	RETAIL CUTS
Hindsaddle	**50%**	
1. Leg	33%	a. Roast
		b. Chops or roast
2. Loin	17%	Loin chops
6. Flank		Stew meat
Foresaddle	**50%**	
3. Rack	11%	Rib chops or roast
4. Chuck	25%	a. Shoulder roast or chops
		b. Neck slices or stew meat
5. Breast, inc. shank	14%	Stew meat

Fig. 2-2. Lamb carcass. (Courtesy, Purdue University)

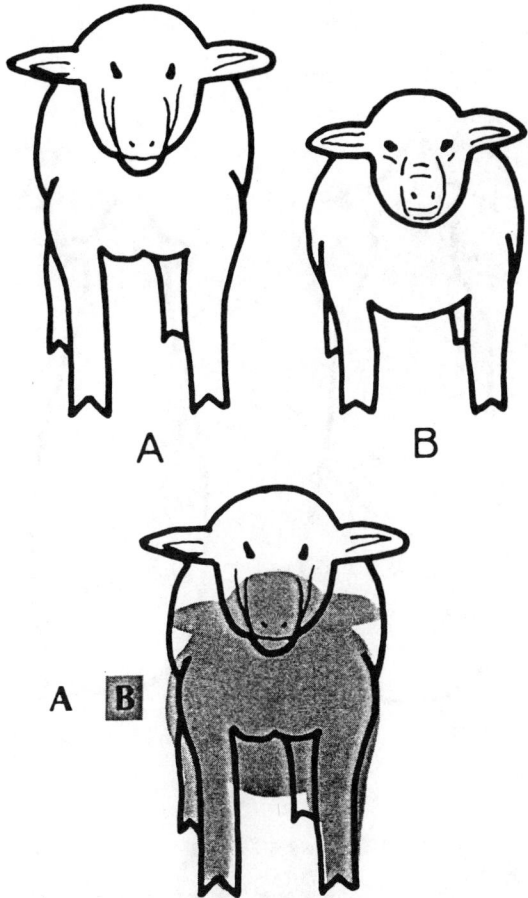

Fig. 2-3. Front view of modern (A) and old-fashioned (B) market lambs, with superimposed comparison. (Courtesy, Purdue University)

should indicate above-average substance of bone. A coarse head and an excessively wide, heavy front are undesirable. However, a narrow chest with the front legs close together indicates a lack of ruggedness or constitution.

Top View

Look for a reasonably slim, clean neck that blends neatly into the shoulders and a slight prominence at the shoulders, which is due to a muscular shoulder, arm and forearm (Fig. 2-4). The lamb should be a shade tight in the heart and free of excessive cover in this area. It should have more spread and spring to the rear rib than the fore-rib, and should carry this extra spread and thickness back into the loin.

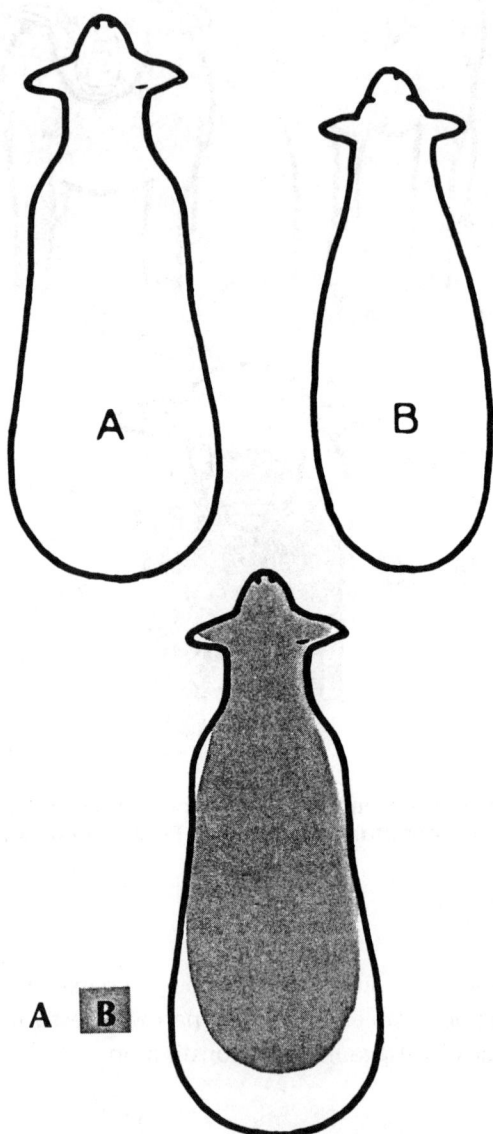

Fig. 2-4. Top view of modern (A) and old-fashioned (B) market lambs, with superimposed comparison. (Courtesy, Purdue University)

A long, level topline; a clean, neat front end; a trim, tight middle, along with a definite degree of cleanness, trimness and firmness in the rear flank and crotch regions — all add to the attractiveness and style of the lamb. A heavy, wasty middle tends to allow more excessive fill thus reducing the yield of the carcass and lowering the dressing percentage. A reasonable amount of middle permits capacity and tends to be associated with good performance and economical production. Smoothness and neat blending of parts also contribute to eye appeal. Coarseness and roughness, denoting a lack of quality, can be seen in the head and in the junctions between the neck and shoulders and fore-rib or rack. Sheep should not show a marked degree of coarseness in the knee, hip or hock joints.

Side View

Length and ruggedness to go along with the desired size and scale are important in yielding carcasses with the largest proportion of the weight in the leg, loin and rack regions (Fig. 2-5). Extremely short-bodied, compact lambs or extremely shallow-bodied, stringy lambs are undesirable. Extra length of rib, loin and especially rump is particularly important in contributing to a larger leg and more weight in the hind-saddle.

The animal should be strong topped, neat and clean in the fore and rear flanks and trim and tight through the throat and breast regions, along the underline (or middle) and in the leg and crotch (twist) areas.

The legs should be at least average in length to be in harmony with growthiness, age and stage of maturity. The feet and legs should be placed squarely on all four corners of the animal's body, and the animal should stand on enough good, clean substance of bone for its size and structure. Even though breeds of sheep vary considerably in size, a "stubby," "rangy" or "spindly" appearance is never desired.

Rear View

The lamb should be extremely long from the hooks to the pins or long in the rump and carry back reasonably full and level out over the dock (Fig. 2-6). This extreme thickness should carry down into deep, thick, full, firm, muscular legs, with the rear legs set wide apart to permit the presence of the muscle volume desired in meat-type lambs. Extreme width at the shoulders and over the fore-ribs, tapering back to a narrow, pinched rump and a narrow, close stance on the rear legs, contributes to a lack of balance or relatively less weight in the hindsaddle.

Condition and Handling: Meatiness

By observing and handling, you can rather quickly detect whether a lamb is

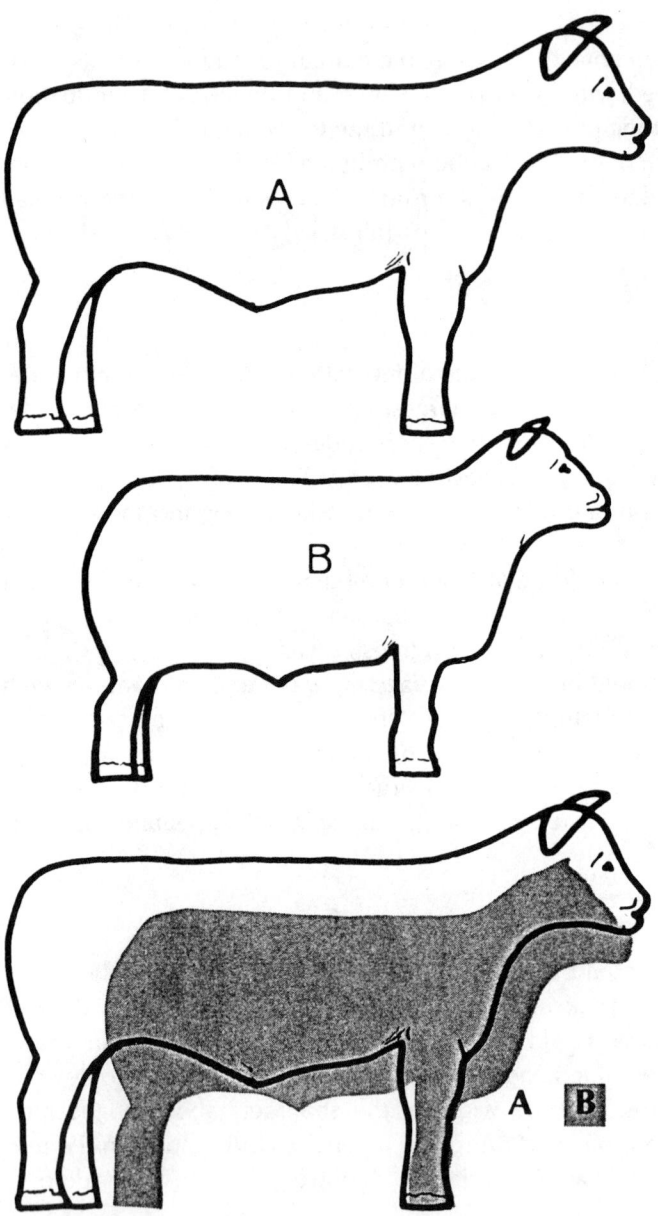

Fig. 2-5. Side view of modern (A) and old-fashioned (B) market lambs, with superimposed comparison. (Courtesy, Purdue University)

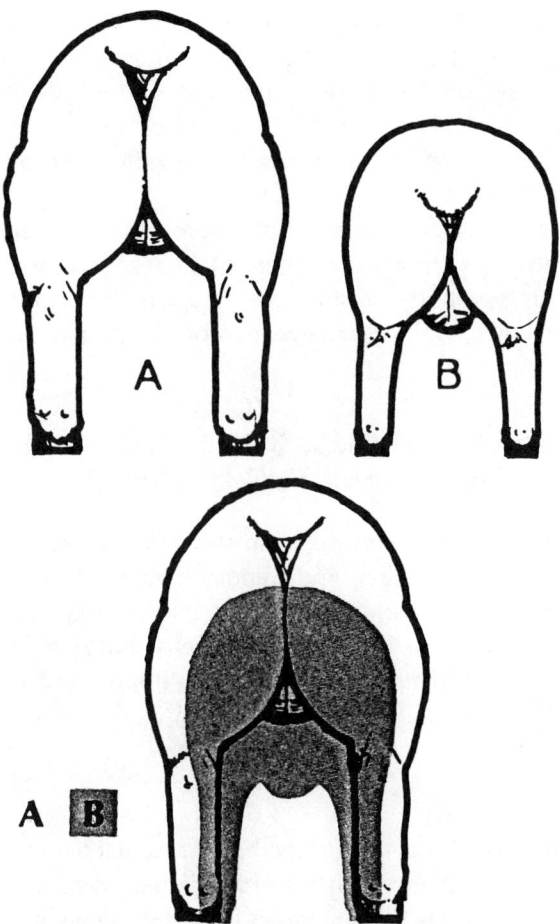

Fig. 2-6. Rear view of modern (A) and old-fashioned (B) market lambs, with superimposed comparison. (Courtesy, Purdue University)

extremely thin or fat. However, small differences of 0.03 – 0.05 of an inch of fat cover are not always accurately detected even after considerable experience.

The handling procedure must be practiced to determine apparent differences in the degree of finish. Lamb buyers and judges vary in their procedure. Parts of the sheep handled are over the backbone and ribs; edge of rack, back and loin; over the hips; and along the joints of the docked tail. The hardness or firmness of the lambs over the back and loin is also used for determining condition, particularly in grading to distinguish between an average lamb and a top-quality market lamb. In handling over the spinal column or ribs, extend the fingers together rather than *spread* apart so that you do not get a separate "feel" from each finger. Also, the fingers should be pressed firmly (through the *fleece* if not shorn) to the flesh or bone. Digging in with the points of the fingers and rubbing or rotating the fingers back and forth is a poor technique that is a waste of time and effort, as well as being hard on the sheep.

In addition to a desirable amount of finish (0.2 – 0.3 inch of fat over the 12th rib, as indicated in the preferred consumer standards), a muscular lamb with particularly good development in the loin and leg is desired. Determining the amount of loin-eye area in the live lamb is merely speculative with the present procedure of observing and handling. However, a lamb that is meaty, firm handling down the top and in the loin and leg area is desirable.

Live animal judges must practice established procedures, pursue new techniques and follow the lambs they appraise to their carcasses, in order to become more competent in detecting properly finished, well-muscled lambs that have 2.5 square inch (3.0 square inch as a goal) of loin-eye area per 50-pound carcass.

CHARACTERISTICS USED IN SELECTING
AND EVALUATING LIVE SHEEP

Generally speaking, all breeders of sheep strive for a common type. To be sure, there are drastic differences in size and minor characteristics, such as head, hair color, wool type and wool distribution. The modern meat-type sheep is characterized by a long, thick, muscular body with symmetry, balance and breed character and a smooth, uniform, firm finish, a strong constitution and outstanding quality (Figs. 2-7, 2-10, 2-12, 2-24, 2-28, 2-30 and 2-33).

Body

The **head** contributes a great deal to the breed characteristics of sheep (Figs. 2-8, 2-11, 2-13, 2-23, 2-27 and 2-38). Each breed is different in appearance. However, in general, a good head is evidenced by a broad forehead, plenty of width between the eyes and ears, a face of moderate length, a well-shaped nose, large nostrils and mouth, with a harmonious blend of the various parts giving a picture of distinctiveness and character. Ears for different breeds vary from gross prominence to relatively small size. Sheep should have ears of proper size and medium texture, which are set on the head in a manner so as to harmonize with the other features of the head. The ears should be free of wool, and the ears and face should possess a distinct color that is characteristic of the breed. The face should be "open" and free of wool down on the nose. The eyes should be clear, well-developed and alert. In observing the head, note especially the appearance of any characteristics, such as scurs, black or brown fibers around the head and particularly in the horn pits, wool blindness, etc., that might disqualify certain breeds. In horned breeds, such as the Horned Dorset and Rambouillet, Debouillet and Merino (rams), the horns should emphasize the character and attractiveness of the head rather than detract from its appearance (Figs. 2-15, 2-16, 2-48 and 2-52). Hereditary unsoundnesses, such as overshot or undershot jaws, scurs, horn stubs, etc., are objectionable in all breeds (Fig. 2-57).

An ideal **neck** is moderately short, strongly muscled, blending smoothly with the head and shoulders (Figs. 2-7, 2-9, 2-19, 2-30 and 2-45). A neck that is properly set on the body adds style and beauty to the animal's features.

Shoulders that possess a minimum covering of flesh and are expressive, heavily muscled, joined neatly at the top and blended smoothly with the neck in front are considered ideal. Modern sheep will not be full and smooth back of the shoulders in the **heart girth,** or fore-rib, area. The skeletal structure and muscle development do not permit an animal to be smooth in its appearance in this region. If the animal is smooth and full in the heart girth, it is due to excessive finish. A muscular animal that is properly finished should not be required to be extremely neat shouldered and full in the heart girth. A certain amount of judgment is involved in determining whether the shoulders are coarse and prominent due to the actual skeletal structure of the animal in question or if the shoulder, arm and forearm areas indicate extreme muscle development. The **top of the shoulder** should be wide and smooth and should contribute its part in carrying out the quality, trimness and muscling desired in the modern sheep type.

In front and below the shoulders lie the **breast** (sometimes called the **brisket**) and **chest** areas. A wide, full chest is indicative of constitution and capacity (Figs. 2-15, 2-16, 2-20, 2-34 and 2-38). Width between the front legs is very desirable for it denotes a wide chest floor with plenty of room for the vital organs embraced by that region. The breast should be neat and trim and should fit clearly and tightly into the neck, shoulder and chest regions. Extreme fullness in the breast area is practically all waste and is highly undesirable in modern sheep. Adequate depth (as deep as the middle) and fullness in the heart girth add materially to the ruggedness, hardiness and appearance of sheep. Too often sheep are faulty in the shoulder region. It is quite common to find open shoulders, a depression immediately behind the shoulders (slack back of shoulders), extremely prominent (coarse) shoulders and narrow chests. Sheep with any of these undesirable points should be discriminated against.

The **back** is indicated by that region between the shoulders and the loin. A strong back with plenty of spread, width, flare, thickness and muscling is an indispensable feature in the selection of a good individual (Figs. 2-20, 2-34 and 2-38). A weak back (top) is a serious fault in sheep and is usually indicative of a light-muscled, overconditioned individual. The back should be broad, straight, strong and heavily muscled, carrying a minimum amount of smooth, firm finish.

The **ribs** join with the backbone to form the width of that region. The rib and back areas of the live animal correspond to the wholesale cut called the "rack" in the carcass. A long, well-sprung (widely curved), close-ribbed sheep usually possesses adequate stamina and digestive capacity. The animal should have more spring (width) to the rear ribs than to the fore-ribs, with a minimum amount of finish covering the entire rib section. The ribs serve as a protective skeleton for the respiratory system and part of the digestive organs and produce valuable cuts of meat when they are heavily muscled. Many sheep are narrow, short-ribbed individuals. Sheep of this conformation lack ruggedness, or constitution, and capacity.

The **loin** is that region between the last rib and the hip bone (Fig. 2-1). The loin is one of the highest-priced cuts on a lamb or mutton carcass; therefore, it plays an important part in its value. The loin should be strong, long, wide, deep, thick, heavily muscled and covered with a minimum amount of hard, firm, smooth finish. You should be able to feel the "buttons" (lateral spinal processes) along the loin edge when you are allowed to inspect and handle the individual in question. The extremely meaty, heavy-muscled lambs will handle very firm over the loin due to the rigid muscle tone resulting from their outstanding muscle structure.

A trim *middle* is especially important in market lambs (Figs. 2-14, 2-21 and 2-32). The meat-type lamb should display a very tight middle and a long, clean underline. The individual should exhibit moderate depth in the foreflank (fore-rib or heart girth area) and should be tucked up a bit in the rear flank, which is quite acceptable under present-day evaluation standards. Research has yielded considerable evidence that extra depth in the foreflank and the rear flank is due to nothing but fat. A neat middle helps to increase the dressing percentage, thus adding to the value of the carcass.

The *flanks* should be neat and clean, thus contributing to the trimness and higher yield of valuable cuts desired in modern-type sheep. A heavy middle, a deep, full foreflank and a deep, full loose rear flank result in a wasty, unattractive lamb that would yield an overfinished, sloppy, undesirable carcass. The flanks and middle should be constructed and correlated so that they add to the symmetry and balance of the entire body. Sheep get their roominess and capacity from thickness and width, not from extra depth in the middle, foreflank and rear flank areas.

Sheep seldom have faulty *hips.* However, keep in mind that the hips should be level, wide and smooth. Prominent hips detract from the balance, symmetry, quality and smoothness desired in the modern sheep type.

The *rump* is the upper portion of the leg of lamb (or mutton) extending from the hips back to the pins, sometimes referred to as "length of rump" or "length from the hooks to the pins." A desirable rump is evidenced by being long, level, muscular, wide, meaty and square at the dock (Figs. 2-22, 2-27 and 2-28). The *dock* is the end of the rump. You should select sheep that are wide and square at the dock and that possess a long, level, muscular rump that will carry down into a large, meaty leg of lamb (or mutton). It is quite common to find sheep that are peaked or drooping at the rump, which is not desired by constructive sheep breeders. The rump and dock should carry a minimum amount of hard, firm, smooth finish. In some market lambs and occasionally in breeding sheep, the dock becomes gobby with fat. This is very objectionable in modern-type sheep.

The *thigh,* commonly referred to as the leg of lamb or leg of mutton, is one of the most valuable wholesale cuts from the lamb or mutton carcass. The leg (wholesale cut) makes up 33 percent of the carcass weight and yields about 40 percent of the value realized from the average carcass. A long, large, thick, plump leg is indicative of a thick, meaty, muscular-type sheep (Figs. 2-7, 2-10, 2-19, 2-23 and 2-27). The leg should be wide, deep, full, heavily muscled and trim. The leg should show evidence of muscling in the inner portion of the leg as well as in the outer portion and should be covered with a minimum amount of hard, firm, smooth finish. When you are allowed to handle and inspect the sheep in question, you should be able to feel the muscle seams in both the inner and the outer leg.

The *twist* is the area between the rear legs. The modern-type sheep will be clean and neat in the twist and cod area and will exhibit an excessive amount of muscle development in this region. In the past, sheep judges selected animals that were deep and full in the twist and cod. Research has demonstrated that this area was loaded with fat, which was highly responsible for the deep, full-looking appearance observed in the twist and cod. The muscle structure in the twist and inner thigh area in no way forms and develops to give the deep, full look sought after by past sheep judges. During the feeding period, fat deposits early in the twist and cod area. The trend is to select sheep that are somewhat "cut-up" in the twist but that possess the muscling and meatiness in this area that is desired in the modern sheep type.

The *legs* should be medium length, set wide apart, with ample bone and strong pasterns (Figs. 2-10, 2-13, 2-24 and 2-30). The front and rear legs should be straight and strong and should set out squarely on the corners of the body. A square-standing, straight-legged sheep will have the width of chest floor, the sturdy constitution and the over-all body capacity to meet any challenge that might confront the future of the sheep industry. It is extremely desirable to select sheep that are straight and clean in the hocks and correct and strong in the pasterns, especially in breeding stock. Rear legs that are set wide apart are indicative of a deep, thick, meaty, muscular rump, thigh and leg. Sheep that are heavily muscled must have wide-set legs and thickness between them to carry the vast amounts of red meat and muscle tissue that develop in this area. Breeding sheep with crooked legs and weak pasterns often become rather useless.

The *bone* from the knee to the ankle joint (cannon bone area) is one of the most accurate and quickest indicators of bone size and substance of an animal. Research data show that heavier-boned animals have more muscle. In other words, there is a positive relationship between bone size and substance and muscling in sheep. Bone size and substance can also be used as an indicator of quality and ruggedness (Figs. 2-16, 2-20, 2-23, 2-27 and 2-38). The size of bone should be in proportion to the size of the body, and the bone should possess *joints* that are clean and strong. In rams, larger, heavier bone is desired to give ruggedness, strength and masculinity to the individual. In female stock, more refinement of bone is sought.

The *feet* should be sound, straight and correctly set under the body, with the weight equally distributed on all toes. A ram, ewe or wether standing up straight on all toes on sound feet and legs portrays style and carriage. This is very desirable. The pasterns should be strong and fairly straight. Rams and ewes that are weak on their pasterns are discriminated against by breeders.

Exercise is important in order for sheep to be sturdy and useful. Therefore, when selecting breeding stock, pay particular attention to the feet, legs, and pasterns, especially when rams are to be used for range-breeding purposes.

Heritability Estimates

Heritability is the fraction of the total variation for a given trait within a population that is due to, or attributed to, the additive effect of the genes. The heritability for a given trait may vary between breeds or sheep populations of different genetic backgrounds. There are several methods of estimating heritabilities, and the results obtained may differ. Estimating heritability helps determining the amount of improvement that might be expected in one year or one generation. In general, heritabilities of less than 20 percent are considered low; between 20 percent and 40 percent, medium; and over 40 percent, high.

The average heritability estimates for various traits in sheep, expressed as percentages, are as follows:

Type score (weaning)	10	Lambing date	37
Dressing percentages	10	Grease fleece weight	38
Weaning weight at 60 days	10	Neck folds (weaning)	39
Carcass grade	12	Staple length (weaning)	39
Number of lambs raised	13	Lean weight	39
Multiple births	15	Mature body weight	40
Condition score	17	Type score (yearling)	40
Carcass weight / day / age	22	Skin folds	40
Fat thickness over loin	23	Clean fleece weight	40
Milk production	26	Clean yield of wool	44
Bone weight	30	Gestation length	45
Birthweight	30	Staple length (yearling)	47
Weaning weight over 100 days	30	Retail cut weight	50
Rate of gain	30	Loin-eye area	53
Carcass length	31	Face cover	56
Fleece grade	35	Fat weight	57

Fleece and Skin

Relatively speaking, fleece and skin are of minor importance in mutton-type sheep. The discussion here will deal only with fleece and skin in the selection of mutton breeds. Every breed of sheep has a type of wool that is typical of that breed. Judging these types or grades of wool can be mastered only by a special study of wool. In judging the fleeces of any breed of sheep, keep in mind that there is a typical fineness that belongs to that kind of sheep. (See "Distinguishing Breed Characteristics of Fine-Wools.") It would not be a typical breed characteristic to find a Hampshire with a fleece as fine as a Southdown. In general, judges like fleeces that are long and dense, with fine, distinct crimp, and sound and uniform in length and fineness, with a medium amount of yolk, which is bright and clean. In examining a fleece, part the wool at the shoulder (finest fleece), the belly (intermediate fleece) and the leg or britch (coarsest fleece). Black fibers and hairs among the wool are undesirable in all white-faced breeds. For further information on judging wool, refer to the information under the heading "Judging a Fleece" later in this part.

A pink skin usually indicates good health, good quality and a good feeder. Breeds with black points often have skin with a bluish tinge. The absence of bright pink skin is *not* always an indicator of poor condition and quality among the black-faced breeds.

A long, clean, bright, dense fleece with a medium amount of crimp and a pink skin is usually acceptable in most medium-wool breeds.

Constitution

Constitution, sometimes referred to as ruggedness, embraces those characteristics that contribute most highly to the economic and practical usefulness of an individual. The ability to resist disease, to convert feed efficiently into muscle, bone and wool and to reproduce strong, vigorous lambs is the result sought. A beautiful individual without enough ruggedness and stamina to produce valuable offspring has a low value.

Constitution is very important in the selection of breeding stock. Constitution is evidenced by (1) a strong head with breed character; (2) a wide, full chest and adequate depth in the heart girth and rib area; (3) tremendous spring of ribs, especially the rear rib area; (4) a strong back and loin, emphasized by extreme thickness, spread, flare and muscling down the top; and (5) four sturdy, well-placed legs (Figs. 2-10, 2-13, 2-24 and 2-30).

Natural Fleshing and Finish

Usually a class of market lambs or wethers can be correctly placed on the fat covering over the ribs, back, loin and rump and on the size (meatiness) of the leg of lamb. Muscling, meatiness and finish (degree of fatness) are of prime importance in judging market classes. Muscling and meatiness are inherited characteristics, more or less, and finish is produced by feeding and management (environment). A lamb should have a deep, thick, meaty covering of muscle and a minimum amount of hard, firm, smooth finish that is evenly distributed over the shoulders, ribs, back, loin, rump and leg. The overfinished lamb will have a crease down its back, while the partly finished or thin lamb will have a ridge down its back. The correctly finished, muscular, modern lamb will have muscle extending up even with the backbone or vertebral column. The muscles will flex upon movement and bulge and express themselves upon stimulation. A keen touch that can determine the differences in the degree of finish and the amount of muscling in close classes can be developed only by persistent practice.

The packer demands not only a thick, meaty, muscular lamb with a minimum amount of fat cover but also a smooth, firm, evenly covered one. A minimum of fat covering, especially in the high-priced cuts, such as the rack, loin or leg, adds firmness to the meat for good shelf life and good shipability of the carcass, as well as to the quality on foot. The muscling, meatiness and finish should be evenly distributed and, figuratively speaking, "as hard as a board." Lambs and wethers that are overfinished (overdone) many times become soft and patchy. Extreme softness, patchiness and flabbiness in a market lamb usually drop that individual to the bottom, provided there are any correctly finished lambs in the class. Firmness is an indispensable requirement of today's sheep and should be carefully noted in selection and evaluation.

In the placing of breeding stock (ewes and rams), muscling, meatiness and finish are also of extreme importance, especially when the animals have been under the same feeding conditions. Muscling is important because it represents the true physical make-up of the animal. Ewes and rams should be meaty and heavily muscled and covered with a minimum amount of evenly distributed firm finish. In breeding classes, it is very important that the fat be firm and smooth. Gobby fat on the lower part of the ribs, over the back, along the loin edge and around the dock is not desirable. A weak back usually indicates a lack of strong muscling in that region. Muscling and meatiness can best be determined by examining the shoulder, arm, forearm, back, loin, rump and leg regions, especially the center of the leg (stifle muscle area). Some of the areas that are indicators of excessive finish are the point of the shoulder, the hip or hook bone and the backbone or vertical column. These are bones that have only a small amount of tendon and skin covering over them. As an animal fattens, a layer of fat forms between the tendon and the skin layer. When the animal is handled in these areas, everything under the hide is fat. The top of the shoulder, the fore-rib, the rear rib, the loin edge, the flank, the elbow pocket, the breast and the twist are all areas that can be used to determine the amount of finish a breeding or market sheep is carrying at one phase or another during the growing and finishing period.

Quality

Quality is desirable in the selection of both breeding and market sheep (Figs. 2-7, 2-10, 2-12, 2-24, 2-28, 2-30 and 2-33). Size without quality leads to a very wasteful and inefficient-type animal. Quality sheep possess (1) a clean-cut, well-shaped head covered with refined wool or hair; (2) ample-size bone and clean joints; (3) a minimum amount of firm, smooth, evenly distributed finish; (4) a good, bright, dense fleece; (5) a pink skin; and (6) a symmetrically balanced body. Quality in market sheep is very important, for it indicates to the packer that the animal will hang up a minimum-fleshed, firm, trim, neat, heavily muscled carcass and that the muscle will be fine textured and very palatable. Quality, size, muscling, degree of finish and age are the main factors that determine the value per hundredweight at any given time.

Sex Character

Sex character deals with those distinguishing features, other than the presence of sex organs, that differentiate one sex from another.

Masculine sex character in a ram is expressed by a strong, bold head and neck; by strong, rugged, heavy bone; by a massive, muscular, powerful appearance; and by a proud movement and bold manner that are characteristic of a prepotent sire (Figs. 2-8, 2-11, 2-23, 2-27, 2-38 and 2-50). More ruggedness is accepted in rams than in ewes.

Feminine sex character is expressed by fineness of features throughout, especially about the head and neck (Figs. 2-7, 2-10, 2-19, 2-28, 2-33 and 2-45). The bone of a ewe is smaller and more refined than the bone of a ram. The general appearance of a ewe gives the impression of those motherly instincts that are necessary for successful reproduction, but the ewe must not lack the size and substance that are inseparable for a sound, good-doing, highly productive individual.

Balance

Balance is extremely important in livestock selection. It is the blending together of all parts of an individual in a harmonious fashion (Figs. 2-7, 2-10, 2-12, 2-24, 2-28, 2-30 and 2-33). A ram with a long body and a short rump lacks balance. Also, an individual that is deep, thick and massive through the shoulders, chest and breast region and narrow through the rump and light in the leg is not balanced. Balance involves uniformity in width, thickness, muscling, finish, character and quality, with each of these factors correlated into one highly functional unit. Balance also adds to the beauty, style and general appearance of an animal.

Style and Smoothness

Style is manifested by an erect and well-set head and neck; by a refined, alert ear; by proud, collective action; by a manner of carriage that attracts attention; and by a responsive and pleasing disposition (Figs. 2-9, 2-19, 2-28 and 2-35). Stylish sheep hold their bodies together, walk smartly and collectively and give the impression of possessing pride in their physical make-up. Style sets off an individual to an advantage, thus quickly and frequently attracting the eye of the buyer or the judge. Style should not be mistaken for good body conformation, for it does not always follow that the most stylish specimen is the most useful, modern-type animal. Style is an asset as long as it is associated with a foundation composed of practical mutton characteristics.

A long, level topline; a clean, neat front end; a trim, tight middle, along with a definite degree of cleanness, trimness and firmness in the rear flank and crotch regions, add to the attractiveness and style of sheep (Figs. 2-10, 2-19, 2-25, 2-28, 2-38 and 2-39). A heavy, wasty middle tends to allow more excessive fill, thus reducing the yield of carcass and dressing percentage. A reasonable amount of middle permits capacity and tends to be associated with good performance and economical production. Smoothness and neat blending of parts also contributes to eye appeal.

Coarseness and roughness denoting a lack of quality can be seen in the head and in the junctions between the neck and shoulders and fore-rib, or rack. Sheep do not show a marked degree of coarseness in the knee, hip or hock joints.

Breed Type

Although the ideal mutton type serves as a common standard for all mutton breeds, there are certain breed differences with which you, as a judge, must be familiar. The characteristics that make up breed type are principally color markings, ear size and carriage, amount of wool or hair covering on the head and legs, shape of the head, presence or absence of horns, size and striking differences in the fineness, length and density of fleece. The distinguishing characteristics of the various breeds are presented under the discussion of breeds of sheep.

Size and Weight

The preferred live weight for market lambs is 85 – 120 pounds (depending on breed). Lambs weighing under 85 or over 120 pounds are seasonably discounted on the market even though they may cut acceptable carcasses. Normally, lambs weighing within this range will yield (average dressing percentage) 49 – 51 percent. Breeding sheep (both rams and ewes) should possess the size and weight that are characteristic of the breed they represent.

The popularity of some breeds may be justified on the basis of size, rather than on the basis of the most ideal mutton conformation. A sheep should be the proper size for its age, sex and breed. For example, a Hampshire ram that is perfect in body conformation and muscling but considerably below the size accepted as standard for the breed cannot be classed as the proper type. Size is a very important factor in the selection of sheep, for many times the greatest difference between breeds of sheep is size. You should note the breeds that emphasize size and ruggedness in their standards and give due consideration to these points in placing these types.

BREEDS OF SHEEP

Distinguishing Breed-Type Characteristics

In this brief review of the most popular breeds of sheep in the United States, those characteristics that will familiarize you with the outstanding breed differences are listed. These distinctive features, known as breed type, are evidenced by differences in body size and shape, head and ears, wool type and covering and color markings. The pictures representing each breed are excellent representatives. Table 2-1 lists the breeds that are discussed in the following sections.

TABLE 2-1. Breeds of Sheep, by Type and Subtype

Breed	Type	Subtype
Southdown	Mutton	Medium-wool
Shropshire	Mutton	Medium-wool
Dorset (Horned and Polled)	Mutton	Medium-wool
Polypay	Mutton	Medium-wool
Hampshire	Mutton	Medium-wool
Suffolk	Mutton	Medium-wool
Oxford	Mutton	Medium-wool
Corriedale	Mutton	Medium-wool
Cheviot	Mutton	Medium-wool
North Country Cheviot	Mutton	Medium-wool
Columbia	Mutton	Medium-wool
Montadale	Mutton	Medium-wool
Panama	Mutton	Medium-wool
Targhee	Mutton	Medium-wool
Lincoln	Mutton	Long-wool
Cotswold	Mutton	Long-wool
Leicester (English and Border)	Mutton	Long-wool
Romney	Mutton	Long-wool
Rambouillet	Wool	Fine-wool
Merino	Wool	Fine-wool
Debouillet	Wool	Fine-wool

Medium-Wool Breeds

Southdown

The Southdown is the smallest and the oldest British breed of sheep. Southdowns' long, low-set and heavy-muscled bodies result in very desirable carcasses. They are a small, early-maturing breed that exhibits exceptional smoothness, style, quality and balance throughout and that possesses outstanding muscle development, especially through the rack, loin and leg of lamb (or mutton) (Figs. 2-7 and 2-8). Wool extends below the eyes. The body is wooled to the knees on the forelegs and to the pasterns on the hindlegs. The hair covering the face and legs varies from steel grey to mouse brown. Bright pink skin and white wool are preferred. Southdowns shear 5 to 8 pounds from a 12 months' wool growth, which grades from $1/2$ to $3/8$ Blood of both combing and clothing grades. According to the American Southdown Breeders' Association, the disqualifications for this breed are as follows: (1) horns or evidence of them; (2) dark poll; (3) speckled markings on face, ears and legs; (4) color of face and legs approaching black; (5) open or coarse wool; (6) only one

Fig. 2-7. Modern Southdown yearling ewe. (Courtesy, *Sheep Breeder and Sheepman*)

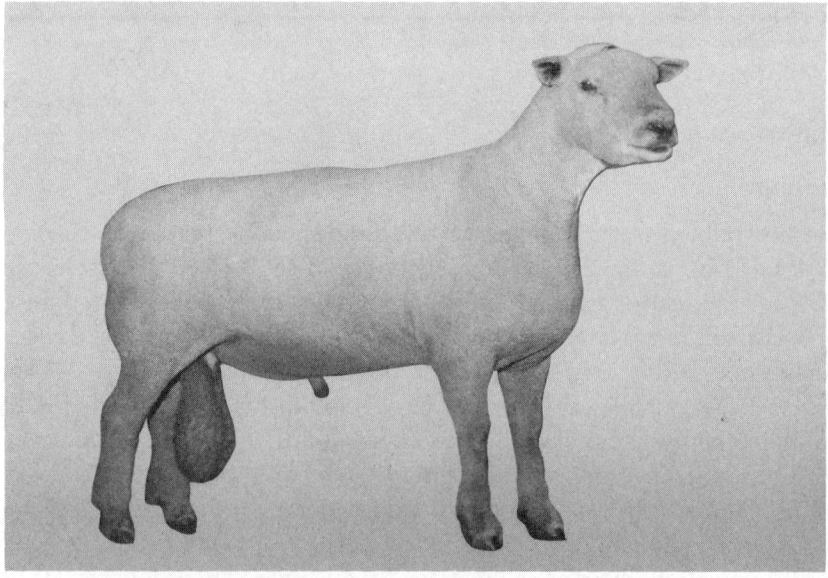

Fig. 2-8. Typical Southdown yearling ram. (Courtesy, *Sheep Breeder and Sheepman*)

testicle down in the scrotum; (7) dark-colored skin; and (8) black or brown fleece. Good mature-type rams in breeding condition weigh from 165 to 220 pounds and ewes from 135 to 160 pounds.

Shropshire

The Shropshire breed is medium in size, distinguished by having a complete covering of wool from the tip of the nose to the feet (Figs. 2-9, 2-10 and 2-11). Less wool over the eyes is preferred; in recent years breeders have been selecting for somewhat less wool on the face, and most at least insist on the sheep being open around the eyes. The ears are short and partly covered with wool. The hair on the nose and legs ranges from very dark grey to deep, soft brown or soft black. The black is not to be confused with the jet black found in some breeds. Shropshires possess an excellent combination of mutton and wool qualities. The typical Shropshire should be low-set, with legs placed well on the corners of a deep body, and should exhibit a lot of style and balance throughout. However, the extreme mutton perfection of the Southdown is usually not attained by the Shropshire. There has been a tendency in the past few years to select for larger, more robust sheep than was formerly popular. Shropshire wool grades from $3/8$ to $1/4$ Blood and grows about $2 1/2$ to 3 inches in 12 months. The average fleece yields between 8 and 10 pounds. Mature rams weigh from 175 to 250 pounds, and mature ewes from 140 to 180 pounds.

Fig. 2-9. Model Shropshire ewe. (Courtesy, American Shropshire Registry Association, Inc.)

Fig. 2-10. Outstanding Shropshire yearling ewe. (Courtesy, *Sheep Breeder and Sheepman*)

Fig. 2-11. Modern Shropshire yearling ram. (Courtesy, *Sheep Breeder and Sheepman*)

Dorset (Horned and Polled)

The Dorset breed has both horned (Figs. 2-12 and 2-13) and polled (Figs. 2-14, 2-15 and 2-16) strains. The horned strain is identified by horns in both sexes and by a white face and legs. It is the only horned mutton breed of sheep in the United States. The polled strain is identical to the horned strain except for the horns. Dorsets possess a compact form with all the mutton characteristics predominating, having good depth of body and a hardy constitution. Their middles are long and roomy. In Horned Dorsets, the horns should spiral close to the side of the face. Rams should possess massive horns and ewes refined horns. Dorsets have wool to the cheeks and to the eyes but seldom have much below the eyes. The face is covered with short, white hair; and hair is the principal covering below the knees and hocks, although within recent years there has been a tendency to get wool in these areas. Most breeders demand wool on the legs, at least to the knees and hocks, and most prefer some below the hocks, as it indicates a heavier wool covering on the body and underline. Breeders do not favor coarse hair on the legs, spots on the skin and in the fleece and horns that tend to grow back. Dorset fleeces are exceptionally white, and the skin is pink. The wool grows about 3 inches in a year and grades $3/8$ to $1/4$ Blood combing. Ewes shear from 6 to 7 pounds per head, and rams from 8 to 12 pounds. The Dorsets are fairly large, with rams weighing 235 pounds or more and ewes 165 pounds or over.

Fig. 2-12. Modern Horned Dorset yearling ewe. (Courtesy, *Sheep Breeder and Sheepman*)

Fig. 2-13. Outstanding Hornet Dorset yearling ram. (Courtesy, *Sheep Breeder and Sheepman*)

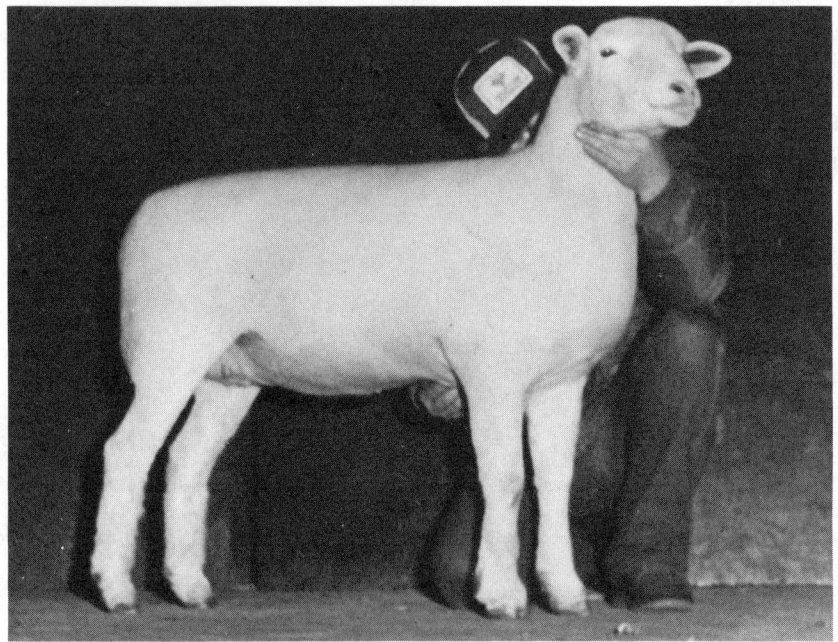

Fig. 2-14. Typical Polled Dorset yearling ewe. (Courtesy, Continental Dorset Club)

Fig. 2-15. Present-day Polled Dorset early fall ram lamb. (Courtesy, Continental Dorset Club)

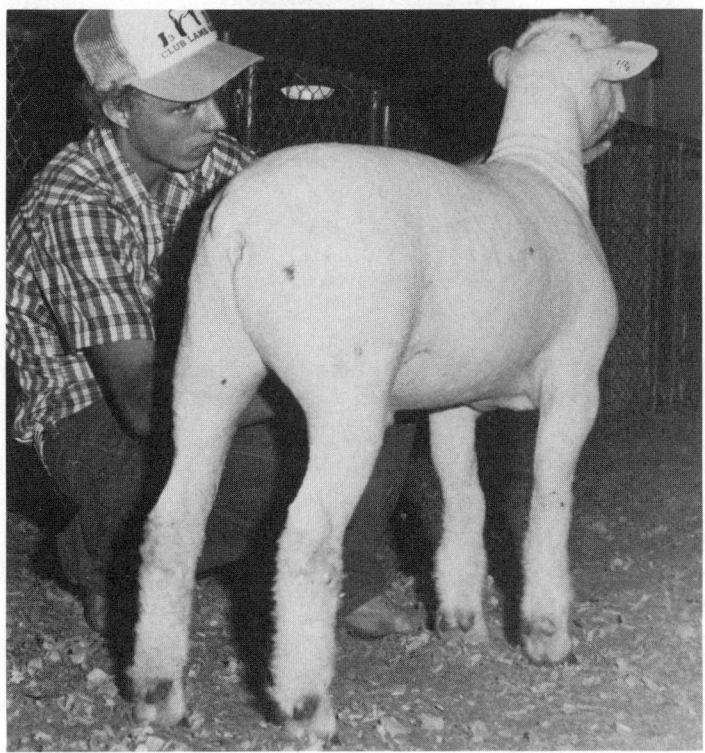

Fig. 2-16. Modern Polled Dorset wether. (Courtesy, Continental Dorset Club)

Polypay

In 1969, the U.S. Sheep Experiment Station at Dubois, Idaho, began initial trials in a series of experiments designed to improve lamb production efficiency and to increase prolificacy per lambing. Reciprocal crosses of (Dorset X Targhee) X (Finn X Rambouillet) resulted in a four-breed cross from which offspring of closed in-line breeding became designated as "Polypay" (Fig. 2-17). After this initial cross, the four-breed crosses were mated interse and were selected along lines for the best lamb production performance. Sires were selected on the basis of the lifetime lamb production records of their dams, plus their own growth rate from birth to weaning.

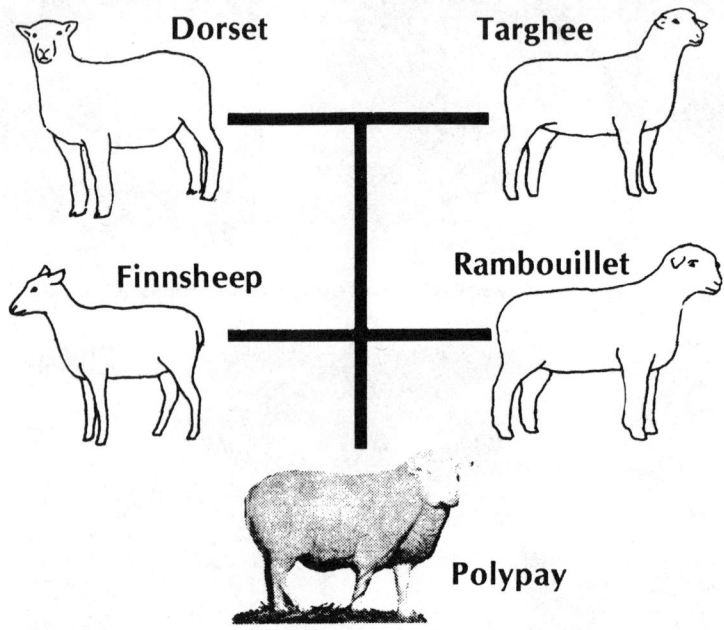

Fig. 2-17. From these four breeds, plus 16 years of research and selection, comes the Polypay — truly the breed of the future. (Courtesy, American Polypay Sheep Association)

In the process a breed developed that has the following traits: (1) high lifetime prolificacy; (2) production of a good lamb crop at one year of age; (3) potential to lamb twice a year; (4) good growth rate of lambs; (5) good carcass quality; and (6) medium wool (Fig. 2-18).

On range conditions, the Polypay can expect to gain as well as the Rambouillet, Targhee and Dorset breeds, but it requires less feed to produce and to maintain its body condition.

The selection criteria set down by the U.S. Sheep Experiment Station remain the foundation for registration and certification by the American Polypay Sheep Association.

Fig. 2-18. These individuals are modern representatives of the Polypay. (Courtesy, American Polypay Sheep Association)

The mature weight of a Polypay ewe is approximately 135 to 150 pounds. A mature ram weighs approximately 180 to 200 pounds.

Hampshire

The Hampshire is one of the largest and most popular breeds of mutton sheep. Hampshire type is evidenced by size, ruggedness and impressive black points. Acceptable face and point color on the Hampshire ranges from dark brown to black, and most breeders prefer the ears to be the same color as the face and to be free from wool. The ears should be thick and medium in length, with the edge of the ear slightly curved, projecting slightly forward and downward. White wool should extend to the knees and hocks. It is of little value to have wool to the pasterns, although this often occurs on the rear legs. Open faces are preferred by sheep breeders. The face and nose profile tends to be Roman and bold; however, extremely Roman noses and very large heads are discriminated against at the present time because of the associated lambing troubles. Hampshires have large bone, strong pasterns, roomy middles, wide tops and strong backs, which are indicative of very strong, muscular, rugged, vigorous sheep (Figs. 2-19, 2-20, 2-21, 2-22, 2-23 and 2-24).

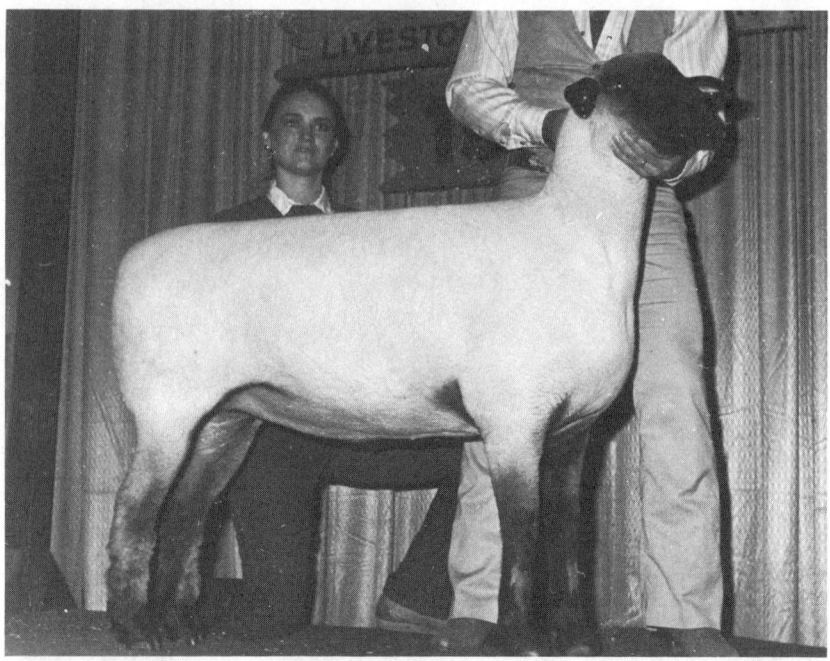

Fig. 2-19. Modern Hampshire yearling ewe. (Courtesy, *Sheep Breeder and Sheepman*)

Fig. 2-20. Outstanding Hampshire yearling ram. (Courtesy, *Sheep Breeder and Sheepman*)

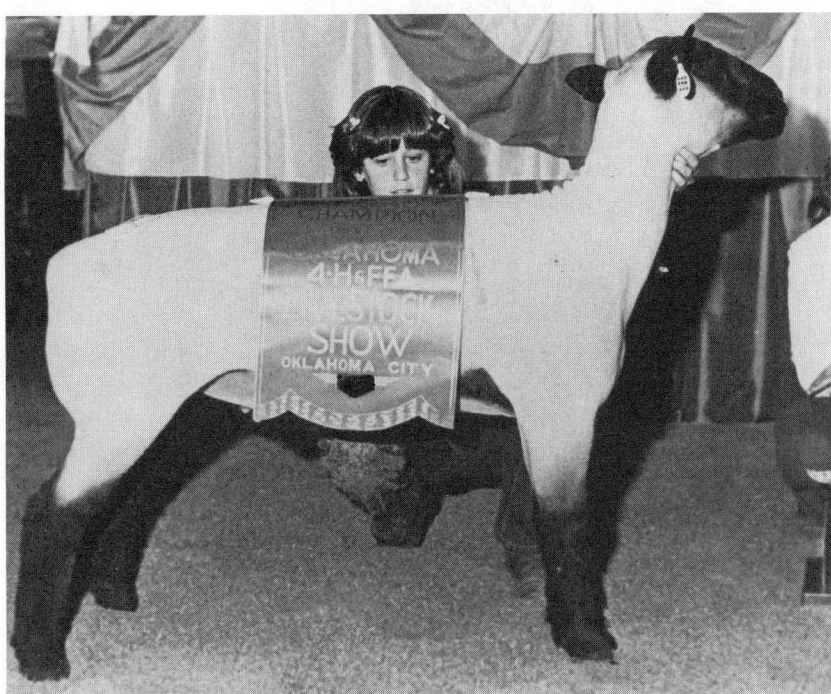

Fig. 2-21. Grand Champion 4-H Hampshire Wether at the 1984 Oklahoma 4-H and FFA Livestock Show, Oklahoma City. (Courtesy, The American Hampshire Sheep Association)

Fig. 2-22. This outstanding group of Hampshires won the Premier Breeder Award at the 1984 Ohio State Fair. (Courtesy, The American Hampshire Sheep Association)

Fig. 2-23. "Perkins," a Grand Champion Hampshire Ram, is an outstanding representative of the Hampshire breed. (Courtesy, The American Hampshire Sheep Association)

Fig. 2-24. Model Hampshire ewe. (Courtesy, *Sheep Breeder and Sheepman*)

Objectionable features are prominent scurs, white specks on face and legs, black wool, wool blindness, coarseness and loose skin under the neck. The skin should preferably be pink; but with the presence of black points, a bluish tinge is evident in most mature sheep of the breed. The wool grades $3/8$ to $1/4$ Blood combing, and a fleece yields on the average of from 7 to 8 pounds. Mature rams in good condition should weigh 275 pounds or over, and mature ewes in good flesh, 190 pounds or over.

Suffolk

The Suffolk is a large down breed, characterized by jet black, polled head and ears and black legs from the hocks and knees down (Figs. 2-25, 2-26, 2-27 and 2-28). The entire face and legs are covered with short, fine, black hair. The fine-textured ears are medium to long in length and are held horizontally or are drooped slightly forward. The face is bold, with a Roman profile and a clean-cut appearance. The striking contrast between white wool and black head, ears and legs never fails to attract attention. Suffolks are very alert and stylish, with plenty of nervous response. Their bodies are large, with a rugged frame, heavy bone, a dense fleece and an exceptionally large, muscular leg of lamb (or mutton). Ruggedness and excellent constitution are very typical of the best individuals. Breeders object to black fibers in the wool and lack of wool on the belly. While many Suffolks have no wool at all on the head, a small quantity of clean, white wool on the forehead is not considered

Fig. 2-25. This painting by Tom Phillips demonstrates a Suffolk ewe with good bone, sufficient length of body, desired body capacity and fine head. (Courtesy, National Suffolk Sheep Association)

Fig. 2-26. This painting by Tom Phillips shows a model Suffolk ram of the desired size, good bone, ample length, tremendous depth of body and excellent head. He is a prime example of the qualities sought in Suffolks. (Courtesy, National Suffolk Sheep Association)

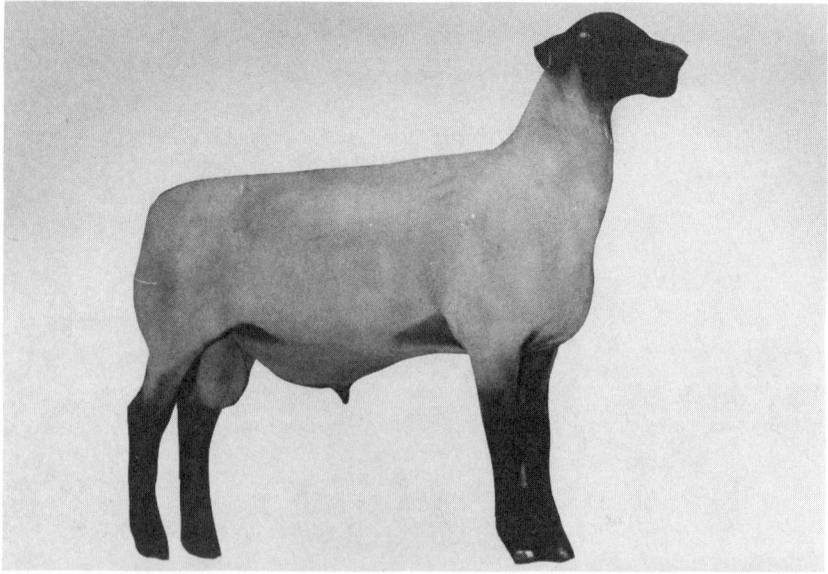

Fig. 2-27. Modern Suffolk yearling ram. (Courtesy, *Sheep Breeder and Sheepman*)

Fig. 2-28. Outstanding Suffolk yearling ewe. (Courtesy, *Sheep Breeder and Sheepman*)

objectionable. Breeders desire pink skin, but there is usually considerable pigmentation, as would be expected in a breed that has such black points. Dark skin is particularly in evidence in older sheep. Dark skins are usually associated with an excess of black fiber in the fleece; consequently, lighter skins are much preferred when they can be obtained without sacrificing merit. The wool is about 2 inches long, shears from 6 to 7 pounds and grades ³/₈ Blood combing or clothing. The body size is very similar to the Hampshire, with mature rams weighing 260 pounds or more and mature ewes 170 pounds or over.

Oxford

The Oxford is manifested by its huge size (usually surpassed in body weight only by the Lincoln) and scale, its abundance of fleece and its dark brown to grey face, ears and legs (Figs. 2-29, 2-30, 2-31 and 2-32). It is a high-headed, stylish breed with a wide, square-lined body and a long, bulky fleece. Wool covers the forehead and cheeks, and trimmed sheep are usually left with a prominent tuft or topknot of wool. This, no doubt, traces back to the Cotswold ancestry. The ears of the Oxford are shorter and wider than those of the Hampshire and often carry a covering of wool. The features of the face are flatter than those of the Hampshire. Bright pink skin is expected. The Oxford is noted for its heavy fleece, averaging from 10 to 15 pounds, with a length between 3¹/₂ and 5 inches. The fineness of the fleece is highly variable, but generally grading ¹/₄ Blood. Full-grown rams should weigh from 275 to 300 pounds, and ewes from 180 to 200 pounds or more.

Fig. 2-29. Champion Oxford Ewe at the Keystone International Livestock Exposition. (Courtesy, American Oxford Association)

Fig. 2-30. Reserve Champion Oxford Ewe at an Illinois State Fair. (Courtesy, *Sheep Breeder and Sheepman*)

Fig. 2-31. Champion Oxford Ram, "Century's Korak 109," at a Keystone International Livestock Exposition. (Courtesy, American Oxford Association)

Fig. 2-32. Champion Oxford Wether and Top-selling Market Lamb at the 1984 Illinois State Fair. (Courtesy, American Oxford Association)

Corriedale

The modern Corriedale is a robust, hardy sheep and a very good producer of both wool and mutton. Probably the breed comes as near to being a dual-purpose breed as any breed in general use in the United States today. The Corriedale has appealed to the range producer because it is open-faced, having little or no wool below the eyes (Figs. 2-33 and 2-34). The face and points are white, but black nostrils are preferred. The poll should be covered with sound wool, but wool below the eyes is not desired, and the Corriedale should be able to see at all times. It is usually well-wooled to the ground on both the forelegs and the rear legs. It stands on more correct feet and legs and has more style than the fine-wool breeds. Dark-colored feet are desirable, and the legs should be free from black or brown wool. Black spots on the face, ears or legs are allowed, but brown spots on any part of the animal are objectionable. Relatively early-maturing, Corriedales have good quality and good mutton conformation and should give the impression of being hardy and having a great constitution. They must carry a dense fleece with plenty of length (5 inches) of even quality, indicating a well-defined crimp with good luster and excellent character. Body wrinkles are objectionable. The wool grades from $3/8$ to $1/4$ Blood. Stud ewes produce from 12 to 18 pounds of wool annually, and stud rams should shear from 15 to 25 pounds or more. Rams in good condition should weigh between 185 and 250 pounds, and mature ewes between 125 and 185 pounds.

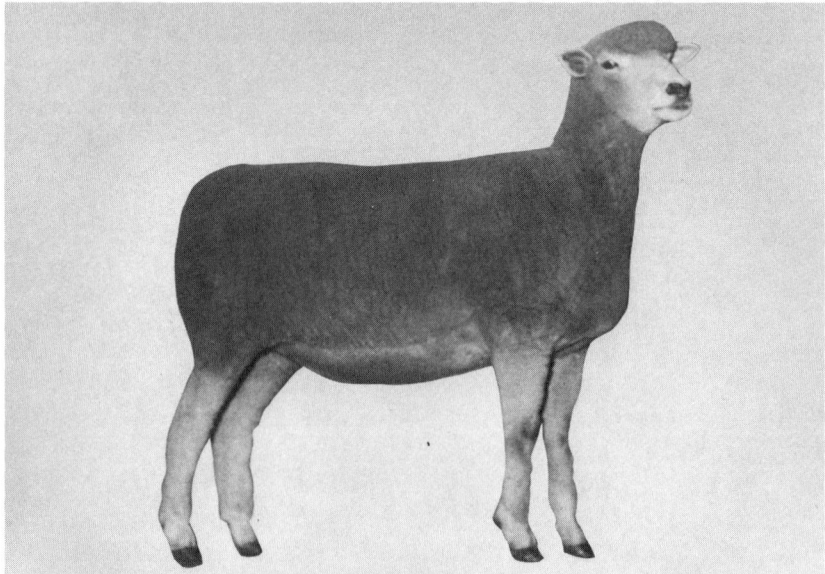

Fig. 2-33. Grand Champion Corriedale Ewe at a North American International Livestock Exposition, Louisville, Kentucky. (Courtesy, *Sheep Breeder and Sheepman*)

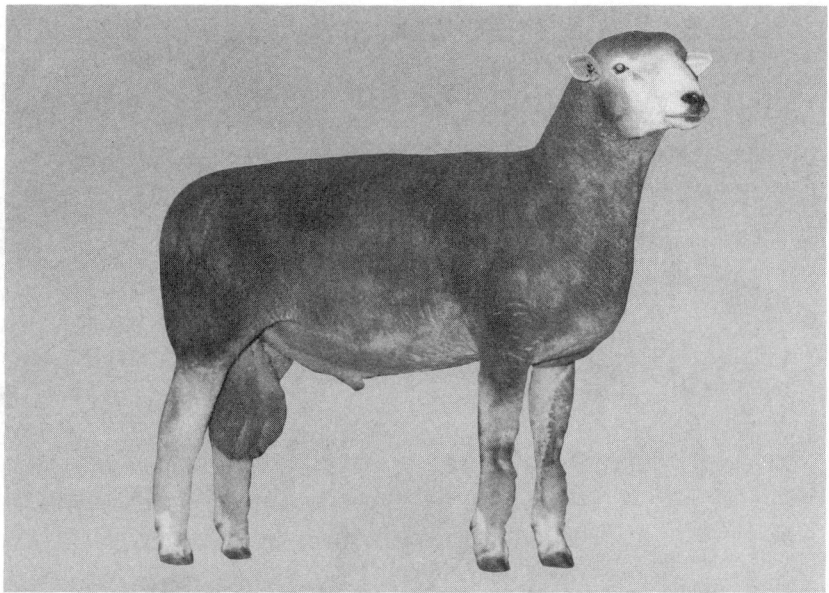

Fig. 2-34. Grand Champion Corriedale Ram at a North American International Livestock Exposition, Louisville, Kentucky. (Courtesy, *Sheep Breeder
and Sheepman*)

Cheviot

The Cheviot is a very stylish and beautiful breed, with the head from the back
part of the ears forward free from wool and covered with white hair. The legs are
covered with white hair below the knees and hocks. The ears are erect and covered
with clean, white hair. The face profile is somewhat Roman. The nostrils and hooves
should be black. The Cheviot is small to medium in size and distinctive by having
fine bone, clean-cut features, a graceful carriage, an unusual quality and a compact
body with large legs (Figs. 2-35 and 2-36). The fleshing qualities and constitutional
vigor are excellent. Objectionable features are flesh-colored nostrils, black spots in
the hair and red or sandy hair on the head or legs. This breed is polled, but slight
scurs do not meet with disfavor. The Cheviot possesses an exceptionally long fleece,
varying from 4 to 5 inches long in one year. Rams clip between 8 and 10 pounds and
ewes 6 and 8 pounds, both falling in the $1/4$ Blood combing class. According to
breed standards, mature rams in good flesh should weigh 185 pounds or more, and
mature ewes, 140 pounds or more.

North Country Cheviot

The North Country Cheviot is similar to the Cheviot except it is a larger, more
rugged, growthier breed that attains greater size at maturity. Because of their small,

Fig. 2-35. Typical Cheviot yearling ewe. (Courtesy, *Sheep Breeder and Sheepman*)

Fig. 2-36. Model Cheviot yearling ram. (Courtesy, *Sheep Breeder and Sheepman*)

narrow heads, the rams can be used on small ewes without risk of difficulty at lambing time. This is a characteristic highly desired in the West, where lambs are dropped on the range without assistance at birth. The wool is springy and white, with a staple length up to 4 inches. Average fleece weight is 8 to 10 pounds, grading 1/4 to 3/8 Blood. Mature rams weigh up to 300 pounds, and mature ewes up to 200 pounds.

Columbia

The Columbia was developed from crosses of Lincoln rams on Rambouillet ewes in an attempt to produce a satisfactory type of range sheep. Columbia sheep have large bodies and rugged constitutions. The hair on the face, ears and legs is white. Although the breed is open-faced, wool extends over the forehead and cheeks, and white hair surrounds the eyes (Figs. 2-37 and 2-38). Developed under strictly range conditions, the breed is now expanding into farm areas. Mutton type is given preference over wool production in selection. The score card for Columbia sheep places 60 percent emphasis on body conformation and 40 percent on wool qualities. The fleece is dense, grows about 3 1/2 inches annually and usually grades 3/8 with some 1/4 Blood. Ewes under range conditions shear an average of 12 pounds, with about a 50 percent shrink. Rams shear 15 to 20 pounds. Range rams weigh from 200 to 250 pounds, and ewes from 135 to 150 pounds.

Fig. 2-37. Extremely modern Columbia yearling ewe. (Courtesy, *Sheep Breeder and Sheepman*)

Fig. 2-38. Modern Columbia yearling ram. (Courtesy, *Sheep Breeder and Sheepman*)

Montadale

The Montadale is the result of crosses of Cheviot rams on Columbia ewes. Montadale breeders insist on sheep with small heads, open faces, bare legs below the knees and white hair on the faces and legs (Fig. 2-39); black spots in the hair are acceptable, but brown in the hair is objectionable. The body conformation should be of acceptable mutton form, and there should be no body wrinkles. Ewes should weigh at least 120 pounds at a year of age and should shear at least 10 to 12 pounds of staple wool that grades from $3/8$ to $1/2$ Blood; rams should be proportionately heavier, weighing approximately 200 pounds, and shear more wool. Montadales have much the same appearance as North Country Cheviots.

Panama

The Panama is the result of crosses of Rambouillet rams on Lincoln ewes and might be considered the counterpart of the Columbia. The Panama was developed by Messrs. Laidlaw and Brockie, in Muldoon, Idaho. It is large in size and produces a good weight of wool. It is very generally distributed throughout the intermountain area of the West and is in demand by range sheep producers who emphasize early lamb and good wool production. Mature rams weigh approximately 200 to 240 pounds, and mature ewes 140 to 160 pounds.

Fig. 2-39. Model Montadale ram. (Courtesy, *Sheep Breeder and Sheep-man*)

Targhee

The Targhee is, in general, the outcome of crossing the first-generation Lincoln-Rambouillet ewes back to Rambouillet rams. Thus, it carries three-fourths fine-wool breeding from the Rambouillet and one-fourth long-wool breeding from the Lincoln. While the general infusion of coarse-wool inheritance has come basically from the Lincoln, some of it has been detoured to the Targhee through the Columbia and Corriedale breeds. Broadly speaking, the Targhee carries the influence of the Rambouillet, Lincoln, Columbia and Corriedale breeds.

The Targhee is a polled, white-faced sheep of intermediate size (Figs. 2-40 and 2-41). This breed has a more desirable mutton conformation than fine-wool sheep, being more blocky, compact and moderately low-set, with good-boned straight legs. The ideal Targhee is free from skin folds and is open-faced. Targhees yield heavy fleeces of long staple wool, which generally grades $1/2$ Blood and which is moderately light in shrinkage. Under range conditions, mature Targhee ewes produce an average per year of about 50 percent of scoured, clean wool. In good range condition, mature rams weigh an average of about 200 pounds, and mature ewes about 130 pounds.

Fig. 2-40. Typical modern Targhee yearling ewe. (Courtesy, *Sheep Breeder and Sheepman*)

Fig. 2-41. Modern Targhee yearling ram. (Courtesy, *Sheep Breeder and Sheepman*)

Long-Wool Breeds

Lincoln

The best Lincoln types have rectangular lines, wide, deep bodies, with a covering of thick flesh, and large legs (Figs. 2-42 and 2-43). Their bone is rugged and large. White hair is found on the legs below the knees and hocks and also on the face in front of the ears, excepting for a tuft of wool on the head. The ears are

Fig. 2-42. Modern yearling Lincoln ewe. (Courtesy, *Sheep Breeder and Sheepman*)

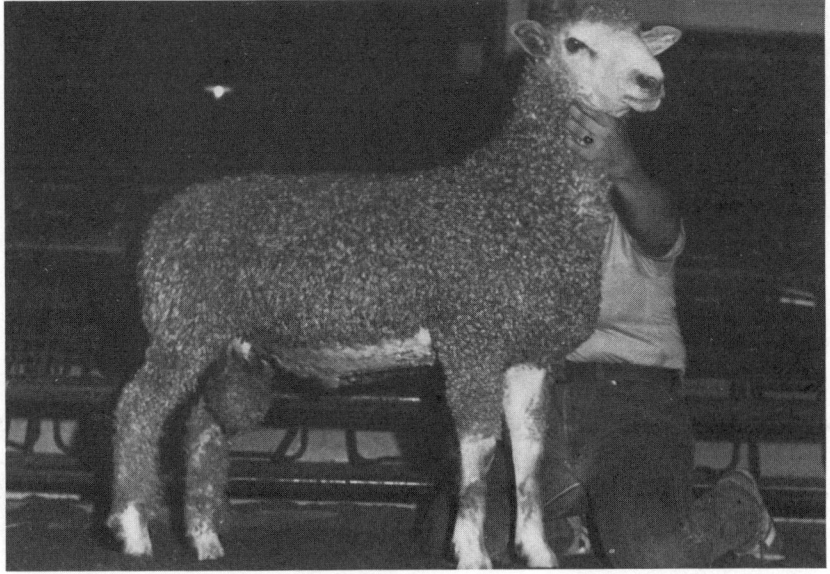

Fig. 2-43. Outstanding modern Lincoln ram. (Courtesy, *Sheep Breeder and Sheepman*)

medium-sized and well-placed. The lips, nostrils and hooves are black, and the skin is pink. The highly lustrous fleece ranges from 10 to 20 inches in length, tends to part in the middle of the back and grades as Braid combing. The wool growth varies from 8 to 12 inches annually. Ewes shear from 12 to 16 pounds annually, and rams, 20 pounds or more. The Lincoln is the largest breed of sheep, with rams in good flesh weighing 300 pounds or more, and ewes between 250 and 270 pounds.

Cotswold

The Cotswold is distinguished from the Lincoln by its excessive growth of wool on the forelock, which hangs over the face and eyes, and by a more bold curl in the fleece (Fig. 2-44). This is especially evident in sheep that have a full year's growth of

Fig. 2-44. Typical Cotswold yearling ewes (shorn once). (Courtesy, American Cotswold Record Association)

wool. On the average, the fleece is somewhat shorter and finer than the Lincoln's. The color of the face and legs may vary from white to greyish white. The face and shanks are white, and some wool below the hocks and knees is favored. Black skin is found around the eyes and on the lips and nostrils. The rest of the skin is pink. The ears are medium fine in quality, are rather long and carried well up. An erect head and compactness and uniformity of width from front to rear are emphasized.

A 12 months' growth of wool measures from 8 to 14 inches, grades Braid and weighs from 10 to 20 pounds. Rams in good flesh weigh from 275 to 300 pounds, and ewes from 180 to 225 pounds.

Leicester

There are two types of Leicesters: the English Leicesters and the Border Leicesters.

The English Leicester carries its head low, has a short neck and is rather rectangular in form with a broad, level top. The bone and head are refined and clean-cut. However, the breed lacks style in comparison to the other long-wool breeds. The face and legs are covered with hair. The head has a tuft of wool, usually the legs have some wool below the knees and hocks. The ears are erect and medium in size. The English Leicester has a coarse Braid wool, 7 to 8 inches long, terminating in a short, twisted curl. On the average, it shears about 10 pounds each year. The English Leicester is the smallest of the long-wool breeds of sheep, with mature rams weighing 250 pounds or more, and ewes, 180 pounds or more.

The Border Leicester is distinguished from the English type in that it is more stylish, has a higher and bolder head, is freer from wool on the forehead and shanks and has whiter hair. The hair on the legs extends well above the knees and hocks. The best types shear from 8 to 12 pounds, and their fleece is somewhat shorter than on the English. Border Leicesters are larger and longer than English Leicesters and usually more upstanding. Mature rams weigh about 280 pounds, and mature ewes 200 pounds.

Romney

The Romney is a low-set, blocky, deep-bodied breed with greater density and fineness of fleece than the other long-wool breeds (Figs. 2-45 and 2-46). Romneys have broad heads; large ears, which are carried straight out from the head; a covering of white hair on the face and legs; and black lips, nostrils and hooves. The legs below the knees and hocks are free from wool, but there is a light tuft of wool on the forehead. In general, the best types are massive and hardy, standing on good, rugged bone.

On the average they yield from 12 to 18 pounds of demiluster wool, 4 to 6 inches long, which usually grades $1/4$ Blood combing. Rams in good condition weigh 225 to 250 pounds, and mature ewes 175 to 200 pounds.

Fig. 2-45. Typical Romney ewe, standing on good, rugged bone. (Courtesy, American Romney Breeders Association)

Fig. 2-46. Present-day Romney ram that possesses the desired characteristics of the Romney breed. (Courtesy, American Romney Breeders Association)

Fine-Wool Breeds

The Rambouillet, Merino and Debouillet are the fine-wool breeds in the United States, with the Rambouillet being the most prevalent.

Rambouillet

According to the American Rambouillet Sheep Breeders' Association, there is no universal standard of excellence for this breed due to the variation in climatic and feed conditions under which Rambouillets are raised. The Rambouillet breed is widely distributed throughout the United States and thrives under a variety of farm and range conditions. In the Northwest, the most popular type of Rambouillet is large; rugged; thick fleshed, with plenty of bone; relatively smooth (few, if any folds); and open-faced. These qualities are being developed and emphasized: 65 percent for meat to 35 percent for fleece qualities. In the Midwest and the East, however, quality of wool gets somewhat more attention, and folds on the neck are not so discriminated against. This is changing, however, to comply with breed association standards.

As a result of this practical difference in type, at some of the livestock shows the Rambouillets are separated into two classes, B and C. Class B includes the type with folds or wrinkles on the neck, tail and flank, while Class C represents the group having smooth bodies with aprons or folds on the neck.

Regardless of these sectional differences, there are certain characteristics that are common to all Rambouillets. Rambouillets should be large, rugged and wide-chested, with ample thickness of flesh to make good mutton (Figs. 2-47, 2-48, 2-49 and 2-50). Acceptable types have superior style, balance, constitution, quality and breed character. They have deep heart girths, bold heads. The wool covers the entire body, head and legs, except for an opening around the eyes and nose. Although some breeders prefer a heavy-wooled head, even to the point of wool blindness, open-faced types are preferred on the western ranges. When compared to wool-blind ewes of comparable quality, open-faced ewes produce approximately 10 pounds more lamb by weaning time per year on the U.S. Sheep Experiment Station ranges near Dubois, Idaho. Consequently, less extreme covering over the head is being encouraged. The parts that are not wooled are covered with white hair. The rams have horns, and the ewes are hornless. Some strains of polled Rambouillets are now in production. The ears are large, white and slightly droopy. A Rambouillet fleece should be between 2¹/₂ and 3 inches long after one year's growth. The fleece should be characterized by (1) a distinct crimp; (2) fibers that are fine and closely packed; (3) a uniform length and fineness over the body; (4) freedom from kemp, hairs and foreign matter; and (5) an even distribution of yolk. On the average, stud ewes should shear 10 to 18 pounds of wool, and rams 15 to 25 pounds. Rambouillets excel all fine-wool breeds in size. Rams in full fleece weigh 225 pounds or more, and ewes 155 pounds or more. Some prominent heavy-fleshed stud rams will weigh over 300 pounds.

Fig. 2-47. Registered Rambouillet ewe at 15 months of age. (Courtesy, American Rambouillet Sheep Breeders' Association)

Fig. 2-48. Registered Rambouillet ram at 18 months of age. (Courtesy, American Rambouillet Sheep Breeders' Association)

Fig. 2-49. Typical modern Rambouillet ewe. (Courtesy, *Sheep Breeder and Sheepman*)

Fig. 2-50. Outstanding polled Rambouillet yearling ram. (Courtesy, *Sheep Breeder and Sheepman*)

Merino

There are many different types of Merino sheep — the American, Delaine, Spanish, Dickinson, Black Top, Standard, Texas, etc. All these types have certain minor differences, which are of little concern to the student of livestock judging. The most popular breeds in the United States are the American Merino and the Delaine Merino. For the sake of brevity and clarity, all the breeds and types of Merino sheep will be discussed according to the classification of the leading state fairs — namely, Class A, Class B and Class C Merinos.

Some characteristics common to all three types of Merinos are as follows: The rams have large spiral horns, and the ewes are hornless. Wool covering should be well-distributed over the body, covering the legs, the body and the head, except the nose and ears. The ears and nose are covered with silky, white hair, and the skin about the nose is often wrinkled. The ears are fine-textured and carried smartly. There is considerable difference in the quality and quantity of fleece on the different Merino types, but they should all have wool that is fine, uniform and pure with ample length, excellent character and even distribution of yolk (Figs. 2-51 and 2-52).

Fig. 2-51. Reserve Grand Champion Merino Ewe at the 1985 North American International Livestock Exposition, Louisville, Kentucky. (Courtesy, *Sheep Breeder and Sheepman*)

Fig. 2-52. Grand Champion Merino Ram at the 1985 North American International Livestock Exposition. (Courtesy, *Sheep Breeder and Sheepman*)

The mutton qualities vary inversely with the wool qualities, but in spite of this fact, body type that most nearly approaches the ideal mutton type is desirable.

Class A Merinos are characterized by having heavy folds and wrinkles in the skin from the head to the tail. They have the finest, densest, shortest and heaviest wool, which is abundant in yolk. The mutton qualities have been sacrificed for the extreme quality and quantity of wool. In comparison to the mutton type, Class A Merinos are rather narrow and leggy and lack thickness and muscling. In 12 months, an average Class A fleece will measure from 1 1/2 to 2 inches. Fleeces grade Fine clothing and are usually outstanding in quality. Exceptional fleeces should be emphasized in the selection of Class A Merinos. Good-type ewes will clip from 15 to 25 pounds of wool, and rams will shear from 25 to 30 pounds of wool. These fleeces carry a heavy percentage of yolk, resulting in a shrinkage of 70 to 75 percent. Mature rams with a full fleece weigh from 130 to 170 pounds, and mature ewes from 90 to 130 pounds.

Class B Merinos are intermediate between Class A and Class C in the number of folds and also in mutton and wool qualities. Class B Merinos have folds on the neck, foreflanks and hindflanks and abut the dock. In body type they are thicker-fleshed, wider, larger-boned and wider-chested than the Class A Merinos. Good rams shear 25 pounds, and ewes 15 pounds. The wool usually grades Fine Delaine, with a staple measuring between 2 1/2 and 3 inches. The wool shrinks about 65 percent. Mature rams weigh from 150 to 175 pounds, and ewes from 90 to 140 pounds.

Class C, or Delaine-type, Merinos are free from folds on the body and have only two or three on the neck. They are the largest of the three types, have more mutton qualities, have shorter legs, are wider through the rib area and are heavier-fleshed with more balance and greater constitution. The wool is not as dense on the legs and head as it is in Class A and Class B. The best individuals have very desirable mutton qualities and make creditable meat producers. Although their wool has less crimp and density than Class A and Class B Merinos, Class C Merinos have a yearly fleece that measures between 3 and 4 inches in length. Rams should yield 18 pounds of wool, ewes 11 pounds. Rams in full fleece should weigh from 150 to 225 pounds, and ewes from 90 to 150 pounds.

Debouillet

The Debouillet is the result of crosses of Ohio Delaine Merino rams on Rambouillet ewes. By selective mating, A. D. Jones of Tatum, New Mexico, produced a long-stapled, fine-wooled sheep with a large, smooth body. He produced this ideal sheep by combining the length of the staple and the character of the Delaine-Merino fleece with the large body of the Rambouillet. The Debouillet is expected to have a uniform fleece that grades 64's or finer and that is at least 3 inches in length on a 12-month basis. The fleece should be bright white with a deep, close crimp and a high density. Debouillet sheep are open-faced below the eyes and over the nose and are deep-sided, with a long, stretchy body, a strong, level back and sturdy, straight legs. Mature rams under average range conditions will weigh 200 pounds and will shear 20 to 25 pounds of grease wool. Mature ewes will weigh approximately 125 to 150 pounds and will shear 15 to 18 pounds of grease wool.

DISTINGUISHING BREED CHARACTERISTICS OF FINE-WOOLS

Although fine-wools differ in size, type of fleece, number of skin folds and body conformation, the following discussion of body type and fleece characteristics can apply to the selection and evaluation of all the fine-wool breeds.

Body Type

In selecting and evaluating fine-wool breeds, roughly emphasize body type and wool equally (in Rambouillets, emphasize conformation 65 percent and wool 35 percent). It is common knowledge that as a whole, the fine-wool breeds of sheep do not have the excellent conformation and muscling possessed by the mutton breeds. The reason is obviously due to the importance placed on wool. However, there are fine-wool individuals that have excellent mutton type and hang up desirable carcasses. In fact, a great majority of feeder and fat lambs carry some fine-wool blood. Breeders of fine-wools are continuously striving to breed to meet the standards of mutton type but yet still retain the fine-wool characteristics. In selecting and evaluating this type, fix in your mind the ideal mutton type and the fine-wool breed characteristics and use this standard for placing fine-wool sheep on body type. You will seldom find the long, thick, meaty, muscular body type in fine-wools that you will find in mutton breeds, but the conformation that most nearly approaches the ideal mutton form is the kind to select. The body type of fine-wools is medium in its development, particularly in muscling, meatiness, tightness of frame, smoothness of body and quality (Figs. 2-47 and 2-51).

Judging Fleece

The factors that should be carefully considered in the evaluation of fleece are fineness, length, density, uniformity, character, yolk, purity, soundness, foreign matter, color and covering. The quantity of wool is determined by density, length and completeness of covering; the quality of wool by fineness, length, soundness and character; and the condition by purity, amount and distribution of yolk and foreign matter.

Fineness refers to the diameter of the wool fiber. The comparative degree of fineness of various fleeces is difficult to determine accurately with the naked eye. Research on wool has shown that in many cases closeness of crimp varies with the degree of fineness. Although strictly speaking, the number of crimps per inch is not absolutely correlated with the degree of fineness; for all practical purposes, the number of crimps per inch may serve as a fair indicator. Where differences in diameter are great, even the variation in the actual size of the fiber may be detected visually. The short wools are usually the finest quality, and the long-staple wools are the coarsest. Because many other factors are just as important as fineness, in judging fleeces, do not spend too much time detecting microscopic differences. The finest fleece is found on the shoulder region, the next in fineness on the side and the coarsest on the thigh or breech. Uniform fineness contributes to the quality of the fleece and therefore to its value.

With regard to fiber diameter, fleeces are classified under the U.S. system from finest to coarsest — Fine, 1/2 Blood, 1/4 Blood, Low 1/4 Blood, Common and Braid. The British system, or Spinning Count, is a numerical rating plan based on the amount of yarn that can be spun from 1 pound of clean wool. The British system provides a more precise designation for diameter of fiber and is used by almost all mill buyers. The spin count values that correspond to or subdivide the various blood grades are shown in Table 2-2.

Length of the fleece influences both the quality and the quantity of wool. All other factors being equal, the sheep with the longest fleece will produce the most pounds of wool. Excessive length in fine wool is of little advantage from a quality standpoint, as long as the majority of the wool is 21/2 inches long. In placing a class, do not attach too much importance to length of staple, unless you know the shearing dates and the ages. When these are not known, extra length of wool is of little importance. As long as the fleece is of sufficient length to qualify for combing wool, extra length adds nothing to the *quality* of the wool, but it does increase the *quantity.* Combing wool is 21/2 inches or more in length in the fine staple grade.

TABLE 2-2. Shrink, U.S. Blood Grades, Spinning Count, Staple Length and Uses of Groups of Wool

Shrink	U.S. Blood Grades	Spinning Count	Staple Length	French Combining	Clothing
			---------------- (in.) ----------------		
56–61	Fine	80's 70's 64's	Over 2³/₄	1¹/₄–2³/₄	Under 1¹/₄
53–59	¹/₂ Blood	62's 60's 58's	Over 3	1¹/₂–3	Under 1¹/₂
47–53	³/₈ Blood	56's	Over 3¹/₄	—¹	Under 3¹/₄
44–50	¹/₄ Blood	54's 50's	Over 3¹/₂	—¹	Under 3¹/₂
42–48	Low ¹/₄ Blood	48's 46's	Over 4	—¹	Under 4

¹Intermediate 1-inch deviations are sometimes classed as "Baby Combing."

The length of wool is determined as the fleece opens naturally without any stretching of the fibers. A fleece that you can part gently with both hands gives a good estimate of the length. A satisfactory place to measure the length of the fleece is directly over the hip bones. If the wool measures 2¹/₂ inches there, you can generally be assured that the majority of the fleece will be combing wool.

Wool classed as "combing" or "staple" is long enough to be handled by regular combing machines. The more expensive worsted cloth for suits, and so forth, is woven from yarn spun from the combed wool fibers. The length of the fiber also contributes to the quantity of the wool and to the loftiness of "fleece volume," which is reflected in clean fleece weight. Thus, length again contributes to the value of wool for the producer.

Wool classed as "clothing" is short and is made into woolen yarn and cloth that is looser woven and less expensive than worsteds. Extremely short wool, usually from old ewes or from lambs with only four to eight months' growth, is classed as "scouring" wool.

There are special combs that will handle length of wool between that classed as "clothing" and "strictly combing." For Fine and ¹/₂ Blood wool, this intermediate length is called "French Combining." In some trade areas "Baby Combing" refers to lengths between combing and clothing for medium wools.

The USDA length specifications for each grade and the expected shrink range for most wool are shown in Table 2-1.

Density is the closeness with which wool fibers are packed together. It can also be expressed as the number of fibers per square inch of body surface, although this last definition is not technically correct or directly applicable when one fleece is compared with another and there is a difference in the size of wool fibers. In the latter case, you should consider the size of the wool fiber, for a coarse fleece will be more closely packed together and denser than a fine fleece with the same number of fibers to the square inch.

In appearance, a loose fleece looks open and thin, and the wool tends to be more ragged on the ends. Grabbing a handful of fleece (with your fingers together) on the side of a sheep is a good way to compare density. You get a greater handful of wool in a dense fleece. Furthermore, a dense fleece has a rather firm, springy touch or recoil, while a loose fleece has less body, giving the impression of being thin. Parting the fleece to observe how closely the wool fibers occur on the skin, as well as the amount of bare skin exposed, is another way to detect density. A thin fleece shows more skin. A fleece that curls up on the ends in every direction lacks density and length. The amount of yolk may influence the appraisal of density.

In wool, *uniformity* refers to the evenness of length and fineness. An excellent fleece approaches the same uniformity of length of wool over the body. However, there is always some variation in the length of fleece. The longest wool is found on the front and side, and the shortest wool on the breech and underline. By parting the fleece on the shoulder, the side and the breech, examine the wool for uniformity of length.

A fleece that varies decidedly in length will not be as valuable on the wool market, for too much short wool (less than 2½ inches) will throw it into a lower price wool class. A fleece that is short in one area and long in another lacks uniformity of length and is undesirable. Occasionally, among poorly bred sheep, the long fibers are coarse wool, and the short fibers are finer. Fleeces like this are severely discounted.

Uniformity within a fleece, for example, a minimum of variation in fiber diameter and length, is desirable. Uniform fleeces require less sorting into different use categories. Thus, they are preferred by the woolen mills.

Uniformity of fineness in different body areas plays an important part in determining the grade of the fleece. An ideal fleece should have the same degree of fineness in every part. Even though this ideal fleece has not been found, the fleece that most closely approaches this ideal is the kind desired. A fleece that is uniform in fineness receives the benefit of a higher grade on the wool market, in contrast to a fleece that may be even finer on the shoulders but much coarser in the rear quarters. The finest wool is found on the shoulders and sides and the coarsest wool on the thighs.

Uniformity of fineness also refers to uniformity of the fibers from base to tip and uniformity among fibers in any chosen lock. These two factors are considered only in the close grading of fleeces and seldom enter into the placing of fine-wool sheep, except in close classes.

Character in wool is those characteristics that denote a distinct and even crimp. A fleece with character possesses a distinct, impressive crimp that is easy to see. Many fleeces are fine but lack depth of crimp — the yolk is unevenly distributed and the tips of the wool have a fuzzy appearance. Because wool of this type lacks character, it is generally discounted. Wool with character is soft and springy to the touch and shows a marked degree of elasticity under pressure. Fleeces that lack character are sometimes designated as "frowsy" or "harsh."

A distinct and even crimp is desirable throughout the entire fiber length. Parallel arrangements of fibers and locks of wool, plus large or "bold" locks, also contribute to the character of wool. These character traits improve the general appearance, or eye appeal, of fleeces and the spinning qualities for manufacturing.

Yolk is yellow wool fat that accumulates on the wool fibers. The shade of yellow varies. The amount and distribution of yolk are factors in determining the condition of a fleece. Yolk acts as a protective agent, making the wool more resistant to water and snow. It protects the individual fibers from injury, which is common to dry fleeces. A medium amount of light yolk evenly distributed from the base to the tip of the wool contributes to the wool character. A reasonable amount of yolk imparts luster and strength to the wool. An excessive amount of yolk is undesirable because it results in a fleece with heavy *shrinkage,* which is the loss in weight when grease wool is scoured.

The clean fleece yield is the percentage of clean wool remaining after all foreign material has been removed by washing and scouring. Shrinkage and price of wool vary with the content of grease, tags or dung locks, dirt, sand, burrs, chaff, straw, seeds, etc.

Shrinkage estimates are reasonably accurate if they are made by a person of considerable experience who visually examines and handles it. Core sampling and scouring tests permit more precise determination of clean-wool yield.

Purity in a fleece is the absence of hair, kemp, black fibers or dark fibers. Kemp is an abnormal, coarse fiber that is white and brittle. Hair is most frequently found on the wrinkles of fine-wool sheep, dark fibers are most often found in the fleece of the dark-faced down breeds and coarse hair occurs repeatedly on long-wool breeds and crossbreds. All these impurities are serious faults, and a fleece that contains any of them should be discounted heavily. Because impure characteristics are inherited, watch closely for them when you are selecting breeding stock.

Soundness in fleece refers to staple wool — wool with strong fibers and no weak spots. Two kinds of defects may cause weak fibers — namely, tender wool, or wool that is weak throughout the fiber, and wool that has a break or a weak spot. When selecting breeding sheep, do not place too much emphasis on a weak spot in the wool, for this weakness may not be the fault of the sheep. A weak place in wool fibers may be caused by a sudden change in weather or feed; lack of feed during severe weather; sickness, especially with fever; and overfeeding. Tender wool, which is due to weakness throughout the fiber, meets with disfavor in the selection of sheep.

Unsoundness and "second cuts" are the principal items contributing to wastiness. Wool fibers should be strong and elastic. If they possess a definite weak, or tender, spot, the fibers will break or pull apart in processing, resulting in excessive noilage. Healthy, properly fed sheep usually produce sound wool. "Second cuts" are the result of improper shearing.

Foreign matter is objectionable. A fleece is in good condition when it is as free as possible from sand, dirt, chaff, heavy tags, burrs, seeds, sweat locks, etc. The condition of a fleece depends primarily on the care of the sheep. Therefore, a lot of foreign matter does not necessarily indicate an inferior covering of wool. However, under the same conditions, a tight, dense fleece will contain the least amount of foreign matter. Wool with an excessive amount of yolk will collect the most sand and dirt.

Good color in a fleece is indicated by bright wool, free from a yellowish tint, and the absence of off-colored fibers, which impart a greyish tinge to the wool. The color of much so-called white wool is actually light cream. Good color is commonly referred to as brightness in fine wools and luster in longer, coarser wools. Both brightness and luster add to the character of the wool and increase its value. A grey, brown or black fleece is not suitable for the manufacture of white and delicate-colored fabrics and, therefore, is shifted into a lower price bracket. As long as a fleece shows brightness or luster with a cream or lighter shade, the color is acceptable.

Bright white- to cream-colored fleeces are most desirable. Grey or dull-colored fleeces are penalized, particularly if they include excessive amounts of black fibers. These dark or colored fibers, along with hair and kemp, affect the "purity" of wool.

Color variation is designated by choice, bright, semi-bright and stained. Bright white fleeces tend to be cleaner as well as more attractive.

The *covering* of fleece over the sheep's body determines in part the pounds of wool produced. There should be a good covering of wool over the belly, with the wool carried down well on the sides and to the knees and hocks. A covering of wool is desired over the head in Rambouillets and Merinos, as long as it does not interfere with their ability to see. Wool blindness is a common occurrence in range sheep. Thus, it is preferable to have enough bareness around the eyes to prevent wool blindness. Wool on the face, ears and below the knees and hocks is of no advantage on sheep intended for the range. Research studies show that open-faced Rambouillet ewes, when compared with wool-blind Rambouillet ewes of the same quality under range conditions, excel in lamb production by an average of about 10 pounds at weaning time.

UNDERSTANDING THE INTERNAL STRUCTURE OF SHEEP

Anatomy

A knowledge of the skeletal structure and internal organs of sheep, coupled with a practical knowledge of ideal types, will give you a fundamental basis for sheep selection and evaluation. An understanding of the internal structure of sheep adds depth and character to your knowledge of livestock.

Skeleton

The skeleton serves as a frame structure for the attachment of muscles and also as a protection for the vital internal organs. The shape, size, number and placement of the bones influence the general appearance or body type of the animal. The flat bones of the head contribute to the shape of the head and serve as a protection for the soft tissues. The bony regions that exert the most influence on the straight, strong, correct lines of the body are the vertebrae. The size and number of the vertebrae vary, including 7 cervical, 13 dorsal or thoracic, 6 or 7 lumbar, 4 or 5 sacral (fused into one bone) and 23 or 24 coccygeal (Fig. 2-53). The number of vertebrae present in each region, as well as the length and placement of the vertebrae, will have an influence, respectively, on the length of the neck, back, loin, rump and tail. Of course, due to the docking practice, the coccygeal, or tail, vertebrae are of little importance in the selection of sheep.

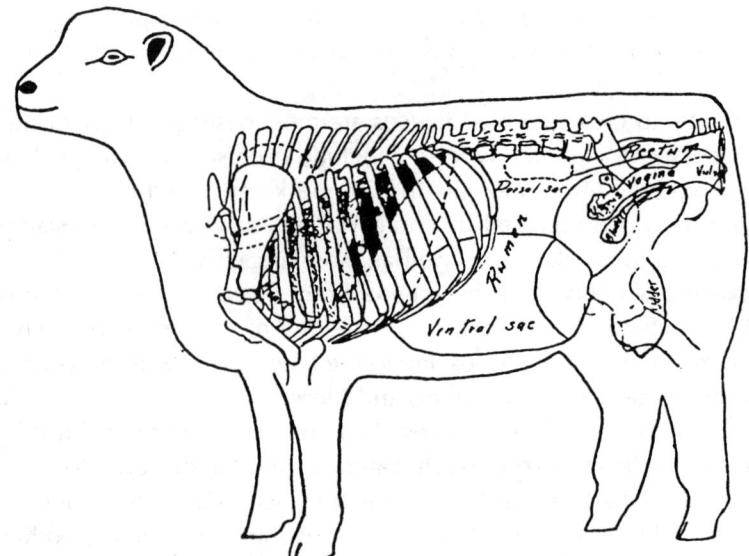

Fig. 2-53. Anatomy of the sheep (left side).

The size, shape and relationship of the sacral, pelvic and coccygeal bones exert considerable influence on the length and levelness of the rump. A short, droopy rump is a serious defect in sheep, for it is usually caused by faulty bone structure and cannot be altered materially by feeding. If the individual is short, droopy and steep in the rump, it will not have the capacity or areas for development of the muscle mass or volume that is desired in this region. Also, ewes that have difficulty in delivering lambs are usually narrow and droopy in the rump. The proper size and articulation of the bones (ilium, ischium, pubis) in the pelvic region aid materially in decreasing trouble at lambing time.

There are 13 pairs of ribs attached to the thoracic vertebrae, 8 pairs connected to the bottom of the chest directly to the sternum or breastbone (sternal ribs) and 4 pairs attached by means of cartilage (asternal). The last pair of ribs are floating with the cartilage unattached. The length, spring and depth of the ribs affect the width, thickness and muscling over the ribs, back and loin and the depth of the fore-ribs and rear ribs. The skeletal structure in the chest and rib regions is a potent factor in determining the constitution of sheep. Extreme weakness in back of the shoulders, a pinched heart girth and a sharp, narrow appearance over the top of the shoulders and ribs are the result of a combination of factors that include skeletal structure and muscle development.

Many bones make up the forelegs and rear legs, but there must be adequate length and size (substance) in the long bones, resulting in an animal that stands off the ground on some length and size of leg. On the average, the bigger, longer, heavier-boned animal will grow faster, utilize feed more efficiently and develop more total muscle in a given period of time than the small, short, light-boned animal. The hock and knee joints should be clean, strong and properly placed. The big, stout, rugged, heavy-muscled sheep has adequate length of leg and ample size bone and legs that are articulated with strong joints and set out on the corners of the body, which makes for very correct underpinning and soundness of structure. The pastern joints should be strong and correctly sloped, and the animal should have sound feet of adequate size to support its mature body weight. Undesirable feet and legs, crooked hock joints and weak pasterns are discriminated against, especially in the selection of rams.

Internal Organs

Sheep are ruminants; therefore, they have the same internal make-up as cattle. In order to provide ample room for the proper development and function of the chest organs, it is very important that the shoulder and chest regions of sheep be spacious and well-developed.

The skeletal system of the forequarter surrounds a group of vital organs — namely, the heart, the lungs, the endocrine glands and the forepart of the digestive system. These organs constitute in part those characteristics that contribute to the vigor and hardiness of a ewe or ram. Broad-chested ewes and rams with adequate spring of fore-ribs and wide chest floors usually have stronger heart and lung systems. Narrow-chested, flat-ribbed sheep usually do not have the mutton, feeding and reproducing qualities that are desired by breeders, feeders and packers.

Breeders insist on roomy middles and wide, straight, long rumps in ewes, for they know that these types digest their feed more efficiently and reproduce more easily. The digestive organs must have room to function. Sheep live primarily on roughages, and their ability to utilize large amounts of roughage feeds and to convert them into lamb or marketable meat products determines their economic value.

A thorough study of Fig. 2-53 shows the definite relationship between the internal and the external structures of sheep.

WHOLESALE CUTS OF THE LAMB CARCASS

Fig. 2-54 shows the location of the wholesale cuts of lamb on the live animal. It is important to have a complete understanding of the location and relative importance of the various cuts from both a packer's and a retailer's viewpoint. This information applies particularly to the *placing of market classes.* The most important wholesale cuts of lamb are the leg, loin (including flank) and rack, corresponding, respectively, on the live animal to the leg, loin, back and ribs. These three cuts include 61 percent of the carcass weight (of the average lamb carcass) and represent about 90 percent of the carcass value (Table 2-3). Therefore, these regions should carry considerable weight in the placing of a class of fat lambs. The wholesale cuts of lesser value are the shoulder, breast and flank, corresponding on the live animal with the neck, shoulder, chest floor, foreshank, foreflank and rear flank. The shoulder and breast are the most valuable of the cheaper cuts; but in relation to the rest of the carcass, they are relatively unimportant. Although the shoulder and breast regions make up 39 percent of the carcass weight, they contribute only 11 percent to the wholesale value.

The values placed on the various parts of the lamb carcass are approximations of the relative importance of various parts of market sheep, but these values are only

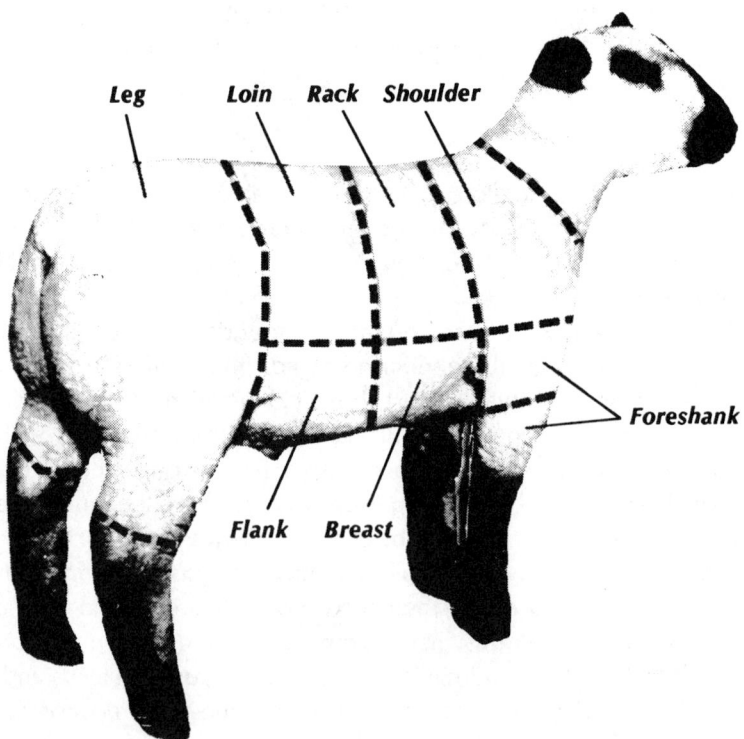

Fig. 2-54. Location of wholesale lamb cuts on live animal.

TABLE 2-3. Relative Importance of Wholesale Cuts on a Lamb Carcass

Cut	Carcass Weight	Wholesale Value
	(%)	*(%)*
Hindsaddle		
Leg	33	38
Loin (including flank)	17	28
Foresaddle		
Rack	11	23
Breast and shoulder (including shanks)	39	11
	100	100

averages and should not be accepted as absolute. On the other hand, the forequarters in breeding stock more nearly approach the value of the rear quarters because constitution, breed character and balance must be given ample consideration when ewes and rams are selected.

EVALUATING LAMB CARCASSES

It is important in judging lambs to be aware of the factors that determine the value of the carcass. Age determines if the carcass is considered lamb or mutton and can be ascertained in the live animal by looking at the teeth. Fig. 2-55 is a guide to the age of sheep. Breeding animals must have sound mouths and have all their teeth for grazing and eating. if some of the teeth are missing, they are referred to as having "broken mouths," and if all the teeth are missing, they are called "gummers."

The carcass merit should be the main concern in the evaluation of live sheep. However, slaughter buyers and producers cannot ignore the contributing value of the pelt and wool. Recently, an industry-wide planning committee outlined the following goals for the "consumer-preferred" lamb carcass.

100 – 120 pounds live in weight lamb[2]
Quality: Choice[3]

Meatiness:
1. Loin eye: Minimum of 3 square inches per 50-pound carcass
2. Leg: Wide, deep and heavily muscled
3. Percentage of trimmed preferred cuts (leg, loin, rack): 70 percent of carcass weight

Fat covering: Minimum of 0.2 and maximum of 0.3 of an inch at the 12th rib.

---

[2]Normally lambs fitting this description should yield 49–51 percent.
[3]"Choice Quality" is defined as:
   a. Quality characteristics of the present Choice grade.
   b. Firmness of meat for good shelf life and good shipability.
   c. Desirable color of lean.

New-born lamb
No teeth may be present, although sometimes the two pinchers and the two first intermediates are pressing through the gums or are even cut through.

12-15 months
Temporary pinchers are replaced by the two permanent ones.

3 months
A full set of completely developed temporary incisor teeth are present.

2 years
First temporary intermediates are replaced by permanent teeth.

3 years
Second temporary intermediates are replaced by permanent teeth.

4 years
Two temporary corner incisors are replaced by permanent teeth. Animal has a "full mouth."

After 4 years
After the sheep has a solid mouth, it is impossible to tell the exact age. With more advanced age, the teeth merely wear down and spread apart, and the degree of wearing and spreading is an indication of age. The normal number of teeth may be retained until the 8th or 9th year; but often some are lost after about the 5th or 6th year, resulting in a "broken mouth." When most of the teeth have disappeared, animals are known as "gummers."

Fig. 2-55. Guide to determining the age of sheep by their teeth. (Courtesy, Purdue University)

The break joint is the most reliable indicator of age. In young lambs, the forefeet are taken off at the break joint or, as it is sometimes called, "the lamb joint." The break joint has four well-defined ridges, which are smooth, moistened and red. As lambs approach the yearling stage, the bones become harder and white. The break joint retains some of the saw-tooth effect but is harder and more porous. Ewes tend to ossify the break joint earlier than do wethers. When the mutton stage is reached, the break joint cannot be made and the forefeet must be taken off at the round or spool joint, which is just below the true break joint (see Fig. 2-56).

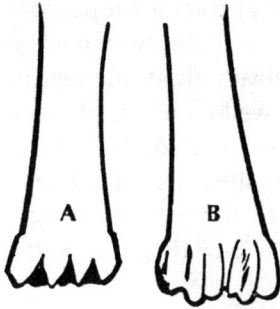

Fig. 2-56. Break-joint (lamb joint) (A) from a lamb and smooth joint (spool joint) (B) from a mature sheep. (Courtesy, Purdue University)

LAMB CARCASS EVALUATION TECHNIQUES

Subjective Carcass Evaluation

Classification

Classification of ovine carcasses is one of the most difficult phases of subjective evaluation, as there is often some overlapping in the development of the class-determining factors. The final judging of class must represent a careful composite evaluation of all factors. The following definitions or descriptions of typical carcasses of each of these three classes may be useful in making evaluations.

1. *Lamb.* — Carcasses always exhibit the characteristic break joint on their front shanks, and these are usually moist and fairly red, with well-defined ridges. They tend to have slightly wide and moderately flat rib bones and a light red color and fine texture of lean (see Fig. 2-56).
2. *Yearling mutton.* — Carcasses may have either break joints or spool joints on their front shanks. They have moderately wide rib bones, which tend to be flat, slightly dark red, and slightly coarse-textured.
3. *Mutton.* — Carcasses always have spool joints on their front shanks. They have wide, flat rib bones and a dark red color and coarse texture of lean (see Fig. 2-56).

Grading

The federal grade standards for the three classes of ovine carcasses are applied without specific sex identification. However, carcasses that show the heavy shoulders and thick necks typical of ram carcasses are discounted in grade in accordance with the extent of the development. Such discounts may vary from less than half a grade in carcasses from young lambs in which the characteristics are barely noticeable to as much as two full grades in carcasses from mature rams in which such characteristics are very pronounced.

Lamb and mutton are graded on a composite evaluation of two major factors — *conformation* and *quality.*

Conformation. — Conformation refers to the general carcass proportions and to the ratio of muscle to bone. The ideal lamb carcass is relatively short, thick, compact and heavily muscled throughout. The shanks are relatively short; the legs deep and full; the loin broad and thick; and the rack full and well-rounded, blending into a smooth but thick, full shoulder. The neck is short and plump. The entire carcass presents a well-balanced, symmetrical unit, with all parts blending together smoothly. Thickness of muscling, width and general development of the leg, loin and rack are of greatest importance, for these are the areas providing the most valuable cuts.

Quality. — Quality in lamb refers particularly to color, texture and firmness of the muscle and to color and firmness of fat. Marbling is also an important point of quality.

Since no freshly cut surface, such as the rib eye in beef, is available for inspection in lamb carcass judging, other points are used as indicators of quality within the lean muscle. Firmness may be judged from the firmness of the legs, back and flank. Firmness and texture of fat are determined by the general firmness in the flank or other areas of heavy fat deposits. Color is indicated by the color of the lean in the flank, the diaphragm and the breast, as well as the general color of the lean tissue in the thoracic region (chest cavity). Obviously the color of fat can be observed over the entire carcass. Marbling is best indicated by the feathering and overflow fat in the thoracic region and the evidence of marbling in the diaphragm, flank and breast.

The high-quality lamb carcass is covered uniformly with a white, firm fat and shows extensive feathering in the thoracic region and strong evidence of marbling in the diaphragm, sternum region and flank. The color of lean in the flank, diaphragm and thoracic region is a light pink. The entire carcass has a general firmness about it as indicated by a firm, dry flank, a hard leg and back and general rigidness. In contrast, the low-quality lamb may possess soft, oily fat, dark, coarse lean and a general lack of substance and rigidness throughout.

There are also distinguishable differences in age among those animals classified as lambs. Young lambs (5 – 6 months) show considerable redness in the break joint and have narrow, round ribs and shank bones that possess a considerable amount of red. In lambs approaching the yearling stage (10 – 12 months), the shank bones are whiter, the ribs are broader and flatter with little red color in them and the break joint itself is whiter and dryer than that found on a young lamb carcass.

There are distinguishable age differences in the characteristics of the muscle of lamb carcasses. In young lambs, the lean is generally bright pink and gives the impression of being moist and tender, whereas, the lean of those approaching the yearling stage is darker in color and generally more mature in appearance. Since younger lambs possess a milder flavor and are more tender than the more mature ones, they are much preferred to those approaching the yearling stage.

Balancing of grade factors. — The federal grades for each class of ovine carcasses are shown in Table 2-4. These grades represent many combinations of the grade factors of conformation and quality. Before this is done, however, it is necessary to understand the minimum requirements for quality indicators in relation to carcass maturity. Table 2-5 shows these requirements for the Prime, Choice and Good grades at two levels of maturity.

The determination of quality is actually made by giving equal consideration to the development of three factors: feathering between the ribs, fat streaking in the inside of the flank muscles and firmness of the fat and lean. All are considered in relation to the apparent evidences of maturity.

The federal grade standards indicate the rate and extent of compensation permitted for variations in the development of conformation and quality. To qualify for the Prime grade, a carcass must possess the minimum qualifications for quality regardless of the extent that its conformation may exceed the minimum requirements for Prime. In all the other grades, including Choice, a carcass that has conformation equal to the midpoint of the grade may have evidence of quality equivalent to the minimum for the upper third of the next lower grade and remain eligible for the higher grade. In all grades, a development of quality that is superior to the minimum specified for the grade may compensate for a development of conformation that is inferior to the minimum specified for the grade. In the Prime and Choice grades, this compensation is on an equal basis. That is, half a grade of superior quality compensates for half a grade of inferior conformation. In the Good and lower grades, half a grade of superior quality compensates for only a third of a grade of inferior conformation.

TABLE 2-4. Grades of Ovine Carcasses Within Classes[1]

Lamb	Yearling Mutton	Mutton
Prime	Prime	Choice
Choice	Choice	Good
Good	Good	Utility
Utility	Utility	Cull
Cull	Cull	

[1]From *USDA Official Yield Grade Standards for Lamb, Yearling Mutton and Mutton.*

TABLE 2-5. Minimum Requirements for Feathering, Fat Streakings in the Flank and Firmness for Lamb Carcasses by Grade and Maturity[1]

Grade	A—Young		
	Feathering	Fat Streaking	Firmness
Prime			
High	Moderate	Modest	Moderately full and firm
Average	Modest	Small	
Low	Modest	Small	Tend to be moderately full and firm
Choice			
High	Small	Slight	Slightly full and firm
Average			
Low	Slight	Traces	Tend to be slightly full and firm
Good			
High			
Average			
Low	Traces	Almost devoid	Slightly thin and soft
Grade	B—Mature		
	Feathering	Fat Streaking	Firmness
Prime			
High	Slightly abundant	Moderate	Tend to be full and firm
Average			
Low	Moderate	Modest	Moderately full and firm
Choice			
High	Modest	Small	Tend to be moderately full and firm
Average			
Low	Small	Slight	Slightly full and firm
Good			
High			
Average			
Low	Slight	Traces	Tend to be slightly full and firm

[1]From *USDA Official Yield Grade Standards for Lamb, Yearling Mutton and Mutton.*

OBJECTIVE CARCASS EVALUATION

Area of Longissimus Dorsi

To observe the loin muscle (longissimus dorsi) in sheep, the carcass must be broken by a transverse cut midway between the 12th and 13th ribs at a right angle to the length (longitudinal axis) of the carcass. A saw-cut is made through any ribs crossing the cut axis diagonally. Only the l. dorsi muscle is included in the loin-eye area measurement.

Depth of Fat

Depth of fat is the average of three measurements over the l. dorsi at one-fourth, one-half, and three-fourths the length from the chine end and perpendicular to an axis through the center of the l. dorsi. In carcasses with a normal distribution of external fat, this factor can be evaluated in terms of the actual thickness of fat over the center of the right and left rib-eye muscles, between the 12th and 13th ribs. This measuring procedure has been readily accepted and is now the most commonly used procedure for calculating fat thickness on lamb carcasses. This measurement may be adjusted, as necessary, to reflect unusual amounts of fat on other parts of the carcass, especially lower body wall or lower rib fat thickness.

In a carcass that is relatively fatter than that indicated by the actual fat thickness over the rib eye, the measurement is adjusted upward. Conversely, in a carcass that is relatively less fat than that indicated by the actual fat thickness over the rib eye, the measurement is adjusted downward. In many carcasses no adjustment is necessary; however, an adjustment of as much as 0.1 inch is not uncommon. Adjustments are made in increments of 0.05 inch of fat thickness. In some cases, a greater adjustment may be necessary. As the amount of external fat increases, the percentage of retail cuts decreases, and each 0.15 inch change in adjusted fat thickness over the rib eye changes the yield grade by one full grade.

Percent of Kidney and Pelvic Fat

The percent of kidney and pelvic fat is the weight of kidney and kidney fat from the loin and pelvic fat from the leg, removed as closely as possible and divided by chilled carcass weight. For carcass contests, percent kidney and pelvic fat may be estimated by qualified individuals.

Cutability or Yield Grades for Lamb Carcasses

The term "cutability of carcass" or "USDA yield grade" refers to the amount of trimmed retail cuts in a lamb carcass (Table 2-6). The USDA yield grade standards were put into effect on March 1, 1969. They are in addition to the quality grading standards of Prime, Choice, Good, etc. Both grade standards are voluntary for the packer.

The consumer-preferred lamb combines thick, high-quality muscling with a min-

TABLE 2-6. Yield Grade Groupings and Percent Boneless Major Retail Cuts for Lamb Carcasses[1]

Yield Grade		Percent Boneless Major Cuts[2]
1 (most desirable)	=	47.3 and over
2	=	45.5 – 47.2
3	=	43.7 – 45.4
4	=	41.9 – 43.6
5 (least desirable)	=	Less than 41.9

[1]From *USDA Official Yield Grade Standards for Lamb, Yearling Mutton and Mutton.*

[2]Includes leg, loin, rack and shoulder.

imum of fat. The identification of such carcasses is needed to more accurately price lamb from producer to consumer and thus determine its true value.

Yesterday's marketing yardsticks of weight, dressing percentage and grade do not accurately determine today's value. Retail cutting tests by several universities and retailers have shown retail value differences of $10 per hundredweight for carcasses of the same grade and weight. The grades of Prime, Choice, etc., tell us the expected eating quality from a carcass, but they do not tell us the cutability (amount of muscling or trimmed retail cuts) expected from a carcass.

In the past, heavyweight lambs were automatically judged as being overfinished, and lightweight animals, correctly finished. Recent breeding, nutrition and management practices have changed this. A recent study sponsored by the America Sheep Producers Council shows that 65 percent of the heavy lambs are correctly finished. Heavy lamb carcasses are 55 pounds and over and are correctly finished with 0.25 inch or less rib-eye fat at the 12th rib.

Today, lambs with the highest dressing percentage usually have excess finish and kidney fat and a lower percentage of trimmed retail cuts.

Bone varies little between carcasses, so it generally follows that the greater the percentage of fat, the lesser the percentage of muscle or trimmed retail cuts.

An accurate measure of the amount of trimmed retail cuts is needed since carcass and live value are determined by (1) the quality of red meat or trimmed retail cuts; and (2) the amount of red meat or trimmed retail cuts.

Cutability provides the yardstick necessary to measure accurately the amount of trimmed retail cuts in a lamb. Up to now, this type of measurement has been missing. This cutability concept separates the trim, heavy-muscled lambs from the wasty, light-muscled lambs and variations between the extremes.

The USDA Lamb Yield Grade standards can be used to measure carcass cutability. This involves estimates of (1) leg conformation (High Prime — 15, Low Prime — 14, etc.); (2) percent kidney and pelvic fat; and (3) adjusted fat thickness over the rib eye in inches.

Leg conformation is an estimate of total carcass muscle. Kidney and pelvic fat and adjusted fat thickness are estimates of total fat trim.

Table 2-7 is an example of a form used for evaluating live sheep and later their carcasses. The information contained in Table 2-8 provides a quick method for determining the USDA yield grade for live sheep and later their carcasses.

TABLE 2-7. Method for Determining USDA Yield Grade for Ovine Carcasses[1]

1. Make the following estimates or measurements on the ovine carcass.

   a. **Fat thickness** — This is a single measurement over the center of the rib-eye muscle, perpendicular to the outside surface between the 12th and 13th ribs. (If the carcass is not ribbed, this must be done by probing.) Measure over both rib eyes and take an average. This measurement may be adjusted, if needed, to reflect unusual fat distribution on the carcass.

   b. **Percent kidney and pelvic fat** — Estimate to the nearest 0.5 percent and express as a percent of carcass weight.

   c. **Leg conformation score** — Estimate the conformation of the leg to the nearest one-third of a grade, e.g., Low Choice, Average Prime, etc.

2. Determine the **preliminary yield grade** based on fat thickness according to the following schedule:

Fat Thickness over Rib Eye	Preliminary Yield Grade
0.00 in.	2
0.15	3
0.30	4
0.45	5
0.60	6

3. Determine the **final yield grade** by adjusting the preliminary yield grade, as necessary, for variations in kidney and pelvic fat from 3.5 percent and variations in leg conformation grade from Average Choice.

   a. **Adjustment for percent kidney and pelvic fat** — For each 1 percent of kidney and pelvic fat more than 3.5 percent, add 0.25 to the preliminary yield grade; for each 1 percent of kidney and pelvic fat less than 3.5 percent, subtract 0.25 from the preliminary yield grade.

   b. **Adjustment for leg conformation grade** — For each one-third grade that conformation exceeds Average Choice, subtract 0.05 from the preliminary yield grade; for each one-third grade that conformation is less than Average Choice, add 0.05 to the preliminary yield grade.

   c. **Adjustment of fat thickness measurement for variations from normal fat distribution** — For determining adjustment of fat thickness for variations from normal distribution, if any, give particular attention to the amount of external fat in such areas as the sirloin, shoulder, breast, flank and cod or udder and to the amount of intermuscular fat in the body wall—all in relation to the thickness of fat over the rib-eye muscle. Body wall thickness may be measured to aid in evaluating intermuscular body wall fatness.

   Body wall thickness is measured 5 inches laterally from the middle of the backbone between the 12th and 13th ribs. If the carcass is fatter over other parts than the fat measurement indicates, adjust the measurement upward. Adjust the measurement downward if less fat is present over the parts than the fat measurement indicates. Normal values are as follows:

Fat Thickness	Body Wall Thickness
0.1 in.	0.5 in.
0.2	0.7
0.3	0.9
0.4	1.1
0.5	1.3

[1]Adapted from *USDA Official Yield Grade Standards for Lamb, Yearling Mutton and Mutton.*

TABLE 2-8. Sheep—Live Animal

Animal Identification	Live Weight		Dressing Percentage		Leg Score		Loin-Eye Area	
	Estimate	Actual	Estimate	Actual	Estimate	Actual	Estimate	Actual

EXAMPLE FOR CALCULATING LIVE PRICE
FOR LAMBS

The following is an example for calculating the live price for lambs:

1. Estimate dressing percentage, carcass quality grade and yield grade (cutability score).

2. For every 0.1 grade difference between estimated and base carcass yield grade, add to or subtract from the current base carcass price per hundredweight $0.25.

3. Provide carcass price per hundredweight, based on carcass grade and weight. For example:

Weight	Prime	Choice	Good	Utility
Under 55 lbs.	$120	$119	$113	$108
Over 55 lbs.	$116	$115	$111	$106

4. Use a standard dressing percentage of 50 percent to calculate the final live price.

and Carcass Evaluation Sheet

Average Backfat Thickness		Percent Kidney and Pelvic Fat		USDA Quality Carcass Grade		USDA Yield Grade or Cutability Percentage	
Estimate	Actual	Estimate	Actual	Estimate	Actual	Estimate	Actual

5. EXAMPLE. If the live weight is 100 pounds and the dressing percentage (standard dressing percentage) is 50, then the carcass weight = 50 pounds. If the estimated yield grade is 3.8 (and the base is 4.0) and the grade is Average Choice, then the estimated carcass value per hundredweight = $120.00 + (0.25) (4.0 − 3.8) = $120.50. Estimated live value per hundredweight = $120.50 (0.50 standard dressing percentage) = $60.20 (rounded to nearest $0.10). If the actual live value per hundredweight = $58.80, then the contestant's score = 60 − ($60.20 − $58.80)/0.05 (1 point of) = 60 − 8 = 52.

6. Assume that the cost of slaughter offsets value of by-products.

7. Express your final estimated price for the live lambs on a $ per hundredweight *live weight basis.*

8. Calculate the official price on a live weight basis.

The on-hoof difference between each yield grade equals approximately $1.25 per hundredweight. The wholesale carcass difference between each yield grade equals approximately $2.50 per hundredweight.

SUMMARY FOR HANDLING, PLACING AND TAKING NOTES ON A CLASS OF SHEEP

Check the fineness, length, density, uniformity, character, yolk, purity, and condition of the fleece on the shoulders, sides and breech. Examine the fleece on the shoulders, sides and breech by using both hands in a flat position to part the wool and keeping the fingers together. Part the wool naturally and easily so the natural crimp in the wool fibers will not be disturbed. Never open the fleece on the top of a sheep or stick your fingers in the wool. On fine-wool sheep, you should spend considerable time studying the fleece, but on medium-wool and long-wool breeds, you should spend very little observing the fleece. When looking at the fleece, look for black fibers and observe the color of the skin. Be sure to check to see that hornless breeds do not have scurs.

In judging market lambs, devote more time to observing the degree of finish, strength of back, width of loin, size of leg and general trimness and firmness of the lamb. Condition is *very important* in market class placings.

The points concerning conformation, meatiness, muscling and finish of market lambs are just as important when you are evaluating sheep for breeding purposes. Additional items to consider or emphasize to a greater extent for breeding sheep are breed and sex character about the head, breed and type, strength and straightness of feet and legs, fleece or wool value, straightness of lines, balance, symmetry and style.

After carefully feeling over all the animals, step back from the class to assemble and weigh all the facts. In your mind segregate the animals into a top pair and a bottom pair or some other possible combination. Spend the remainder of the time checking on the points that are "hazy" in your mind and in placing the close pairs. Mentally outline the outstanding good and bad points of each animal and take a complete set of notes that will help you recall the class in great detail. Before you leave the class, review the points on each animal in both a descriptive and a comparative manner to determine if you are capable of presenting clear-cut, exact and effective reasons for your final placing.

SHEEP JUDGING TERMINOLOGY

Being able to use a wide variety of terms is an indispensable factor in describing sheep accurately and effectively. The following may serve as a stimulus for you to improve your vocabulary. Instead of using these items in a memorized fashion, develop your own, with the aid of the following list, for discussing sheep.

General Expressions

1. A big, growthy, stretchy, straight-legged, heavy-boned lamb carrying more muscling down the top and through the leg than any other lamb in the class.

2. Longer-bodied, stronger-topped and more correctly finished over the shoulders, ribs, back and loin than No. 2.
3. A tighter-framed, cleaner-fronted, trimmer-middled, harder, firmer-fleshed wether that will hang the trimmest, most correctly finished, meatiest carcass in the class.
4. A typey, well-balanced, high-quality, roomy-middled ewe.
5. A thick-topped, broody, roomy, full-chested individual with excellent bone and ruggedness.
6. A ewe lamb showing outstanding breed character, balance and tremendous quality.
7. A heavy-muscled ewe that has more natural thickness down her top and a longer, leveler rump than No. 4.
8. A big, rugged, long-bodied, square-docked, dense-fleeced ewe.
9. A very heavy-boned, strong-headed ram, with more breed type and sex character than any other animal in the class.
10. A big, stout, rugged, heavy-boned ram.
11. A smooth, high-quality, well-balanced ewe that has the finest, densest fleece in the class.
12. A big, growthy ram with good substance but a little too coarse about the head and shoulders.
13. A square-ended ram, exhibiting more style and breed character than No. 3.
14. Structurally the most correct, best-balanced, meatiest, thickest-topped lamb in the class.
15. A breedy-headed lamb that lacks the natural thickness and muscling of the other animals in the class.
16. A narrow, upstanding, easy-topped, peaked-rumped wether.
17. A coarse, upstanding, narrow-chested, open-fleeced, light-legged ewe that is definitely lacking in constitution and breed type.
18. A short-bodied, deep-sided, overdone, wasty-middled, light-boned lamb.
19. A short, dumpy lamb.
20. A straight-lined ewe with excellent breed type.
21. A ram of excellent type that possesses more balance than any other animal in the class.
22. A wether with more spread and thickness down his top, being longer and fuller at the dock and thicker, plumper, meatier through the leg than No. 1.
23. The lowest-set, lightest-boned ewe in the class.
24. More spread over the top with more spring to the rear rib and more length of rump than No. 1.
25. An "off-type," open-shouldered, narrow-docked ewe.
26. A wide-chested, rugged-fronted ram with his legs set squarely out of the corners.
27. A yearling ram with great promise.
28. A ram that carries a good stylish head, but lacks character of fleece and muscling and meatiness through the leg.

29. Superior muscling and thickness.
30. Tremendous balance and excellent quality.
31. A ram that has plenty of ruggedness but lacks the type and quality of the ram placed above him.
32. Very stylish, with a dense fleece and a smooth, firm fat cover.
33. A massive individual.
34. A clean-fronted, trim-middled, firm-fleshed lamb.
35. Up off the ground and smoother than No. 1, but he placed second because he lacks the thickness, muscle development and ruggedness of No. 1.
36. Excellent length but lacks substance of bone.
37. Extremely close top pair of ewes.
38. A short-bodied, tightly wound, shoved-together, overfinished lamb.
39. Lacking size but is smooth and correct.
40. A ram that is more symmetrical throughout than any other individual in the class.
41. A big, rugged wether that is up in the air, heavily muscled, and firmly covered.
42. Outstanding fleece and breed type easily surpasses the other rams in the class.
43. Excellent breed character but lacks the fleece quality of the other ewes in the class.
44. Large, roomy middle and wide chest that indicate an excellent breeding ewe.
45. A stud ram that is the biggest, most rugged, most massive individual in the class.
46. Heavily muscled leg, lines and excellent fleece character.
47. A loose-framed, coarse-shouldered ram.
48. A short-rumped, light-legged wether that lacks the stretch, balance and trimness to place any higher in this class today.
49. A large, roomy ewe with a strong back and good bone.
50. More breed type and sex character about the head and neck than any other ewe in the class.
51. A meatier-topped, nicer-handling, trimmer-middled wether.
52. A U-necked, open-shouldered ewe who is weak in the loin and lacks strength of the pasterns when compared to the other animals in the class.
53. An overshot jaw or mouth (parrot-mouth or parrot-beaked) (Fig. 2-57).
54. An undershot jaw or mouth (Fig. 2-57).
55. "Off" in the mouth (Fig. 2-57).
56. Good, sound mouth. True, even bite.
57. More genetic potential to produce with larger, meatier offspring.
58. Bolder-chested, wider-sprung, deeper-ribbed individual with more total reproductive capacity.
59. Lacking the production potential of the top three individuals in terms of broodiness and capacity.
60. More growth potential to develop into a bigger, more rugged, productive flock matron.
61. More genetic potential to produce the right kind.

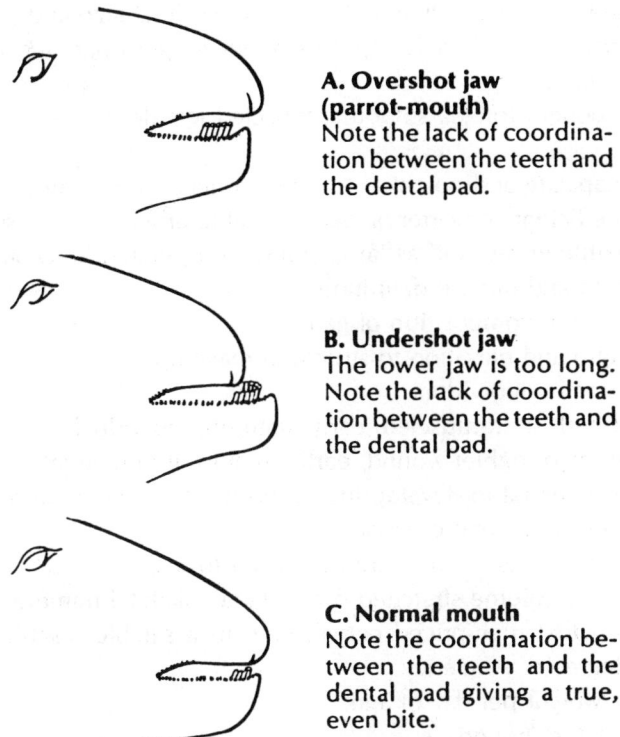

A. Overshot jaw (parrot-mouth)
Note the lack of coordination between the teeth and the dental pad.

B. Undershot jaw
The lower jaw is too long. Note the lack of coordination between the teeth and the dental pad.

C. Normal mouth
Note the coordination between the teeth and the dental pad giving a true, even bite.

Fig. 2-57. Sheep jaw configuration (Courtesy, Purdue University)

62. A ram that would add more size, scale and profit potential to that commercial flock.
63. Greater genetic potential to stamp his offspring with more growth, size and scale.
64. A ram that shows that extra growth potential in terms of developing into a more rugged, stouter, more modern individual that would be more apt to push forward that conventional herd.
65. Lacking size, scale and growth potential; would be the least useful when mature in terms of siring those larger, faster-growing offspring for that profit-minded breeder.
66. More genetic potential to produce the right kind.
67. Greater promise and future outcome.
68. Lacking the production potential of the top females.
69. Taller, longer-skeletoned, more structurally correct.
70. A thicker-topped, broodier, roomier, fuller-chested ewe.
71. A large, roomy-middled, high-capacity ewe.
72. A more stylish, cleaner-conditioned, better balanced individual that puts more good things together than any other individual in the class.
73. A taller, stretchier, framier individual that is jacked up taller on both ends on more length of leg.

74. Larger, growthier individual that has more skeletal size and framework.
75. A more progressive female in terms of her longer bone and longer, smoother muscle structure.
76. A longer-boned, longer, smoother-muscled female that is more progressive in her type.
77. Greater capacity and potential to produce larger offspring.
78. The least efficient converter of feed to salable product with less profit potential.
79. More flexible in his kind as far as going on to higher and / or heavier weights.
80. More width and muscle definition.
81. More desirable composition of gain.
82. More length and elevation to structural make-up.
83. A larger-framed individual.
84. Round-boned, short-muscled, early-maturing individual.
85. Smaller-framed, tighter-wound, earlier-maturing individual.
86. The least potential to develop into a productive flock matron (ewe).
87. (Least or most) genetic promise.
88. Longer, looser muscle structure from front to rear.
89. More muscle volume stretched over a larger skeletal framework.
90. Converting that high-priced cereal grain to a salable, usable product with a greater degree of efficiency.
91. A greater weight per day of age.
92. Taller from the ground up.
93. Lacking growability, doability and profitability to all segments of the industry to merit a higher placing in the class today.
94. Possessing size, scale, frame and all growth factors.
95. A ewe that will lamb those higher-selling offspring for the profit-minded shepherd.
96. Performance and growability.
97. Faster-growing.
98. If the market so demanded.
99. Frame and skeleton.
100. Bigger or larger as he / she reaches maturity.
101. Greater muscle volume.
102. Thicker through the lower one-third of the body.
103. Overpowers contemporaries.
104. Profit potential.
105. More profitable to the progressive breeder.
106. The most flock improvement potential of any ewe in the class.
107. The most breed improvement capabilities of any ram in the class.

Head and Character

1. A lot of breed type and sex character about the head and ear.

2. A bold-headed, strong-jawed ram.
3. A bold-faced, strong-headed ram.
4. Has more Hampshire (or any other breed) type and character about the head and ear.
5. A breedier-headed ewe.
6. A broad-faced ram with outstanding sex character.
7. Blacker about the points than any other ewe in the class (applies to Hampshires and Suffolks).
8. A head and neck that blend smoothly into the shoulder.
9. More open-faced.
10. More feminine head (ewes).
11. Stronger head (especially rams).
12. Wool blind.
13. Coarse-headed.
14. A plain head.
15. A breedy head.
16. A narrow-faced ewe, lacking strength and character about the head.
17. An attractive head.
18. A stylish head and neck.
19. More refinement and quality through the head and neck.
20. A clean-cut but overrefined head.
21. A Roman nose.
22. A long, thin-necked ewe with a plain head.
23. A strong-jawed ram that is cleaner through the throat than No. 2.
24. A long, coarse, heavy ear that is not characteristic of the breed.
25. A sweet-headed ewe with a bright, alert eye and a small, sharp, correctly placed ear.
26. More breed type and character about the head and ear.
27. A darker, more pendulous set to the ear.
28. More strength and broodiness about the head.
29. Low-headed or heavy-headed.
30. Mature-headed.

Forequarters (Neck, Shoulders and Chest)

1. A shorter-necked ewe with more width through the chest floor and more spring through the fore-rib than any other ewe in the class.
2. A neck that blends in smoothly with the shoulder.
3. Smoother shoulders (more neatly laid in).
4. A long-necked ewe that is narrow-chested and very prominent at the point of the shoulders.
5. Smoother blending of neck and shoulders.
6. Rough (coarse) shoulders.

7. A U-necked, open-shouldered ewe.
8. A bolder, wider-fronted wether that is fuller behind the shoulders than any other lamb in the class.
9. A slim-necked lamb that is narrow behind the shoulders (pinched in the heart girth).
10. A bolder front and a wider set to his front legs, which would indicate that he has more constitution and capacity than any other individual in the class.
11. A wide-chested lamb that has more evidence of muscling through his shoulder and down into his arm and forearm region than any other lamb in the class.
12. More prominent in the muscling through his forearm than No. 2.
13. A wide chest floor.
14. A deep, wide-chested individual indicating constitution.
15. A wide-chested, bold-fronted ram.
16. Neater through his neck, wider through the chest and neater through the breast area than any other lamb in the class.
17. Prominent shoulders.
18. Breaks behind the shoulders.
19. Slack behind the shoulders.
20. Low behind the shoulders.
21. Too sharp over the shoulders.
22. Out at the shoulders, or wing-shouldered.
23. A well-sprung fore-rib.
24. Narrow in the fore-rib.
25. Pinched in the fore-rib.
26. The widest-fronted lamb in the class but too wasty through the throat and breast.
27. A narrow-made, open-shouldered, narrow-docked individual.
28. Unbalanced, U-necked individual.
29. A bolder-chested, wider-sprung, deeper-ribbed individual.
30. Tighter-shouldered.
31. More width, length and depth to her chest chamber and reproductive tract.
32. Powerful-fronted.
33. Short, thick neck.
34. Heavy-fronted.

Ribs, Back and Loin

1. A narrow-chested, flat-ribbed, weak-topped lamb.
2. A tighter-framed ewe with a bolder spring to the fore-rib and the rear rib and a thicker, meatier, heavier-muscled loin.
3. A wider-chested ewe that is especially meatier through the rack (ribs) and has more natural spread and thickness over the back than any other ewe in the class.
4. A lamb that is narrow through the rack (ribs), weak-topped and flat-loined.

5. A slack-framed, light-muscled lamb that is narrow over the back and through the loin.
6. A bigger, stretchier lamb that has greater arch (spring) of rib and that is especially wider and meatier down his top than No. 2.
7. Strength of top.
8. A lamb that is low in the top, is narrow through the fore-rib (pinched in the heart girth) and lacks the natural thickness and muscling exhibited by the other lambs in the class.
9. A straighter-topped lamb that has a bolder arch of rib than No. 3.
10. A ewe that is especially low in front and stands downhill.
11. Easy in the topline.
12. A strong back.
13. Low back of the shoulders and high in the loin, being uneven in the topline.
14. A straight topline.
15. Great spring of rib and strength to back.
16. More spread and thickness over the back and through the loin.
17. A great top and an outstanding spring of rib.
18. Shallow-ribbed.
19. Tucked up in the flanks and clean along the underline.
20. Lacking thickness and muscling through the loin.
21. Deep-ribbed.
22. Pinched in the loin.
23. A firm-topped, nice-handling lamb.
24. Cleaner over the top and tighter through the middle than No. 2.
25. Fish-backed (narrow-topped) lamb.
26. A symmetrical, well-balanced lamb that is nicely turned over the top and correct in finish.
27. A square-topped wether that shows evidence of excess condition over the top of the ribs, back and loin edge.
28. More correct in finish over the ribs, back and loin.
29. A lamb that is rough and not uniform in the finish over the ribs and is starting to roll at the loin edge.
30. Soft over the fore-rib and bare over the lower rear rib.
31. A hard, firm, correctly finished, good-handling lamb over the ribs, back and loin.
32. A wide-ribbed, harsh-handling lamb that lacks uniformity of thickness and strength of top.
33. Stretchier individual that has more length through that rib, loin and rump.
34. A leveler-topped, leveler-rumped individual.
35. Squarer-ribbed.

Rump and Leg

1. A longer-rumped lamb with a thicker, more bulging leg than No. 2.

2. Wider and meatier through the rump and especially thicker through the center portion of the leg.
3. A deeper, plumper, shapelier leg.
4. Wider and squarer at the dock, dropping down into the deepest, thickest, heaviest-muscled leg in the class.
5. Longer, leveler-rumped lamb with the pin bones set higher and wider than No. 3.
6. A short-rumped, narrow-docked, light-legged lamb.
7. A lamb that is narrow and tapering at the rump and low at the pins and stands narrow (close behind) on the rear legs, which would truly indicate a definite lack of muscle development.
8. A ewe that is droopy at the rump and has the narrowest, shallowest leg in the class.
9. The largest leg of lamb (or mutton) in the class.
10. A plump leg. A thick leg. A large leg. A bulging leg. A light leg.
11. A plump leg that lacks the firmness of the other lambs in the class.
12. Thick and firmly muscled in the leg.
13. A square-ended ram.
14. More level at the rump.
15. Peaked at the dock.
16. Pinched at the dock.
17. Long, level rump.
18. Patchy around the tail.
19. Firmer in the leg and cleaner and neater through the crotch.
20. A lamb that displays more meatiness and muscling through both the inner and outer legs, and is free of any excess finish in the crotch.
21. When viewed from the rear, a lamb that is especially heavier-muscled through the center of the leg (stifle muscle region).
22. A lamb that displays a deeper, thicker, plumper, more bulging leg as viewed from the side.
23. A lamb that is loose, soft and flabby through the leg.
24. Deep and full in the crotch (twist), indicating excessive fat deposits in this area.
25. A steep, short-rumped wether.
26. A short-rumped lamb that is patchy around the tailhead and carries an excessive amount of finish in the crotch area and at the base of the leg.
27. A deep, full flank that shakes and wiggles when he walks, which would indicate that the flank region is filled with large amounts of waste fat.
28. A wether that is tucked up a bit in the flank but is especially clean and trim in this area.
29. A thick, bulging, muscular stifle that flexes and expresses itself more prominently when the animal moves.
30. More evidence of muscling through the leg.

31. A lamb that is harder, firmer and trimmer about the base of the leg and through the crotch.
32. A lamb that is loose, soft and flabby through the leg and in the crotch.
33. The lightest-muscled individual that possesses the shallowest leg in the class.
34. More volume and dimension to the leg.
35. More muscle dimension through the upper, lower and center portion of the leg.
36. Leveler in the rump and squarer at the dock.
37. Higher at the dock, giving more depth to the leg.
38. Thick in inner and outer lower leg.
39. Higher set to the dock, thus giving more depth of leg.
40. Longer through the stifle and longer from hip to hock, higher tail setting, thus more total dimension to the leg.
41. Wider and squarer through the top of the rump and dock.
42. Muscling through the biceps and over the top of the rump.

Feet, Legs and Substance of Bone

1. A heavier-boned, straighter-legged, squarer-standing lamb than No. 2.
2. A rugged-boned lamb that stands straighter, squarer and stronger on his feet and legs than any other animal in the class.
3. A bigger, stouter, longer-legged, more upstanding lamb.
4. A short-legged lamb that is too close to the ground; needs more leg under her.
5. More ruggedly made.
6. Straighter (front or back) legs.
7. A light-boned lamb with a narrow set to the rear (front or both) legs.
8. A ewe that is weak on her legs and lacks strength of pasterns (weak on pasterns).
9. A small, fine-boned, shoved-together, light-muscled ewe.
10. A low-set, short-legged, light-boned ewe.
11. A rugged-boned ram.
12. Ample quantity and quality of bone for size.
13. Lacking quality and refinement of bone.
14. Narrow-set legs. Stands too close on (front, rear or both) legs.
15. Too coarse in the joints. Puffy and swollen hocks (joints).
16. Plenty of substance.
17. A smaller-boned ewe.
18. Cow-hocked (narrow at the hocks, or close at the hocks).
19. Sickle-hocked (too much angle or set to the hock).
20. Toed-out or toed-in — either front or rear legs, but usually front.
21. Close at knees (knock-kneed).
22. Back at the knees.
23. Over at the knees (buck knees or bucked knees).

24. Post-legged (straight-hocked or too straight-legged).
25. Stands up on toes.
26. A good set of feet and legs.
27. Crooked hind legs (or front). Straight hind legs (or front).
28. Stands on adequate bone for size.
29. Set too wide at the hocks.
30. A ram that rolls over on his front feet, which is especially noticeable at the walk.
31. Lacks substance of bone and is the most crooked-legged ewe in the class.
32. "Off" (incorrect) on her feet and legs.
33. The soundest feet and legs in the class.
34. Unsound set of feet and legs.
35. Plenty of clean-cut, hard bone with strong pasterns and a free and active walk.
36. Stands up on more rugged substance of bone.
37. A ram that stands wider on both ends, indicating more total dimension to the leg.
38. Incorrect in underpinning.
39. A ewe that stands down wider at her hocks.
40. Splay-fronted.
41. Splay-footed.

Finish

1. A trim-fronted lamb that is more correct in finish over the ribs, back and loin and displays more firmness through the leg than any other lamb in the class.
2. Harder, firmer-finished lamb that would hang the cleanest, trimmest carcass in the class.
3. A nicer-handling lamb that is more uniformly covered down the top and firmer in the leg.
4. A higher-quality, firmer, more uniformly finished lamb.
5. An overdone (overfinished), heavily covered lamb that is rough and rolling over the loin edge.
6. A lamb that is extremely soft in his finish down the top, being very soft through the leg and crotch.
7. A heavy-fronted, excessively finished ewe.
8. A thin, underfinished lamb.
9. A trimmer-finished (handling) lamb that is carrying a correct kind of finish, but is extremely wasty-fronted and lacks the quality and refinement of the other lambs in the class.
10. A neat, clean-fronted lamb that is bare (lacks cover) over the ribs, back and loin.
11. The fattest, most undesirable lamb in the class.
12. A correctly conditioned wether that is quite smooth in his finish.

13. A lamb with extraordinary handling qualities.
14. Too light in natural covering. A thin-fleshed lamb.
15. Roughly finished.
16. The thinnest lamb in the class.
17. Too soft in his finish.
18. Soft, flabby, overdone wether.
19. Not in as high condition (flesh) as the other ewe.
20. A smoothness of finish that is unsurpassed.
21. A firm touch to the back.
22. Neatly covered with quality finish.
23. An even distribution of finish.
24. Cleanest, hardest, firmest-handling, strongest-topped lamb in the class.
25. Firmer, harder-handling individual that possesses more thickness, spread and muscle tone all down the top and through all portions of the leg.
26. Less finish on lower fore-ribs and rear ribs.
27. Harder, firmer handling with more even distribution of fat cover (condition).
28. In a more desirable degree of condition for the breeding pasture.
29. More fit and true in her breeding condition.
30. Rough handling over the loin edge.
31. Soft handling over the lower ribs.
32. More uniform over the rib and loin edge (referring to fat cover).
33. More market-ready in conditioning.
34. Freer of waste from end to end (head to tail).

Quality and Style

1. Ample quality and style.
2. A more stylish, higher-quality, nicer-balanced ewe.
3. A plain, "off-type," low-quality ewe.
4. A typier ewe with more flash and eye appeal than any other animal in the class.
5. A ewe that is coarse about the head, ear, neck and bone, displaying the least quality of any ewe in the class.
6. A ewe that possesses more quality and refinement about the head, ear, pelt (hide) and bone than any other individual in the class.
7. A ewe whose refined head and neat joints indicate quality.
8. A stylish lamb with plenty of smoothness and quality.
9. Even, firm fleshing and trimness, which indicates quality.
10. Smooth, clean-cut and stylish.
11. A tight-framed, stylish, well-balanced, high-quality ewe.
12. A ram whose coarseness about the head, neck and bone indicates a lack of quality.
13. A quality ewe with a thin, pliable hide and pink skin.
14. A ram whose softness of flesh and blue skin show a lack of quality.
15. Refined head and quality wool.

16. Feminine refinement and unusual style.
17. A ewe whose alert eye, refined ear and clean-cut face denote extreme quality.
18. A bold-headed ram but somewhat coarse in his features.
19. A typier, more stylish ewe that is somewhat over-refined and lacking in ruggedness.
20. A refined horn that adds quality and style to his head.
21. A U-necked, open-shouldered ewe that has the least desirable side-view profile.
22. More reproductively sound in his testicular development and descension.
23. Balance and tightness of frame.
24. Higher quality.
25. Slack-framed.

Carcass

1. A lamb that will hang a meatier, more shapely carcass with a greater percentage of carcass weight in the leg, loin and rack than No. 2.
2. A lamb that will hang up the heaviest-muscled, most correctly finished carcass in the class.
3. A higher-yielding (dressing or killing) lamb that will hang a neater, trimmer carcass than No. 4.
4. A lamb that will hang a long, narrow carcass with the lightest leg in the class.
5. A lamb that will be an extremely low-yielding (dressing or killing) lamb that will hang a wasty, overfinished carcass.
6. A lamb that will hang the thickest-topped, trimmest-middled carcass in the class.
7. A lamb that will hang a more angular, narrower, lighter-muscled carcass than any other lamb in the class.
8. A lamb that will yield an overfinished carcass that will be extremely wasty and heavy through the middle.
9. A lamb that will hang a neater, firmer, more uniformly covered carcass than No. 1.
10. With his firmness of finish and tidy middle, a lamb that will yield the cleanest, trimmest, meatiest carcass in the class.
11. A high-killing lamb.
12. A lamb that will hang a carcass with more weight through the high-priced (most valuable) cuts than any other lamb in the class.
13. A lamb that will produce a more desirable carcass, in that it will be firmer, trimmer and heavier-muscled.
14. A lamb that will hang a carcass that will be meatier down the top and trimmer through the breast, along the underline and in the flank and crotch region than any other lamb in the class.
15. A lamb that will yield the most muscular carcass in the class.
16. Most predisposed to fat (waste).

17. Least flexible in going to heavier weights.
18. More growability and doability for the efficient-minded lamb feeder.
19. A cleaner-conditioned individual that has a more desirable composition of gain.
20. More flexible for that profit-minded feedlot producer in terms of going to that heavier market weight and maintaining a desirable composition of gain.
21. Best chance of reaching the self-service meat counter with the highest-quality grade.
22. Lacking the size, scale, and frame to go to heavier weights.
23. Trimmest, tightest-framed, most nicely balanced.
24. More potential to reach marketing weight at a younger age.
25. A larger, growthier individual that appears to be more profitable in terms of feed efficiency and growability.
26. More growability and more profit potential for the producer, by more efficiently converting feed into a more salable product.
27. A lamb that will hang a cleaner, trimmer, more nicely balanced carcass with more consumer appeal.
28. A lamb that will hang a thicker, heavier-muscled carcass with more total pounds of trimmed retail cuts, thus a higher merchandising value.
29. A lamb that will yield a more massively muscled carcass with more total lean red meat and muscle mass that will be worth more dollars and cents to both the packer and the producer.
30. A lamb that should hang the lightest, most angular-muscled carcass with the least consumer appeal and merchandising value.
31. Most total volume through the leg, loin and rack, thus a greater merchandising value.
32. A thicker-muscled individual that should hang a meatier carcass with the highest leg score.
33. A lamb that should hang up a heavier-muscled carcass with a longer hindsaddle, meaning more profitability to the packer and producer alike.
34. A lamb that should hang a thicker, heavier-muscled carcass with a higher percent of its weight, in the leg, loin and rack.
35. A lamb that should hang a longer, trimmer carcass with a higher cutability score.
36. A lamb that should hang a carcass with the least merchandising value and profit potential.
37. A lamb that should hang a cleaner, more expressively muscled, and thus more easily merchandisable, carcass.
38. More total red meat and muscle volume wrapped into one package.
39. When killed, carcassed and hung on the rail, a lamb that should yield a thicker, more shapely, more eye-appealing carcass.
40. Before being placed in the self-service meat counter, a lamb that should have less fat trim.
41. A lamb that should hang a carcass that could split a larger loin chop.

42. When killed and carcassed, a lamb that should have more total muscle volume from rail to floor.
43. A lamb that should yield a carcass with a greater percent lean.
44. A lamb that should yield a higher cutability carcass, thus giving a greater return to packer and producer alike.
45. A lamb that should yield a greater amount of retail lean per day of age.
46. Before entering those merchandising channels, a lamb that should require less fat trim.
47. Conventional, light-muscled lamb.
48. A lamb that handles with more muscling today (need to say where).

Fleece, Wool and Skin

1. A longer, denser fleece with more character (crimp) than No. 2.
2. The finest, highest-quality fleece in the class but a little open.
3. A fleece with character, length and density but lacking uniformity.
4. A long staple with good density, but the wool lacks fineness and character.
5. Outstanding length of staple, character and uniformity.
6. More crimp and elasticity to the fleece than any other ewe in the class.
7. A ewe that is extraordinary in meatiness and muscling, but her fleece is too short, rather coarse and open.
8. A strong fleece throughout with no weak spots.
9. A bright, clean fleece.
10. Too many black fibers (Hampshires and Suffolks).
11. Extremely black about the horn pits and numerous black fibers throughout the fleece.
12. The poorest fleece in the class, due to her dark skin and black fibers.
13. A fleece with a lot of kemp fibers present.
14. Harsh fleece, lacking luster and character.
15. A fleece that contains too much foreign matter, lacks uniformity and is very hairy on the breech.
16. Rather dull and listless wool on the 2 ewe, which "runs out" on the leg.
17. A heavier-shearing fleece that is more uniform in character and has more evidence of yolk and grease throughout than any other fleece in the class.
18. A loose, open, uneven fleece.
19. Dark-skinned ewe that will have a very light-shearing, dirty fleece.
20. A finer, denser, cleaner, higher-quality fleece that has more character than any other fleece in the class.
21. A denser, brighter, cleaner fleece than No. 3, but lacking the uniformity of character displayed by the fleece on No. 3.
22. Burry wool (wool that contains burrs).
23. Braid wool (coarse grade of wool).
24. Too much breech wool.
25. Cotted fleece (matted or tangled fibers).

26. Defective wool (damaged fleece).
27. Dingy wool.
28. Stained wool.
29. Frowsy wool (dry, wasty wool lacking character).
30. Very little difference between the shoulder and breech wool.
31. A pure fleece containing no hair, black or brown fibers or kemp.
32. A fleece that is free of stains.
33. Too many heavy tags and sweat locks.
34. Dog hair (coarse wool).
35. A fleece that shows life.
36. Good color — bright.
37. Excessive yolk.
38. Poor condition.
39. Lofty wool (springy wool, full of life).
40. Excellent-conditioned skin and fleece.
41. Evenly distributed yolk.
42. A good distribution of yolk and moderate in quantity.
43. A flesh-pink skin.
44. Too many folds on the neck.
45. Close and distinct crimp.
46. A ewe's fleece that is in excellent condition.
47. Too coarse wool on the hindquarters, in comparison to the wool on the fore-quarters (lacks uniformity).
48. A ewe's fleece showing a uniform and prominent crimp from base to tip.
49. Pencil-fleeced individual that possesses the poorest-quality fleece in the class.
50. A tighter, denser fleece that should have a higher spinning count.
51. Freer from black fibers.
52. Brighter, higher-quality fleece free from black fibers.

REASONS FOR PLACING A CLASS
OF MARKET LAMBS

It is important for you to have a definite system of giving reasons and to be able to use it in an effective manner. In brief, the form used here is to state the outstanding good points of the top animal first, and then in a comparative manner, tell why you placed this animal over the one below. Then, grant or admit the outstanding good points of the lower-placing animal in the pair. Follow the same order of describing and comparing for each pair of animals in the class, adjusting the relative amounts of description and comparison to fit the class at hand. The following set of reasons for a class of market lambs (Figs. 2-58a, 2-58b, 2-59a, 2-59b, 2-60a, 2-60b, 2-61a and 2-61b) is presented as a guide to help you describe and compare animals that you will be placing in future judging classes.

"I placed this class of market lambs 4-1-3-2, feeling that this was definitely a two-pair class, with 4 and 1 being the top pair and 3 and 2 being the bottom pair. I started this class with 4 and placed it over 1 in a rather close placing because it is a longer, stretchier, meatier, harder, firmer, straighter-lined, more correctly balanced, higher-quality wether than 1. The No. 4 wether is longer through the rib, has more spring to the fore-rib, is longer in the loin, is longer and more nearly level in the rump and is squarer at the dock than 1. He stands wider, with a thicker, meatier, more heavily muscled leg that has more muscle expression through the stifle and is especially thicker through both the inner and outer portions of the lower leg. He also is straighter and stronger-topped and neater and trimmer through the breast and chest area, in the rear flank and especially through the crotch (twist) area. He is a firmer-handling, more correctly finished lamb over the ribs, back and loin and possesses a harder, firmer leg. He should hang a harder, firmer, trimmer, more correctly finished, nicer-balanced carcass and should yield a higher percent of leg, loin and rack than 1. However, I grant that 1 is a larger, growthier, more rugged wether. He is wider through the chest floor and exhibits more constitution and capacity when viewed from the front than 4. No. 1 shows more muscle development through the arm and forearm than 4. He has more spring to the rear rib, is wider through the loin, is thicker through the top part of the leg and is deeper in the twist than 4. He is also a bit trimmer and tighter along the underline than 4.

"For my middle pair I placed 1 over 3 in the easiest placing in the class. No. 1 is bigger, more rugged, stouter and has more size and scale, length of body, capacity and ruggedness than 3. He is wider through the chest floor, shows more spread and thickness all down the top, being thicker over the shoulders with more spring to the fore-rib and rear rib, and wider in the loin and exhibits more muscling and meatiness through the top part of the leg and through the stifle. He is firmer and more correctly finished over the ribs, back and through the leg. When viewed from the rear, No. 1 is trimmer-middled than 3 and should hang a carcass with a greater percent of its weight in the hindsaddle, thus having greater merchandising value. However, I grant that 3 is a bit stronger-topped than 1. He is more level at the rump, squarer at the dock and thicker through the lower leg, especially through the inner portion, than 1.

"Coming to my bottom pair, I placed 3 over 2 in a very close placing. No. 3 is a thicker, meatier, more heavily-muscled wether that stands wider on both the front and rear legs than 2. He is thicker over the fore-rib and rear rib and wider at the loin, showing more overall spread, spring, expression, and muscling down the top than 2. No. 3 is squarer at the dock, which drops down into a deeper leg that is thicker and meatier through the top and through the stifle muscle area than 2. He should hang a thicker, meatier, more heavily muscled carcass than 2. However, I grant that 2 is stronger-topped, longer through the rib, loin and rump and harder, firmer-handling, more correctly finished than 3. He is neater through the breast and chest, cleaner and tighter through the middle and particularly trimmer through the twist than 3. He is also thicker on the outside of the lower leg than 3.

"I placed 2 at the bottom of the class because it is a narrow-topped, light-muscled lamb that lacks size, scale and total muscle volume and would hang the least desirable carcass of any lamb in the class."

Fig. 2-58a. Side view of market lamb No. 1. (Courtesy, Purdue University)

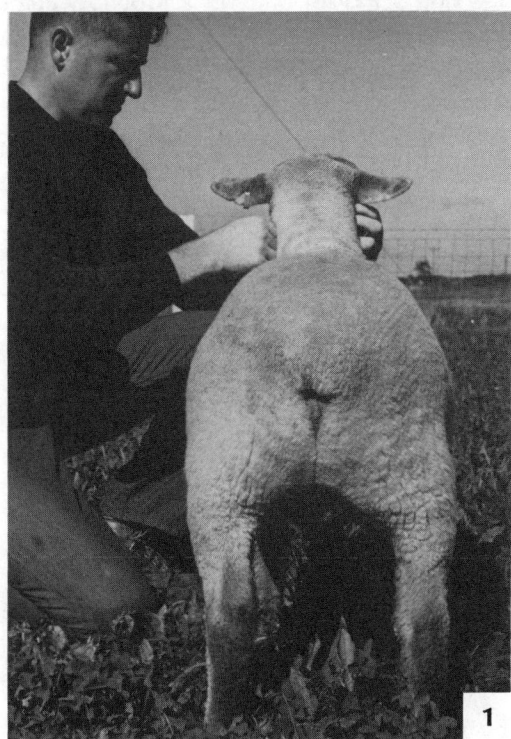

Fig. 2-58b. Rear view of market lamb No. 1. (Courtesy, Purdue University)

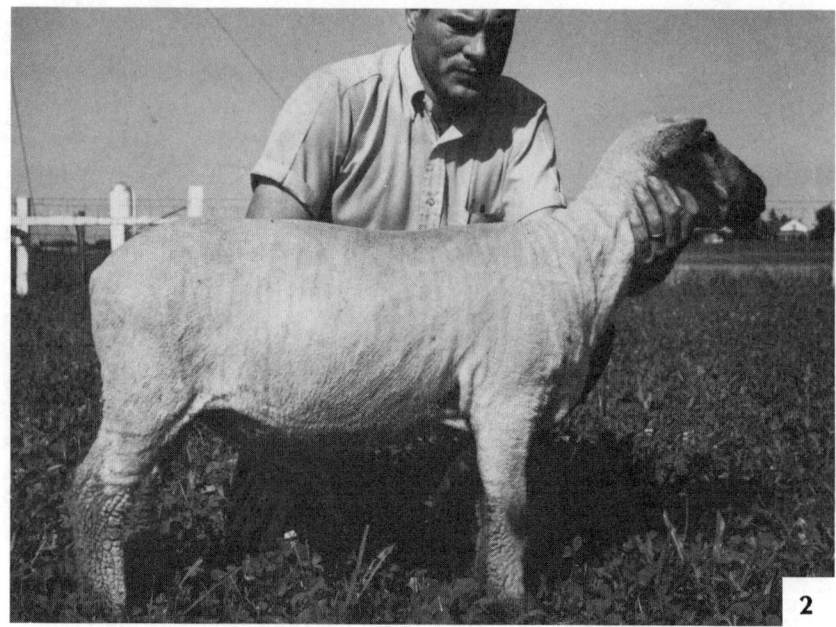

Fig. 2-59a. Side view of market lamb No. 2. (Courtesy, Purdue University)

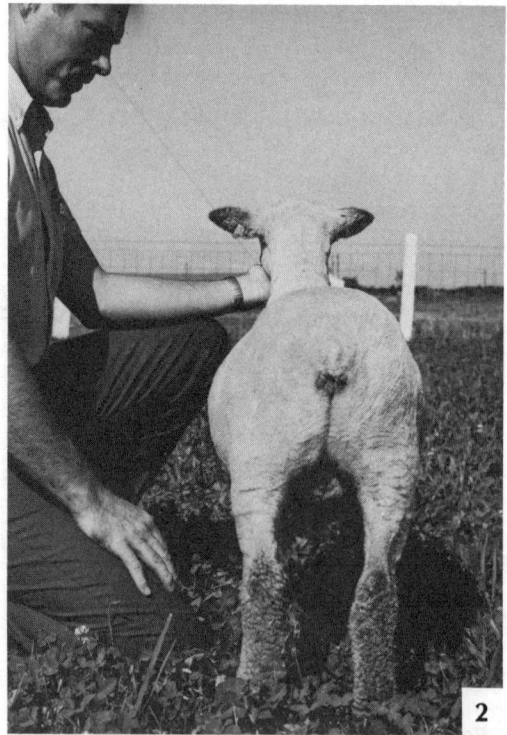

Fig. 2-59b. Rear view of market lamb No. 2. (Courtesy, Purdue University)

Fig. 2-60a. Side view of market lamb No. 3. (Courtesy, Purdue University)

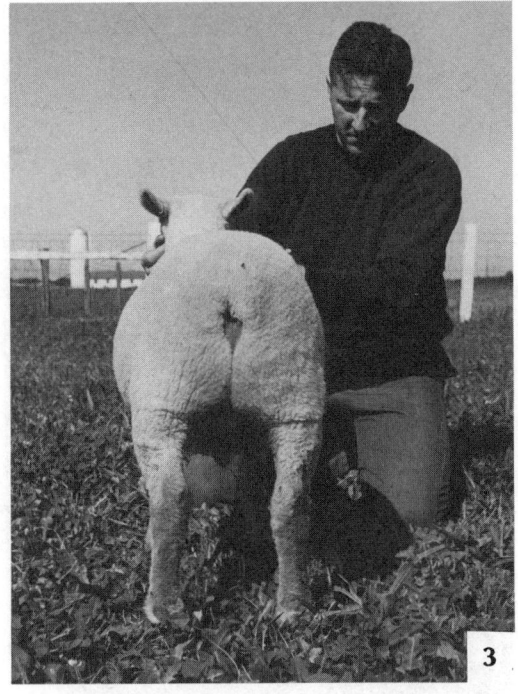

Fig. 2-60b. Rear view of market lamb No. 3. (Courtesy, Purdue University)

Fig. 2-61a. Side view of market lamb No. 4. (Courtesy, Purdue University)

Fig. 2-61b. Rear view of market lamb No. 4. (Courtesy, Purdue University)

BREED REGISTRY ASSOCIATIONS

Breed	Association	Officer and Address
Cheviot	American Cheviot Sheep Society, Inc.	Larry E. Davis Secretary Department of Agriculture Missouri Western State College St. Joseph, Missouri 64507
Columbia	Columbia Sheep Breeders' Association	Richard L. Gerber Secretary P.O. Box 272E State Route 182E Upper Sandusky, Ohio 43351
Corriedale	American Corriedale Association, Inc.	Russell Jackson Secretary Box 29C Seneca, Illinois 61360
Cotswold	American Cotswold Record Association	Virgil A. Bortel 282 Meadowboro Road Rochester, New Hampshire 03867
Debouillet	Debouillet Sheep Breeders' Association	A. D. Jones Secretary 300 S. Kentucky Avenue Roswell, New Mexico 88201
Delaine-Merino	America Delaine-Merino Record Association	Elaine Clouser Secretary-Treasurer 1193 Twp. Road 346 Nova, Ohio 44859
Dorset	Continental Dorset Club	Marion A. Meno Secretary Box 577 Hudson, Iowa 50644
Finnsheep	Finnsheep Breeders' Association, Inc.	Claire H. Carter P.O. Box 34303 Indianapolis, Indiana 46234–0303
Hampshire	The American Hampshire Sheep Association	Jim Cretcher Executive Secretary P.O. Box 345 Ashland, Missouri 65010–0345
Lincoln	National Lincoln Sheep Breeders' Association	Ralph O. Shaffer Secretary 5284 S. Albaugh Road West Milton, Ohio 45383

(Continued)

BREED REGISTRY ASSOCIATIONS (Continued)

Breed	Association	Officer and Address
Montadale	Montadale Sheep Breeders' Association, Inc.	Mildred Brown Secretary P.O. Box 44300 Indianapolis, Indiana 46244–0303
North Country Cheviot	American North Country Cheviot Association	John C. Goater Secretary Hitchner Hall University of Maine Orono, Maine 04473
Oxford	American Oxford Association	W. W. Watts Secretary Route 4 Ottawa, Illinois 61350
Panama	American Panama Registry Association	W. G. Priest Secretary-Treasurer Route 2 Jerome, Idaho 83338
Polled Dorset	Continental Dorset Club	Marion A. Meno Secretary Box 577 Hudson, Iowa 50644
Polypay	American Polypay Sheep Association	Belva Hansen Executive Secretary P.O. Box 371 Rexburg, Idaho 83440
Rambouillet	American Rambouillet Sheep Breeders' Association	LaVerne McDonald Secretary 2709 Sherwood Way San Angelo, Texas 76901
Romney	American Romney Breeders' Association	John H. Landers, Jr. Secretary 4375 N.E. Weslinn Drive Corvallis, Oregon 97333
Shropshire	American Shropshire Registry Association, Inc.	Elizabeth Glasgow Secretary P.O. Box 1970 Monticello, Illinois 61856
Southdown	American Southdown Breeders' Association	W. L. Henning Secretary 212 S. Allen Street State College, Pennsylvania 16801

(Continued)

BREED REGISTRY ASSOCIATIONS (Continued)

Breed	Association	Officer and Address
Suffolk	American Suffolk Sheep Society	VeNeal Jenkins Secretary 1115 N. Main Logan, Utah 84321
	National Suffolk Sheep Association	Betty J. Biellier Executive Secretary P.O. Box 324 Columbia, Missouri 65205
Targhee	U.S. Targhee Sheep Association	Vicki Arnold Secretary Box 40 Absarokee, Montana 59001
Tunis	National Tunis Sheep Registry	Leona Fitzpatrick Secretary Wayland, New York 14572

3

DAIRY CATTLE

The primary function of dairy cows is the economical production of milk and butterfat. The cows with correct dairy-type possess characteristics indicating the ability to perform this function. These characteristics include a large number of factors that influence the production and usefulness of the dairy cows, not only for one or two locations but also for a long lifetime performance.

Each of the national breed associations has a type classification plan whereby cows are rated for type according to the official breed score card. Each also has a herd test plan whereby the milk and butterfat production is recorded. This correlation of type and production is not infallible, as sometimes good-type cows are not high producers, and all high-producing cows are not good type. In general, however, a good judge of dairy cattle can select high-producing cows from low-producing cows. Since a large percentage of dairy cows are evaluated commercially by their type characteristics, all dairy producers should be familiar with the type standard adopted for their particular breed.

Every judge of dairy cattle should be familiar with the names, locations and correct form of the different parts of a dairy cow, as presented in Fig. 3-1. In order to receive proper benefits from instruction in judging, you must be able to correlate names and form.

DAIRY TYPE

The official breed score card describes desirable type in dairy cows under four main divisions: general appearance (including breed characteristics), dairy character, body capacity and mammary system (Figs. 3-2, 3-3, 3-4, 3-5, 3-6, 3-7, 3-8, 3-9 and 3-10).

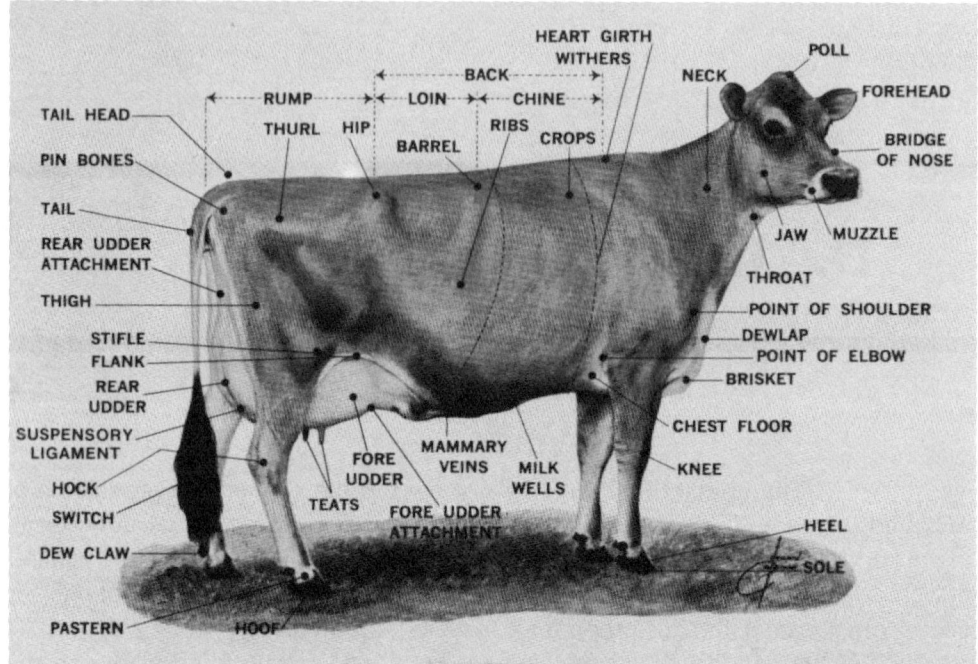

Fig. 3-1. Names and locations of the parts of a dairy cow. (Courtesy, Purebred Dairy Cattle Association)

General Appearance

General appearance includes the smooth blending of all parts, which should result in symmetry and balance, combined with style, carriage and alertness. A cow in milk should give a general impression of milkiness and display evidence of feminine refinement and outstanding dairy character. In addition to balance of the mammary system, capacity, constitution and conformation, the back and topline should be strong and straight, the loin broad and level, the rump long, level and extending to and including the tailhead. Pin bones should be set wide and high (almost level with the hip bones), and the tailhead should be slightly above and neatly set between the pin bones without roughness.

Breed type refers to the characteristics of a particular breed as related to head, color and size. A dairy cow should conform to the characteristics of her particular breed (discussed later in this chapter).

Dairy Character

Dairy character embodies those characteristics denoting angularity of body, animation, general trimness and refinement, lack of beefiness and general openness throughout. The ideal dairy cow has a rather long, lean neck blending smoothly into the shoulders and brisket, with a clean-cut throat and dewlap. The withers are prominent and wedge-shaped, with the shoulder blades slightly lower than the vertebrae. The ribs are wide apart, and the hips and pin bones are prominent, denoting a general lack of beefiness and appearing just the opposite from the beef cow in being active and alert. The skin is pliable, loose, of medium thickness and with soft, fine hair, indicating excellent quality and refinement throughout. The bone is hard, clean, flat and ample in amount and high in quality. The feet and legs are straight and strong and correctly placed on all four corners of the cow's body.

Dairy character does not mean that a cow should be thinly fleshed. It is expressed by dairy producers as "milky" in appearance. A dairy cow should carry sufficient flesh to indicate that she is thrifty (Figs. 3-2, 3-3, 3-4, 3-5, 3-6, 3-7, 3-8, 3-9 and 3-10).

Body Capacity

Body capacity includes barrel and heart girth. Good body capacity requires a long, wide, deep barrel, strongly supported with ribs well sprung and wide apart, appearing slightly wedge-shaped with greater depth and width toward the rear of the barrel.

The size of the muzzle is closely correlated with feed capacity, so the muzzle should be broad with open nostrils. A high-producing cow must consume and digest large quantities of feed (Figs. 3-2, 3-3, 3-4, 3-5, 3-6, 3-7, 3-8, 3-9 and 3-10).

Size is an important factor. Other characteristics being equal, the larger the cow within the breed, the better. However, coarse-boned cows lacking quality, refinement and dairy temperament are not desired. Small cows within the breed may be too refined and may lack substance, capacity and constitution.

Heart girth indicates constitution and vigor and is determined by the circumference of the body just back of the shoulders. In judging the size of the heart girth, consider both the depth of the cow just behind the front legs and the width between the front legs. A strong heart girth is indicative of the ability of a cow to stand up under the strain of heavy milk production. This requires a large heart and lung capacity, which is indicated by a deep, full heart girth, resulting from long, well-sprung fore-ribs and a deep, wide chest floor. The crops should be full behind the shoulders, and the chest floor wide and full just behind the front legs and between the front legs.

Mammary System

The mammary system includes the udder, teats and milk veins. A cow with continued high production for a long period of years must have a large, strongly attached, well-carried high-quality udder (Figs. 3-2, 3-4, 3-5 and 3-7). The udder should be large, as the size indicates the quantity of milk-producing tissues present — a small udder lacks sufficient milk-producing tissue. Also, the udder is a storage reservoir for milk between milkings. When the udder becomes full of milk and pressure builds up, the rate of milk secretion is decreased until the cow is milked and the pressure relieved. The udder should be long and wide and of moderate depth, extending well forward and high behind. The quarters should be even and uniform in size. A funnel-shaped or tilted udder is objectionable. The floor of the udder should be reasonably level and not deeply cut up between the quarters. The udder should be attached well forward with a neat, firm junction at the body wall, with no evidence of breaking at the front, rear or bottom. A pendulous udder is very objectionable because it is easily injured and often becomes infected. The udder should be soft, pliable and elastic, indicating active milk glands that take the milk constituents from the blood. A hard, meaty udder or a lumpy one is undesirable. A quality udder shrinks greatly when it is milked, while a meaty udder remains hard and does not contract very much.

Teats should be wide apart, squarely placed, of convenient size and free from obstructions. Funnel-shaped teats are objectionable, as are extremely long/short, large/small teats.

Blood vessels on the underside of the body are called *milk veins*. These milk veins should be long, crooked and prominent and should branch to enter large or numerous milk wells. Veins on the udder should be numerous and clearly defined.

JUDGING DAIRY CATTLE

Purpose and Value of the Score Card

Each national breed association formerly published an official score card for its breed. Score cards of the different breeds were somewhat similar but varied in certain details. One breed score card emphasized one part of the anatomy, while another breed score card emphasized a different part. This led to confusion. Consequently, the five leading national dairy cattle breeders organizations joined together to organize the Purebred Dairy Cattle Association, which published a simplified uniform score card for all breeds. A numerical value was assigned to the various parts of the animal in accordance with the relative importance of each part. This uniform score card for all breeds replaced the individual breed score cards. The score card describes the ideal type of dairy cow in general conformation of all breeds.

Many breeders and some judges of all breeds have maintained that all dairy cows have the same general conformation regardless of the breed and that all breeds should be judged by the same standards, with the exception of strictly breed characteristics, which for the most part, refer to color, size and the head, including horns. These three characteristics distinguish representatives of one breed from those of another.

When comparing the weak and strong points of two animals, you should remember that while the score card is a guide to the relative importance of the different parts, a greater adjustment should be made than appears on the score card if marked deficiencies should occur. For example, an extremely sloping rump calls for a deduction for a defective rump and also for a deficiency in "general appearance." Likewise, a very pendulous udder or an extremely funnel-shaped udder calls for a deduction greater than most score cards specify. Since dairy cattle are judged in the show ring by comparison, use the score card only to aid in acquiring knowledge of the relative values of the different parts.

DAIRY COW UNIFIED SCORE CARD[1]

(Breed characteristics should be considered in the application of this score card.)

	Perfect Score
General Appearance	30

Attractive, individuality with femininity, vigor, stretch, scale, harmonious blending of all parts and impressive style and carriage; all parts should be considered in evaluating a cow's general appearance. (Use pictures throughout text freely in this study.)   10

Breed Characteristics

1. *Head* — Clean-cut, proportionate to body; muzzle broad with large, open nostrils; jaws strong; eyes large and bright; forehead broad and moderately dished; bridge of nose straight; ears medium size and alertly carried.

2. *Shoulder blades* — Set smoothly and tightly against the body.   10

3. *Back* — Straight and strong.

4. *Loin* — Broad and nearly level.

5. *Rump* — Long, wide and nearly level from *hook bones* to *pin bones;* clean-cut and free from patchiness; *thurls* high and wide apart; *tail-head* set level with backline and free from coarseness; *tail* slender.

6. *Legs and feet* — Bone flat and strong, pasterns short and strong, hocks cleanly molded; *feet* short, compact and well-rounded with deep heel and level sole; *forelegs* medium in length, straight, wide apart and squarely placed; *hind legs* nearly perpendicular from hock to pastern from the side view and straight from the rear view.   10

	Perfect Score
Dairy Character	20

Evidence of milking ability, angularity and general openness, without weakness; freedom from coarseness, giving due regard to period of lactation.

1. *Neck* — Long, lean and blending smoothly into the shoulders; throat, dewlap and brisket clean-cut.

2. *Withers* — Sharp; *ribs* wide apart, rib bones wide, flat and long; *flanks* deep and refined; *thighs* incurving to flat and wide apart from the rear view, providing ample room for the udder and its rear attachment; *skin* loose and pliable.   20

	Perfect Score
Body Capacity	20

Relatively large in proportion to size of animal, providing ample capacity, strength and vigor.

1. *Barrel* — Strongly supported, long and deep; ribs highly and widely sprung; depth and width of barrel tending to increase toward rear.   10

2. *Heart girth* — Large and deep, with well-sprung foreribs blending into the shoulders; crops full; full at elbows; chest floor wide.   10

(Continued)

DAIRY COW UNIFIED SCORE CARD (Continued)

	Perfect Score
Mammary System	30

A strongly attached, well-balanced, fine-textured, capacious udder, indicating heavy production and a long period of usefulness.

1. ***Udder*** — Symmetrical, moderately long, wide and deep, strongly attached, showing moderate cleavage between halves, no quartering on sides; soft, pliable and well collapsed after milking; quarters evenly balanced. ... 10

2. ***Foreudder*** — Moderate length, uniform width from front to rear and strong attachment. ... 6

3. ***Rear udder*** — High, wide, slightly rounded, fairly uniform width from top to floor and strongly attached. ... 7

4. ***Teats*** — Uniform size, medium length and diameter, cylindrical, squarely placed under each quarter, plump and well-spaced from side and rear views. ... 5

5. ***Mammary veins*** — Large, long, tortuous, branching. "Because of the natural undeveloped mammary system in heifer calves and yearlings, less emphasis is placed on mammary system and more on general appearance, dairy character and body capacity. A slight to serious discrimination applies to overdeveloped, fatty udders in heifer calves and yearlings." ... 2

TOTAL ... 100

Note: Subscores are not used in breed-type classification.

---

[1]Copyright © Purebred Dairy Cattle Association, 1957, approved, American Dairy Science Association, 1957. Courtesy, Purebred Dairy Cattle Association.

Judging Young Dairy Animals

Many heifers and young bulls are selected each year to build up the herds of the future; however, the evaluation of heifers and young bulls that will develop into the correct type of mature animals is less dependable than when mature animals are considered. As young animals develop, their conformation often changes. Yearlings, especially, tend to change in balance and body proportion. The proper evaluation of udders on heifers is particularly uncertain, and yet in milking cows, the udders are exceedingly important. Some of the more serious defects of milking udders can be detected in the undeveloped heifer udder. Defects such as an udder tilted forward, teats too close together, quarters not uniform in size, teats unequal in length or size, double teats, a hole in the side of a teat and an undeveloped udder may be detected. Marked defects in undeveloped udders seldom improve. They are more apt to become very pronounced as the udders develop. The problem in judging heifer udders is to detect marked defects and to evaluate underdeveloped udders in accordance with their resemblance to the desirable milking udders. If defects are not found, the heifer mammary system should not be criticized.

Often a bred heifer starts making up an udder that is not uniform in the size of quarters. It is difficult to evaluate properly such an udder; however, this condition should not be severely criticized if the teats are correct in size and shape and properly spaced and there is sufficient rear attachment.

The same characteristics are desired in heifers as in mature animals, except that less emphasis is given to mammary development. The same breed type and general appearance, dairy conformation, capacity and constitution are desired. Care should be taken to differentiate between beefiness and good condition, as young animals are frequently in good condition. The clean-cut, angular, dairy-like heifer is preferred to the close-coupled, short, meaty, thick-necked type, which lacks quality, finish and refinement.

In general appearance, the smooth, straight-topped, level-rumped, deep-bodied kind is selected to develop into the desirable type of cow.

Judging Dairy Bulls

With the exception of the mammary system and femininity, you should look for the same essentials when you are judging bulls as when you are judging cows. The angular, open, smooth "dairy-like" bull is preferred to the short, thick, coarse type. In addition, the bull should possess masculinity. This is indicated by massiveness, heavy front quarters and a well-developed crest. The head is broader, and the horns coarser and thicker than those of a cow.

The bull should have a long, clean neck and open ribs. He should be smooth and well-balanced, without excess flesh. Breed character and alertness are very important. Crooked legs and weak pasterns, especially hind legs, are very objectionable. Because sickle-hocked, crooked hind legs lack strength and durability, they are severely criticized. A dairy bull should possess smooth withers, a straight, strong top and level rump; he should be well cut up between the rear legs and have thin, angular thighs. Thick, beefy rear quarters; heavy, open shoulders; and short, low-set frames are discriminated against.

Judging Dry Cows

Every cow should have a dry period between the milking and the freshening periods. The dry cow is apt to be in good flesh and appear thick. She may be heavy and rounded over the shoulders and may lack the leanness and angularity of the milking cow. Even so, she should retain breed type, quality and feminine refinement. The udder should show strong attachments, should be evenly spaced and should have uniformly sized teats and numerous veins. It should also be soft and pliable. Other conditions being equal, milking cows are favored over dry cows in the show ring, as the judge cannot be sure of the quality of the udder, the strength of attachment or the symmetry of the udder while the cow is dry.

Judging Dairy Cattle by Comparison

Dairy cattle are judged by comparison. In comparative judging, animals are ranked according to trueness to type, so the first step is for you to form a picture of the ideal animal in your mind. This can be achieved by a careful study of the correct kind of living individuals possessing desirable type and of pictures of representative animals of the different breeds presented in this chapter and in breed magazines.

Mentally compare the animals in the ring to this ideal. The first impression frequently gives the correct evaluation of an animal unless you later discover defects. First appraise an animal on general appearance. You should look for a pleasing individual, one that most nearly resembles the true-type model. Look for an individual that is symmetrical and well-balanced, showing quality, refinement, smoothness and outstanding dairy character, combined with breed character, conforming most nearly to the standards established for the particular breed.

Usually four animals are used in judging practice and judging contests and are assigned numbers or letters from right to left (exactly opposite from the numbering system used for beef cattle, horse, sheep and swine classes).

The four main divisions of the score card can be used in practice judging. Cards similar to the following — general appearance, dairy character, body capacity and mammary system — can be prepared for this purpose.

The ranking on divisions should be used only as a guide to teaching the fundamentals of judging dairy cattle.

DAIRY CATTLE PLACING CARD

Class _____ Date _____

Placing: 1st _____ 2nd _____ 3rd _____ 4th _____

		1st	2nd	3rd	4th	Grade
General appearance	(30)					
Dairy character	(20)					
Body capacity	(20)					
Mammary system	(30)					

Name _____

Due to extreme variation, the final placing cannot always be arrived at by adding the rankings. For example, two cows may be extremely close in score on body capacity but differ widely in score on mammary system. In practice, a cow might rank first in general appearance, dairy character and body capacity but be deficient enough in mammary system to place last in the class of four cows.

First, make a comparison by standing 20 to 30 feet from the animals and get a general impression of the group, preferably as they walk. It is always desirable to see the animals walk as this may bring out defects, such as a sloping rump, a weak loin or lameness, which may not be observed while the animal is standing. This view gives a general conception of the type, breed character and balance of the individuals in the class.

While the animals are walking at a distance, they should be observed and compared from the side for major characteristics. General appearance, breed type, size, angularity, straightness of back, levelness of rump and the presence of outstanding defects or weaknesses, if any, should be noted. This view also shows the general balance and the relative capacity and constitution, as indicated by the length and depth of barrel and the depth through the heart girth. Attachments, shapes and sizes of the udders should be compared at this time.

Often there is an outstanding cow that is superior in practically all characteristics or strong in several and not unduly weak in any. There may be one animal much inferior to the others; she may have an udder broken away, or she may lack sufficient barrel or be extremely pinched in the heart or perhaps have some other weakness that makes her an easy bottom for the class.

Second, the animals should be lined up side by side, all facing in one direction. The view from the rear permits observation and comparison of the rumps, width of pin bones, size and attachment of rear udders, evenness of rear quarters, placement of rear teats and straightness of hind legs. The position also affords a good opportunity to compare the spring of rib, width of barrel and smoothness of withers and crops of the individuals in the class. The front view permits comparison of character of heads and the width of chests.

Third, the animals are again lined up, one behind the other, all facing the same direction and close together. This view presents a final opportunity to compare the topline, the relative depth and size of the barrels and udders, the placement of teats and the levelness of the udders. It is necessary to be at least 20 feet away to make the best comparison from the side.

Fourth, if close examination of the best-appearing cow fails to disclose any defects, place her first in your mind; and if there is an extremely poor animal in the class, disposed of her mentally by placing her fourth. The two that are more nearly equal should be studied from the side, rear and front. The better of the two should be given second place and the other third.

Sometimes the class naturally divides itself into two pairs. Analyze each pair and decide which pair should go at the top and which at the bottom. Then, select the superior animal from each pair.

The remainder of the time should be used to check the entire class to determine that some weakness has not been overlooked and that there is a sound reason for each placing. If reasons are to be given, sufficient notes on each animal should be made to recall each animal vividly. Notes should be made in the order in which the animals are placed. The notes should include the outstanding as well as the undesirable points of each animal and the characteristics by which one animal excels another.

Handling the Animals

In judging contests, handling or feeling the animals is often prohibited. In such cases, it is assumed that all animals are sound except for such defects as can be seen. This restriction requires observation of the attachment of the udder on both sides, in the front and at the rear, also observation regarding the evenness of all quarters, the length and the size of the veins and the silkiness of the hair.

The purpose of handling is merely to verify the previous impression as to the preceding points. If handling is permitted, you should feel the udder to determine whether it is soft and pliable or whether it is hard and lumpy. The front attachment of the udder, the size and crookedness of the milk veins, and the size and number of milk wells should be examined. Also, you should feel the hide to ascertain its thickness and mellowness and the softness of the hair.

BREEDS OF DAIRY CATTLE

Jersey

The Jersey is the smallest of the leading dairy breeds. Jerseys should have angularity and strength, indicating productive efficiency. Jerseys vary greatly in color, but they usually are some shade of fawn, with or without white markings. The muzzle is black, encircled by a light-colored ring, and the tongue and switch may be black, white or mixed. The Jersey head is clean-cut, well-proportioned, moderate in length, with a moderately dished face. Jerseys are noted for well-shaped udders and strong udder attachments.

The Jersey cow's weight is influenced by her condition and stage of lactation. The ideal mature weight for a Jersey cow in milk is about 1,000 pounds. The height at withers for a mature Jersey cow should be 50 inches. Jerseys usually calve for the first time at 23 to 25 months of age. They reach full maturity when they are about six years old. The Jersey bull weighs between 1,200 and 1,800 pounds.

Jerseys are usually considered more angular, leaner, free of excess flesh, with more uniform refinement and finish, than the larger dairy breeds (Figs. 3-2, 3-3 and 3-4).

Fig. 3-2. "Generators Topsy," Excellent - 97 percent, was the 1973 National Grand Champion Jersey and produced several records over 20,000 pounds of milk and 1,000 pounds of butterfat. She won the 1985 *Jersey Journal* Great Cow Contest. (Courtesy, The American Jersey Cattle Club)

Fig. 3-3. An artist's conception of the ideal Jersey cow. (Courtesy, The American Jersey Cattle Club)

Fig. 3-4. "Empire Crusader Heidi" is the highest classified living Jersey cow at Excellent - 97 percent. She was Grand Champion at the 1979 and 1981 All-American Jersey shows and has completed three consecutive lactations exceeding 1,000 pounds of butterfat. (Courtesy, The American Jersey Cattle Club)

Guernsey

Size, strength, quality and character are desired in Guernsey cows. Cows weigh about 1,100 pounds, and bulls about 1,700 pounds. Breeders stress dairy character, smoothness, symmetrical udders and freedom from excessive flesh (Fig. 3-5). Both bulls and cows should be stately and majestic in appearance. Their color may vary from almost red to a light fawn with various sizes of white markings. The Guernsey head is moderately dished. The horns curve forward and are yellow at the base. A clear or buff muzzle is preferred. There is no discrimination for the absence of horns.

Guernseys produce a superior milk in total milk components, flavor and color. Because the milk is naturally high in total milk solids, it is a market leader, whether used for fluid purposes or for high yields of manufactured products, such as cheese and butter. Guernsey milk is adaptable to all pricing procedures, including multiple component pricing.

The unique qualities of Guernsey milk, especially flavor and color, have always been recognized. Guernsey breeders took steps to protect and emphasize these qualities as early as 1923 when the Golden Guernsey Trademark was developed. In 1933 Golden Guernsey Incorporated, a subsidiary of the AGCC, was established.

Under the Golden Guernsey trademark program, dairies are franchised to distribute Golden Guernsey Milk Products. The usual Golden Guernsey product has the following advantages for wise consumers.

	Average Market Milk Contains . . .	Golden Guernsey Milk Contains . . .
Solids-not-fat	8.71%	9.32%
Protein	3.19%	3.69%
Lactose	4.65%	4.91%
Butterfat	3.68%	4.40%
Ash	0.69%	0.72%
Total nutritive energy (units)	625	716

Comparison of the chemical composition of average milk and natural Guernsey milk by Dr. G. A. Richardson, Oregon State College, Corvallis.

Courtesy, American Guernsey Association.

The golden color of Guernsey milk, which comes from unusually high contents of beta-carotene, has the potential of adding even more value for producers and consumers.

Fig. 3-5. This outstanding Guernsey was recently nominated for All-American honors for the fourth consecutive year. "Chedco TH Maxine" is also the world Guernsey butterfat production champion. She has been designated as the breed's highest-selling cow ever at public auction. At the October 1984 World Premiere Sale in Madison, Wisconsin, she sold for $51,000. (Courtesy, The American Guernsey Cattle Club)

Holstein-Friesian

The Holstein-Friesian, commonly called Holstein, is the largest of all the breeds of dairy cattle (Figs. 3-6 and 3-7). Cows weigh from 1,300 to 1,700 pounds and bulls 2,000 pounds and up. Larger size, when accompanied with finish and quality, is not a discrimination. Color markings are black and white, varying from nearly all black

Fig. 3-6. Ideal young Holstein-Friesian female. (Courtesy, Holstein-Friesian Association of America)

Fig. 3-7. Ideal mature Holstein-Friesian milking female. (Courtesy, Holstein-Friesian Association of America)

to nearly all white. The switch is white, and there is some white extending on each leg to the hoof. The Holsteins are rugged, have large barrels and udders and are more thickly fleshed than the Jerseys or the Guernseys.

Brown Swiss

The Brown Swiss is strong and vigorous but not coarse. Size and ruggedness with quality are desired. Extreme refinement is undesirable. Cows weigh from 1,200 to 1,600 pounds, and bulls from 1,800 to 2,500 pounds. The Brown Swiss are usually heavier-boned than other breeds. They are, as a rule, more muscular and show more evidence of beefiness than other breeds. The skin appears loose, often hanging in folds about the neck and brisket. The color is solid brown, varying from very light to dark. White or off-colored spots are objectionable. Females with any white or off-color markings above the underside of the belly or with a white core in the switch do not meet the color standards of the Brown Swiss and should be so designated when they are registered. Mouse color is very common, with a light-colored strip encircling a black nose. The switch and horn tips are black. A light-colored strip generally extends along the backbone. The horns are incurving, inclining slightly upward, of medium length, showing quality and tapering to the tips. Polled animals are not barred from registry, and no discrimination is given for absence of horns (Figs. 3-8 and 3-9).

Fig. 3-8. Modern Brown Swiss female. (Courtesy, Brown Swiss Cattle Breeders' Association of America)

Fig. 3-9. Typical Brown Swiss cow and calf. (Courtesy, Brown Swiss Cattle Breeders' Association of America)

Ayrshire

The Ayrshire is strong and robust, showing constitution and vigor, symmetry, style and balance throughout. The udders are strongly attached, evenly balanced and well carried. The shape of the udders, the levelness of the rump and the straightness of the topline have been uniformly stressed by the breeders until these qualities are outstanding in the breed. Mature cows should weigh between 1,050 and 1,400 pounds, and mature bulls between 1,700 and 2,200 pounds. Mature cows in milk should weigh at least 1,150 pounds. Red and white coloring is preferable, though any shade of red, brown or mahogany, brown with white spotting or almost all white is acceptable. Each color should be clearly defined. Black or brindle markings are objectionable. The horns incline upward, show refinement, are of medium length and are tapered. There is no discrimination for absence of horns (Fig. 3-10).

EVALUATING DEFECTS AND DISQUALIFICATIONS

Occasionally, you will be confronted with an animal showing a defect or one that is at a disadvantage, such as a dry cow in a milking class, a lame cow or a dehorned cow. The following guide to the evaluation of defects has been approved by the Purebred Dairy Cattle Association and should be followed in the judging of all breeds.

In adapting these rules in judging practice to a class of four animals, place an

Fig. 3-10. Modern Ayrshire cow, "Oak Ridge Flashy Starlet 656179" 2E —
90.9 feet and 92.3 feet. DIHR: 11 - 0 years, 302 days; 18,640 pounds milk, 3.9
percent 734 pounds butterfat lifetime production; 108,310 pounds milk, 5.0
percent 5,463 pounds butterfat. Twice All-American Aged Cow. (Courtesy,
Ayrshire Breeders' Association)

animal with a defect termed "slight discrimination" second if the two animals are
otherwise equal. An animal with a defect termed "serious discrimination" should be
placed third or fourth, depending on the degree of seriousness, if the animals are
otherwise equal. The term "disqualification" means the animal should be placed last
in a class of four.

GUIDE TO EVALUATION OF DEFECTS
IN JUDGING COWS[1]

NOTE: In a show ring, disqualification means that the animal is not eligible to win a prize. Any disqualified animal is not eligible to be shown in the
group classes. In slight to serious discrimination, the degree of seriousness
shall be determined by the judge.

Eyes

1. Total blindness: Disqualification.
2. Blindness in one eye: Slight discrimination.
3. Crossed eyes: Slight discrimination.

Wry Face

Slight to serious discrimination.

Cropped Ears

Slight discrimination.

(Continued)

GUIDE TO EVALUATION OF DEFECTS (Continued)

Parrot Jaw
Slight to serious discrimination.

Shoulders
Winged: Slight to serious discrimination.

Tail Setting
Wry tail or other abnormal tail settings: Slight to serious discrimination.

Legs and Feet
1. Lameness — Apparently permanent and interfering with normal function: Disqualification.
2. Lameness — Apparently temporary and not affecting normal function: Slight discrimination.
3. Bucked knees: Slight to serious discrimination.
4. Evidence of arthritis, crampy hind leg: Serious discrimination.
5. Boggy hocks: Slight to serious discrimination.

Absence of Horns
No discrimination.

Lack of Size
Slight to serious discrimination.

Udder
1. Blind quarter: Disqualification.
2. Abnormal milk (bloody, clotted, watery): Possible disqualification.
3. Udder definitely broken away in attachment: Serious discrimination.
4. A weak udder attachment: Slight to serious discrimination.
5. One or more light quarters, hard spots in udder, obstruction in teat (spider): Slight to serious discrimination.
6. Side leak: Slight discrimination.

Dry Cows
Among cows of apparently equal merit: Give strong preference to cows in milk.

Freemartin Heifers
Disqualification unless proved pregnant.

Over-conditioned
Slight to serious discrimination.

Temporary or Minor Injuries
Blemishes or injuries of a temporary character not affecting the animal's usefulness: Slight discrimination.

Evidence of Sharp Practices
1. Animal showing signs of having been operated upon or tampered with for the purpose of concealing faults in conformations, or with intent to deceive relative to the animal's soundness: Disqualification.
2. Uncalved heifers showing evidence of having been milked: Serious discrimination.

---

[1]Courtesy, Purebred Dairy Cattle Association.

Types of Defects

Side View of Udders

Figs. 3-11 and 3-12 illustrate desirable udders as they are viewed from the side. These udders are capacious, extending far forward and well out and up behind. They are held firmly to the body, with the floor of each udder higher than the hocks. The quarters are evenly balanced, and the teats are uniform in size and evenly spaced. The rear attachment is high, wide and strong. Veins on each udder are numerous and well-defined, and the veins on the abdomen are large and prominent. These characteristics are associated with high milk production and many years of usefulness.

In contrast, Figs. 3-13, 3-14, 3-16, 3-17 and 3-18 illustrate weakness, as viewed from the side, which reduce milk production and shorten the useful life of a cow. Figs. 3-13, 3-14 and 3-17 show short, low-hanging, weakly attached udders that are easily injured. The usefulness of these udders is limited. Fig. 3-13 is already pendulous, with a broken front attachment, and the udder in Fig. 3-17 shows weakness in the front attachment. These udders lack capacity in both the rear and the front, which greatly limits production. Weak attachment of the udders, either front or rear, is much more serious in young cows than in old ones. Two- and three-year-old heifers with weak attachment of the udders are seriously discriminated against, as this weakness becomes worse as the cows get older. Note that the floor of these udders is much lower than the hocks, which makes the udders easily bruised as the cows walk or lie down.

Figs. 3-16 and 3-17 illustrate tilted udders with weak and blind front quarters that greatly reduce production from the front quarters as compared to the rear quarters. These udders are also small and lack capacity.

Fig. 3-16 shows a weak front quarter with the rear quarters larger, making the udder unbalanced. The rear quarters hang lower than the front ones. The capacity of this udder is reduced due to the small size of the front quarters.

These defects and deficiencies in udders greatly reduce the usefulness of the cow and are seriously discriminated against in the show ring. A defect that reduces capacity of the udder decreases production and the value of the cow (Figs. 3-17 and 3-18).

Rear View of Udders

Figs. 3-21, 3-23, 3-24 and 3-25 illustrate desirable rear udders. The udders show depth, width and capacity, with teats well-placed. The rear attachment of these udders is strong, high and wide. The quarters are large and even, resulting in capacious, evenly balanced rear quarters.

In contrast to the desirable rear udders, note the defective rear udders pictured in Figs. 3-19, 3-20 and 3-22.

The udders in Figs. 3-19 and 3-23 are approximately equal in size, but Fig. 3-23 shows the superior udder, as it has greater capacity at the top. Fig. 3-19 shows a narrow, weak attachment. Note also that the udder in Fig. 3-23 is well held up, the

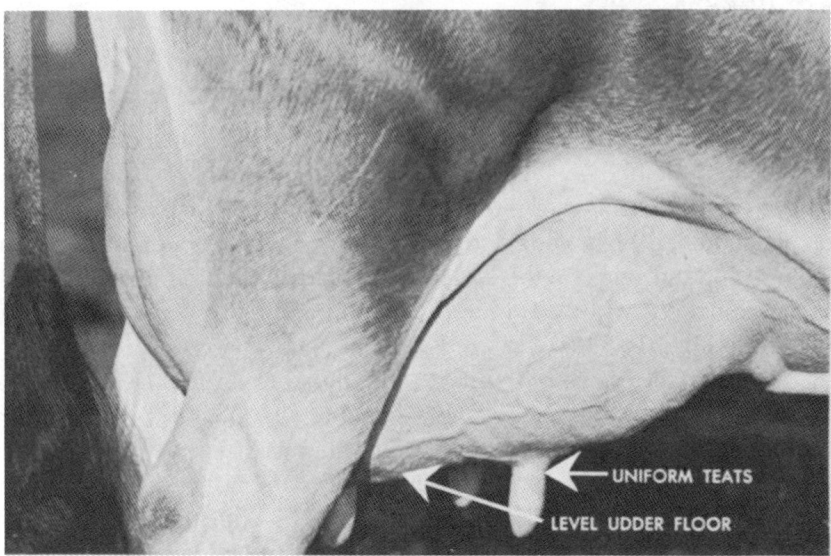

Fig. 3-11. Teats and udder floor. Look under the udder. Four evenly placed teats of uniform size and shape are required. The four quarters of the udder should be equal and ample in size and should blend together without large crevices between them. The udder floor should be level.

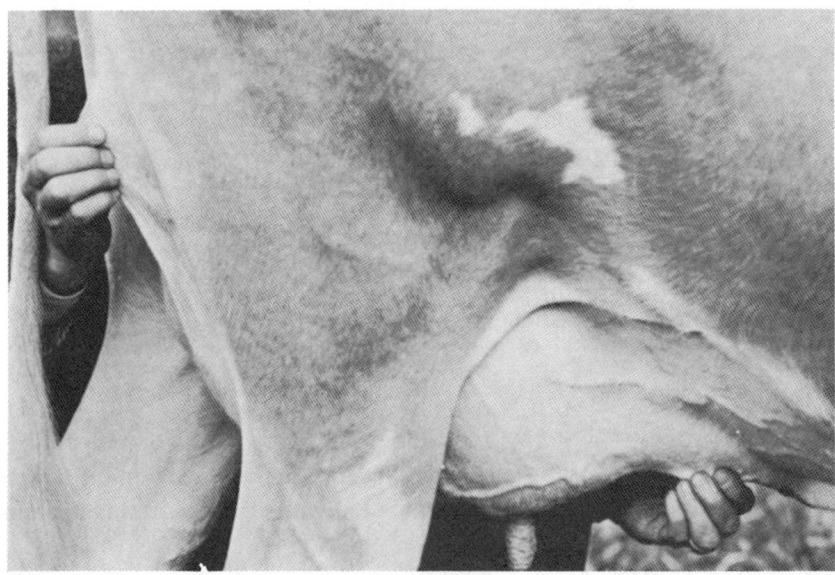

Fig. 3-12. Udder attachments. Front and rear, the udder should blend into the body with no sign of "breaking away." A high attachment at the rear and a front attachment that carries well forward are desirable. An udder of good texture is soft, pliable and free of excess meat and coarse hair.

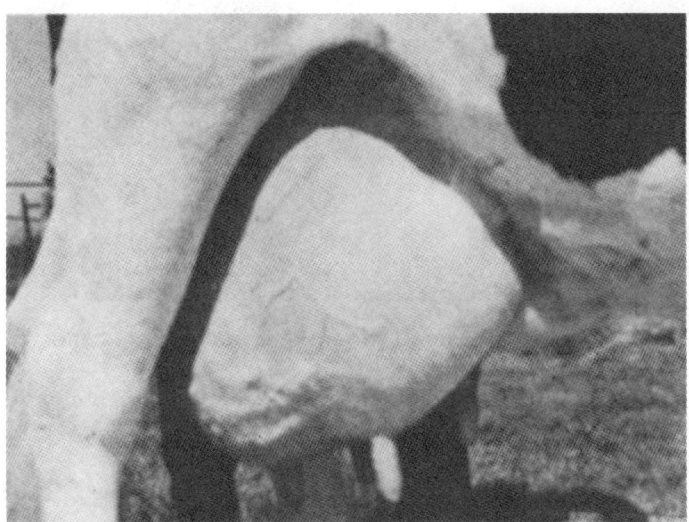

Fig. 3-13. A broken and / or very faulty udder.

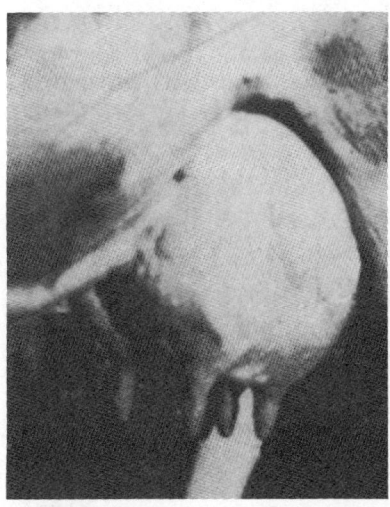

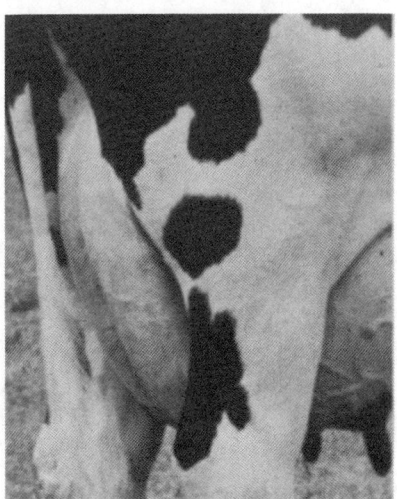

Fig. 3-14. The foreudder is too short.

Fig. 3-15. This is an ideal rear udder — high, wide and strongly attached.

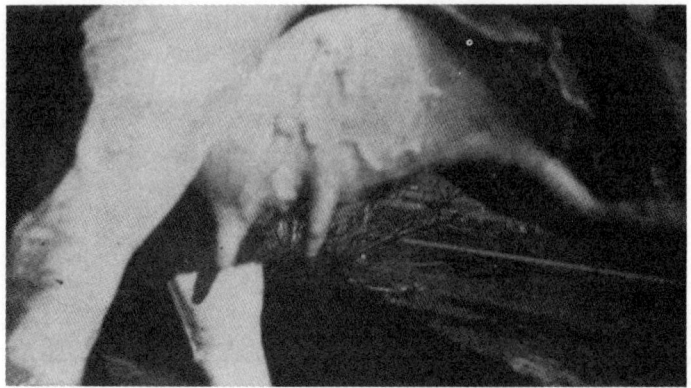

Fig. 3-16. An unbalanced udder with a weak front quarter.

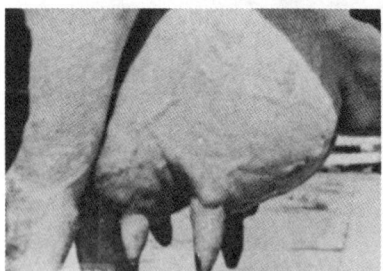

Fig. 3-17. The foreudder is short, bulgy and loose, and the teats are undesirable in size and shape.

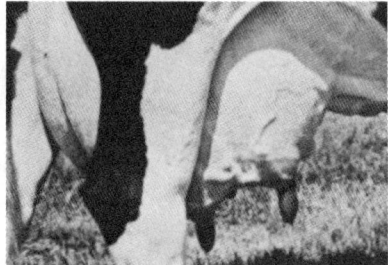

Fig. 3-18. The rear udder is low, and the foreudder attachment has given way. So, the udder is too low, and the teats are not squarely placed.

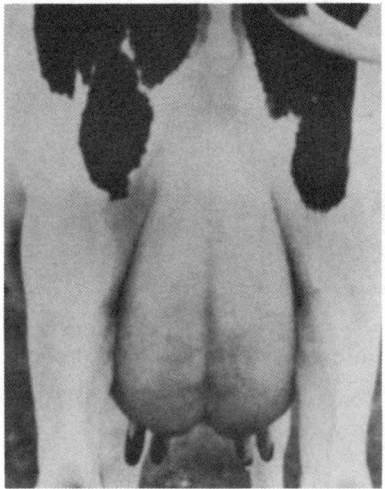

Fig. 3-19. The rear udder is too low.

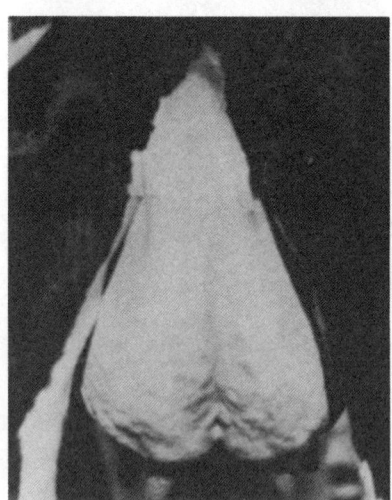

Fig. 3-20. The rear udder is low, narrow and pinched.

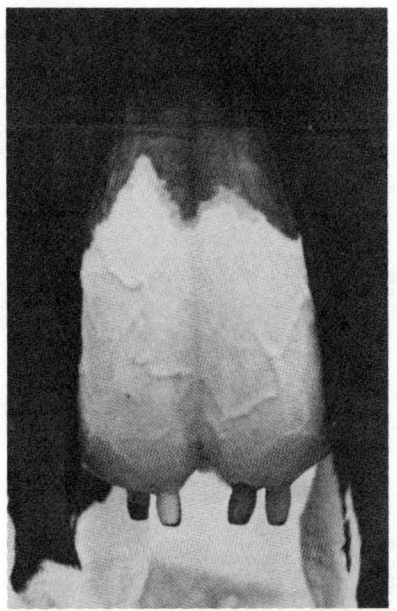

Fig. 3-21. The median suspensory ligament is very strong, and the teats are squarely placed.

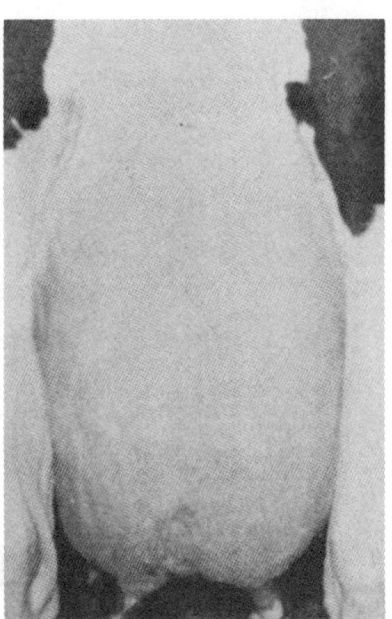

Fig. 3-22. A weak suspensory ligament causing the teats to strut and the udder attachments to weaken.

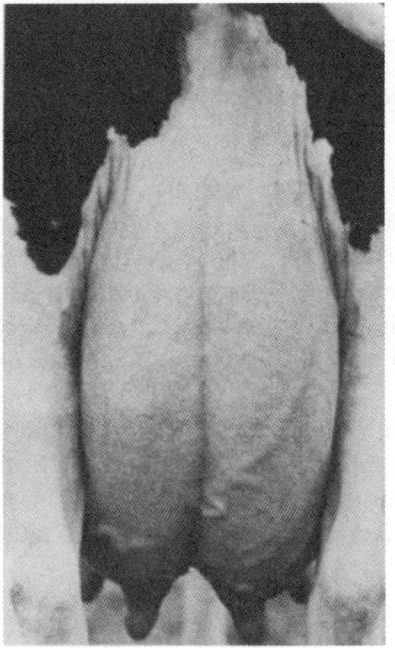

Fig. 3-23. A desirable udder in milk. (See the same cow in Fig. 3-24.)

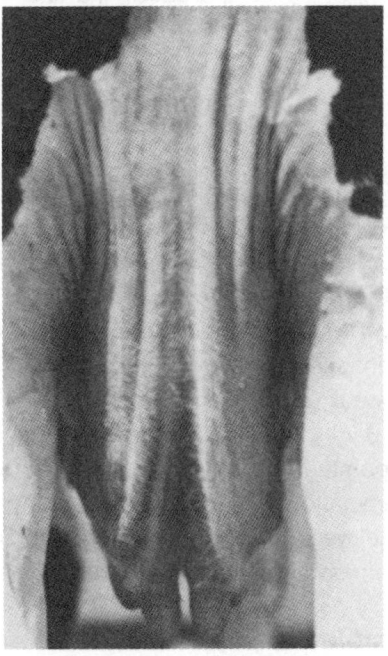

Fig. 3-24. A desirable udder when dry. (See the same cow in Fig. 3-23.)

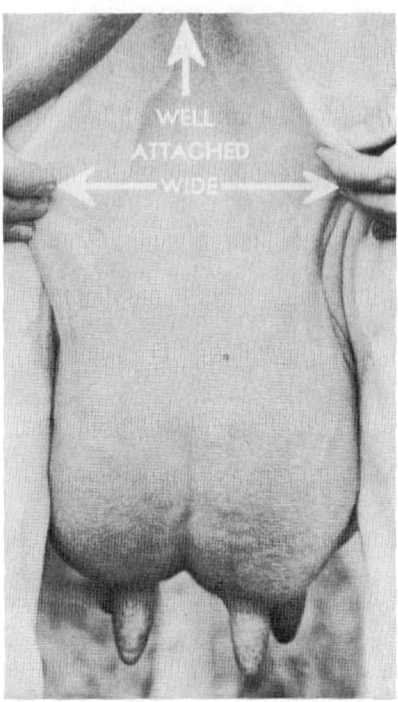

Fig. 3-25. Rear udder. Note the width of the udder, the firm, high, wide attachment, the pliability of the skin and the uniform teat placement. Deep, swinging udders are hard to milk and are more easily injured and contaminated than those that are well attached.

floor being higher than the hocks. The poorly attached, low-hanging udder shown in Fig. 3-19 indicates that its usefulness is limited.

Figs. 3-20 and 3-22 illustrate rear udders lacking in depth, width and capacity. Fig. 3-20 shows a narrow, weak attachment with the udder cut up between the quarters. The capacity of these udders is limited by the thick thighs, which leave insufficient space, the weak and undersized quarters, the division between quarters and the weak, narrow attachments.

Fig. 3-21 illustrates a narrow udder with rear teats too close together and a narrow, pointed attachment. Fig. 3-21 shows capacity further reduced by the division of quarters, and Fig. 3-22 shows strutting rear teats.

These weaknesses limit production by reducing the capacity of the udder and the usefulness and value of the cow.

Undesirable Body Type

The short, thick neck; weak back; and rounding, meaty type are seriously discriminated against in the show ring, even though the cow does possess good barrel capacity, a strong constitution and a reasonably well-shaped udder. This type of cow is criticized for lack of dairy temperament, or the ability to convert feed into milk, quality, refinement and general appearance.

The meaty, thick type is usually associated with low production. Such cows become fat and patchy, lacking the leanness associated with high production.

Head, Neck and Shoulders

The head, neck and shoulders indicate to a great extent the quality, refinement and breed character possessed by a dairy cow. A short, thick head and neck portray poor dairy quality and are usually associated with a rounding, close-knit, meaty body. These characteristics are severely criticized by judges. Another type of defect is broad, open shoulders. The top of the shoulders should be thin and wedge-shaped and should blend smoothly into the neck and body, with the top of the shoulder blades lower than the chine. Loose shoulders, especially severely winged shoulders, are criticized, depending upon the seriousness of the weakness. The dry cow is expected to be more rounded over the shoulders than the milking cow but it should not be open in the shoulders.

An undesirable head is the narrow head with a small eye and a narrow muzzle. The narrow muzzle is often associated with a weak constitution.

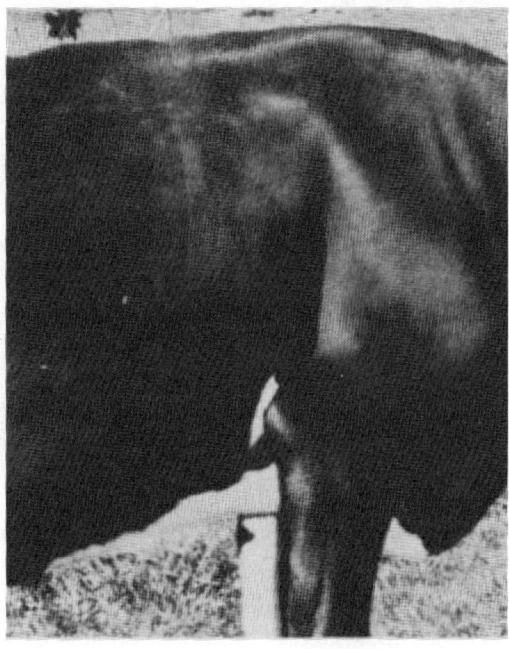

Fig. 3-26. Winged shoulders.

Legs and Feet

Legs and feet affect not only the general appearance of a dairy cow but also her lifetime usefulness. The legs and feet should be set wide and squarely under the body. The front legs should be straight. The hind legs should be nearly straight when viewed from the rear and should not weave or cross when the animal walks. Pasterns should be strong and springy. As more emphasis is placed on lifetime production, strong, straight legs set squarely under the cow become more important.

The Official Dairy Cow Score Card gives the following description of desirable legs and feet:

> Legs wide apart, squarely set, clean-cut and strong with forelegs straight.
> Hind legs nearly perpendicular from hock to pastern, when viewed from behind, legs wide apart and nearly straight. Bone, flat and flinty, tendons well defined, pasterns of medium length, strong and springy. Hocks cleanly molded [Fig. 3-27].
> Feet short and well rounded, with deep heel and level sole [Fig. 3-28].

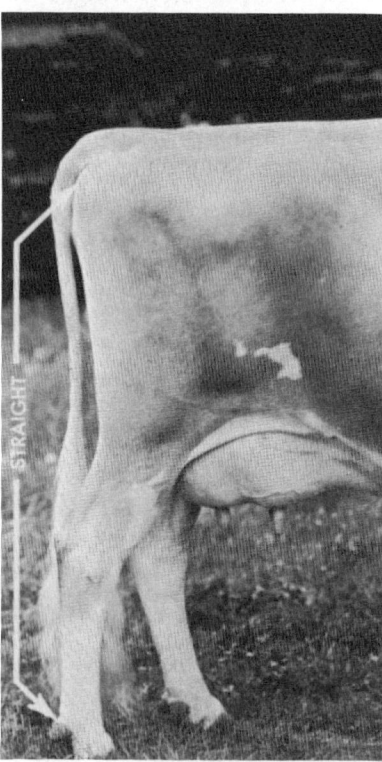 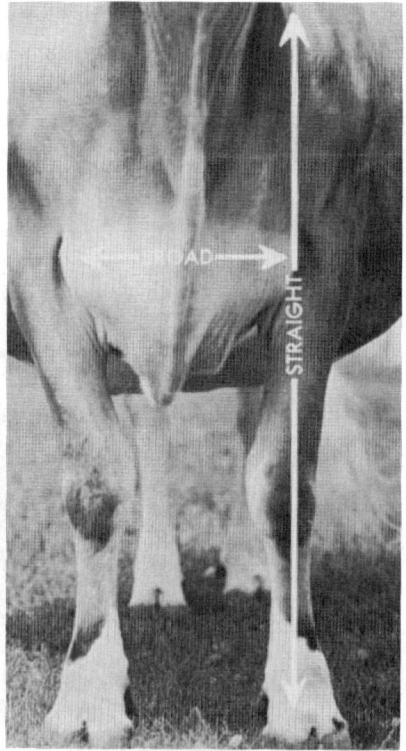

Fig. 3-27. Hind legs. Posed as in this picture, the hind legs of the cow should be straight. A line dropped from the pin bone should just touch the hock and descend, paralleling the lower leg, to the dewclaw. Beware of sickle-hocked animals. Dairy cattle need strong, straight legs.

Fig. 3-28. Forelegs. From a front view, a straight vertical line from shoulder points to hooves is desirable. Legs should be set squarely to support the heavy body and well apart so that there is ample room for the chest. Note how the body starts to widen behind the front legs.

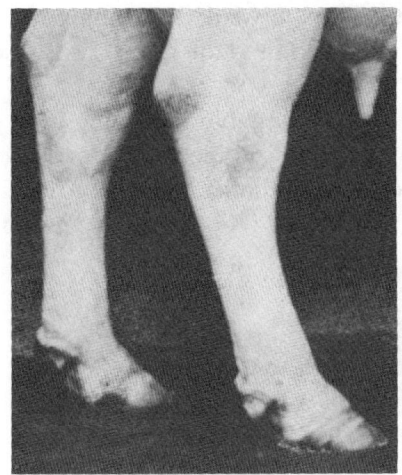

Fig. 3-29. Strong, clean, flat bone; legs squarely placed.

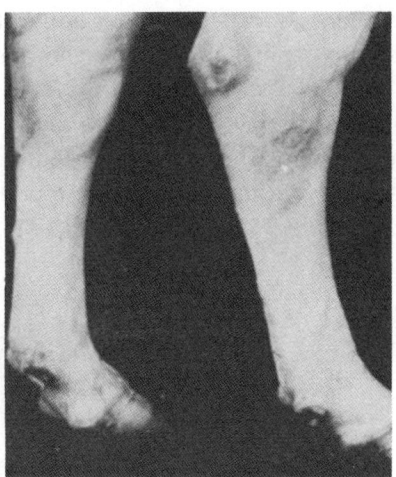

Fig. 3-30. Weak pasterns and shallow heel.

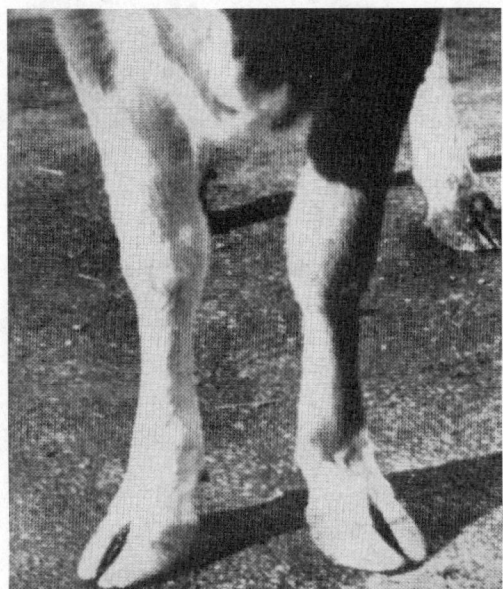

Fig. 3-31. The front legs of a heifer that toe out noticeably. This is usually associated with a narrow, pinched heart girth and an early tendency toward winged shoulders.

Undesirable Rumps

Fig. 3-32 shows a rounding rump with low pin bones. This undesirable type of rump is often associated with an udder that is tilted forward. A rounding rump also detracts from general appearance and does not conform to the rump of the true-type cow.

Fig. 3-35 portrays another common kind of undesirable rump, the high pelvic arch and tailhead. This shape contributes to roughness and detracts from the general appearance. Note the contrast in smoothness of the rumps featured in Figs. 3-32, 3-33, 3-34 and 3-35 with those in Figs. 3-2, 3-3, 3-4, 3-5, 3-6, 3-7, 3-8, 3-9 and 3-10.

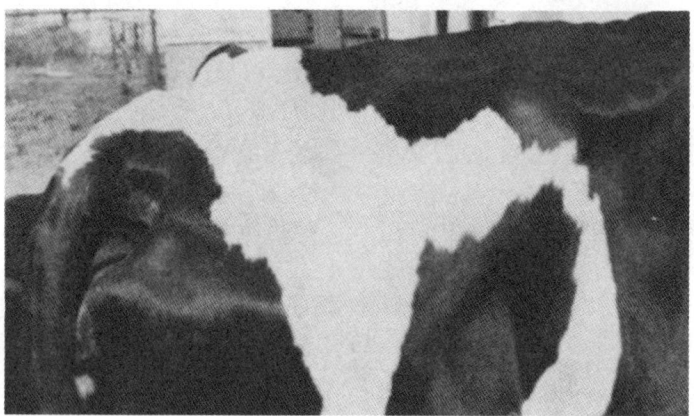

Fig. 3-32. Sloping rump.

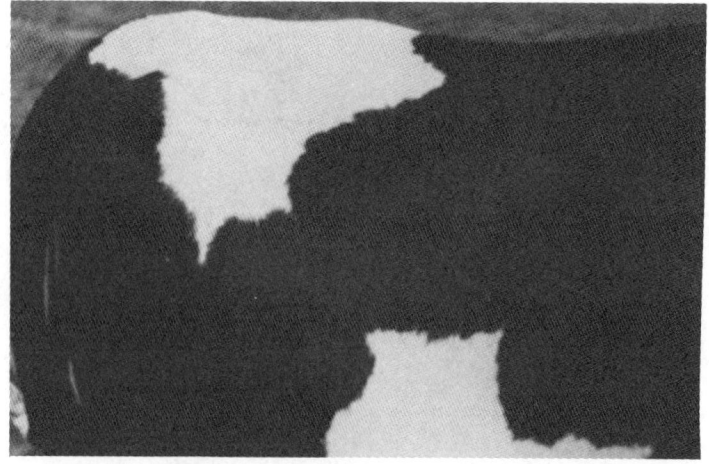

Fig. 3-33. Rump with medium width, length and levelness.

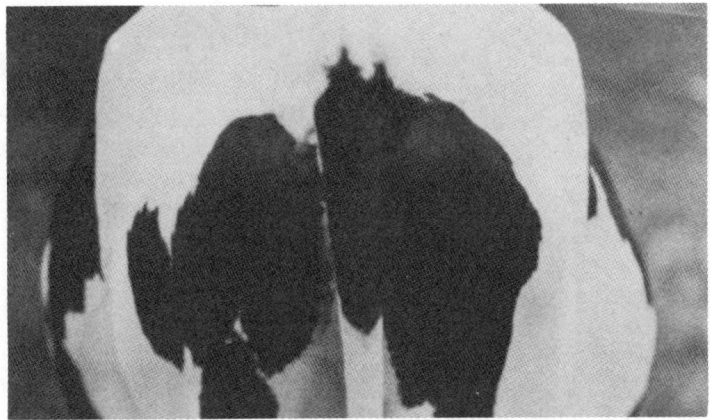

Fig. 3-34. Narrow rump, especially at the pins.

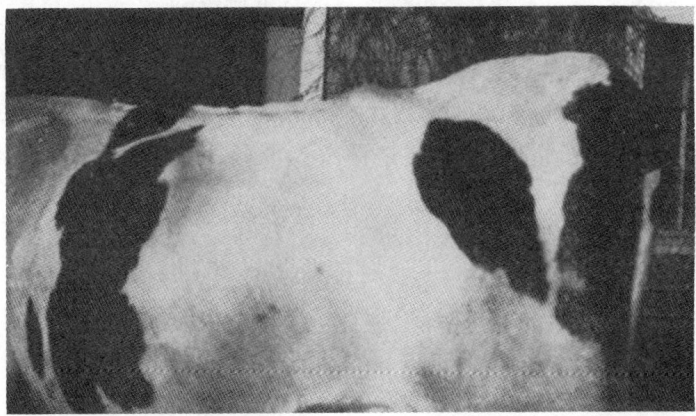

Fig. 3-35. Undesirable rump that is rough, with a high pelvic arch.

Constitution

Figs. 3-36 and 3-37 show a contrast in constitution. Fig. 3-36 illustrates a chest of good width and strength, indicative of a hearty constitution; while Fig. 3-37 portrays a very narrow chest, indicative of a weak constitution and a general lack of thriftiness and "doing ability." Note the closeness of the front legs in Fig. 3-37 as compared to those in Fig. 3-36. A dairy cow must possess a rugged constitution in order to maintain high production year after year. A cow with a poor constitution is difficult to keep on feed and does not have the ability to maintain high production. Many cows with the necessary mammary system, dairy temperament and other characteristics associated with high production fail to produce satisfactorily because they do not have a strong constitution and are not able to hold up under the strain of heavy milk production.

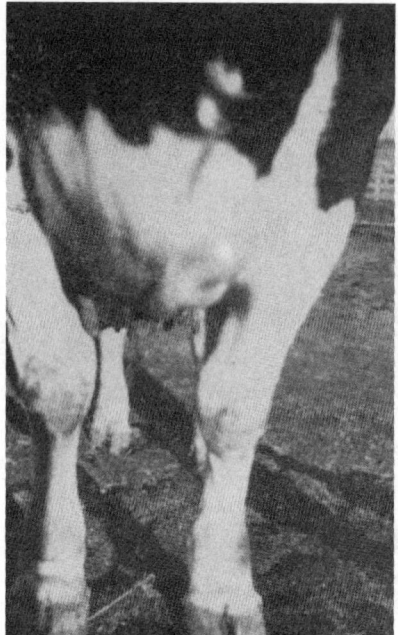 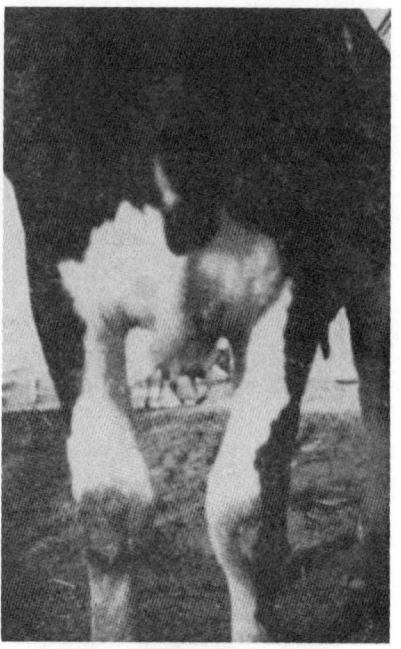

Fig. 3-36. Strong constitution, indicating strength and stability.   Fig. 3-37. Weak constitution, indicating lack of vigor and stamina.

REASONS FOR PLACING CLASSES OF DAIRY CATTLE

Guernsey

"I placed this class of Guernsey cows 1-2-3-4 (Figs. 3-38, 3-39, 3-40 and 3-41). I placed 1 over 2 primarily because of general appearance and mammary system. No. 1 excels 2 in smoothness of shoulders, both at the point of the shoulder and at the elbow. She has greater capacity, being deeper in the ribs with a greater spring of rib. She is also more level over the topline and rump. Both these cows show outstanding dairy character, as noted by their clean-cutness and their freedom from excessive fleshing. No. 1 has a more desirable mammary system than 2, as she has a longer, fuller foreudder and a wider rear udder. She also has a more desirable teat placement and is free of halving; whereas, the udder on 2 shows evidence of halving.

"I placed 2 over 3 and consider this an easy placing. No. 2 has more size and capacity than 3, as noted by her greater depth of heart and depth and spring of rib. She excels 3 in general appearance and breed character, as is shown by her strength of head and her overall strength and balance. No. 2 has a higher rear udder attachment and is stronger throughout the rear udder area. I will grant that 3 has a more desirable foreudder and teat placement.

Fig. 3-38. Guernsey cow No. 1.

Fig. 3-39. Guernsey cow No. 2.

Fig. 3-40. Guernsey cow No. 3.

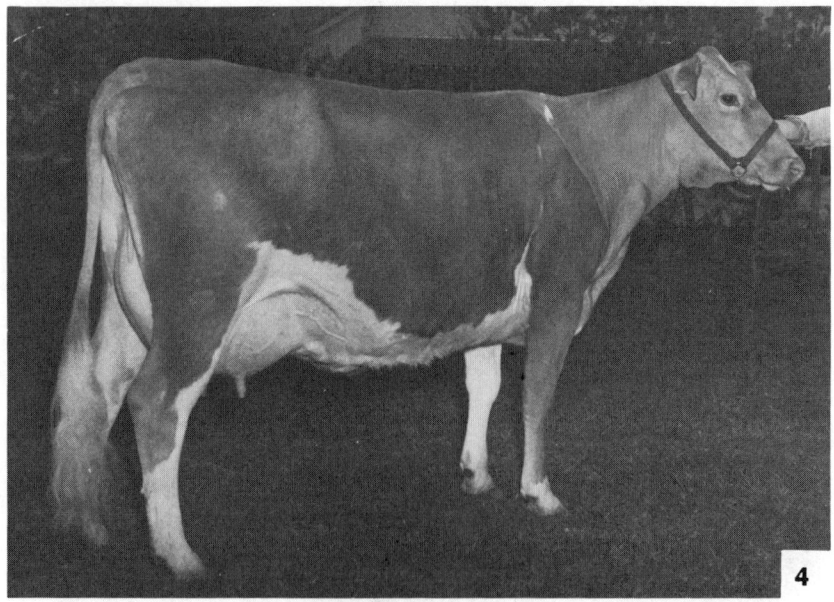

Fig. 3-41. Guernsey cow No. 4.

"I placed 3 over 4 because 3 has more dairy character and possesses a more desirable mammary system. No. 3 surpasses 4 in general appearance and overall smoothness. She is more level over the topline and rump with a neater tail setting. She has a more desirably shaped udder with more fullness to the foreudder. She is more correct in teat placement and foreudder attachments. However, I will admit that 4 is a larger cow than 3 with a deeper rear rib."

Holstein

"I placed this class of Holstein cows 1-2-3-4 (Figs. 3-42, 3-43, 3-44 and 3-45). I placed 1 over 2 because 1 exhibits more Holstein breed character, is more correctly balanced and possesses a more desirable mammary system than 2. She is more correct than 2 in her general appearance, as is indicated by the levelness over her topline and rump and her smoothness throughout. She also expresses more Holstein breed type and sex character about her head and neck. She excels 2 in dairy character, as is shown by the sharpness over the withers, the leanness of the neck and the freedom from excessive fleshing over the body. No. 1 is standing on the most desirable set of feet and legs in the class, as noted by the straightness of her legs and the strength of her pasterns. No. 1 has a more desirable mammary system than 2, as is evident by the more correct shape, length and attachment of the foreudder and the height and the width of the rear udder.

"I placed 2 over 3 considering this the closest placing in this class. No. 2 has more breed character than 3, as is evident by her more desirable type head and general alertness. She is also superior to 3 in smoothness of shoulder both at the point of the shoulder and at the elbow. Both these cows have ample size and capacity, but 2 is deeper in the heart, longer from the withers to the hips and longer-rumped (longer from the hooks to the pins). No. 2 greatly excels 3 in shape and fullness of both the foreudder and the rear udder.

"I placed 3 over 4 because 3 is more correct in her general appearance and possesses more breed character than 4. She has a more desirable type head and neck; she is deeper in the heart and ribs, which signifies greater capacity. She is higher in the thurls, making a more level rump. In comparing the mammary systems of these two cows, I noted that 3 has more length and exhibits a more desirable shape of the foreudder. I will admit, however, that 4 has a superior rear udder, as is indicated by the fullness of the lower rear udder."

Fig. 3-42. Holstein cow No. 1. (Courtesy, Holstein-Friesian Association of America; photo by Strohmeyer & Carpenter)

Fig. 3-43. Holstein cow No. 2. (Courtesy, Holstein-Friesian Association of America; photo by Strohmeyer & Carpenter)

3

Fig. 3-44. Holstein cow No. 3. (Courtesy, Holstein-Friesian Association of America; photo by Strohmeyer & Carpenter)

4

Fig. 3-45. Holstein cow No. 4. (Courtesy, Holstein-Friesian Association of America; photo by Strohmeyer & Carpenter)

Jersey

"I placed this class of Jersey cows 3-4-1-2 (Figs. 3-46, 3-47, 3-48 and 3-49). No. 3 is a large, well-balanced, deep-bodied cow, showing a great deal of breed type and dairy character.

"No. 3 possesses more Jersey breed type and character in that she has a shorter head with more dish to the face than 4. She is smoother over the withers and at the junction of the neck than 4. She is deeper than 4, with more spring of rib and depth through the heart girth and smoother over the rump, with a more evenly shaped udder and stronger hind legs than 4.

"Granting that 4 is a longer cow with a slightly higher and wider rear udder, the extreme smoothness, breed type and body capacity definitely warrant placing 3 over 4.

"For my middle pair, I placed 4 over 1. This is an extremely close pair. Both cows can be criticized for roughness over the withers and at the junction of the neck and over the rump and tailhead. However, 4 is longer than 1, having just as much depth of body and spring of rib. No. 4 shows more dairy character, as is indicated by more openness of ribs and a cleaner-cut neck.

"Coming to my bottom pair, I placed 1 over 2 based on general appearance, mammary system and dairy character. No. 1 possesses a larger udder, with the rear udder being attached much higher and wider than on 2. She is deeper and longer, showing more depth of body and spring of rib. She also is cleaner-cut in the neck and withers.

"I placed 2 at the bottom of the class because she lacks body capacity, mammary system and dairy character."

Fig. 3-46. Jersey cow No. 1. (Courtesy, The American Jersey Cattle Club; photo by Serpa)

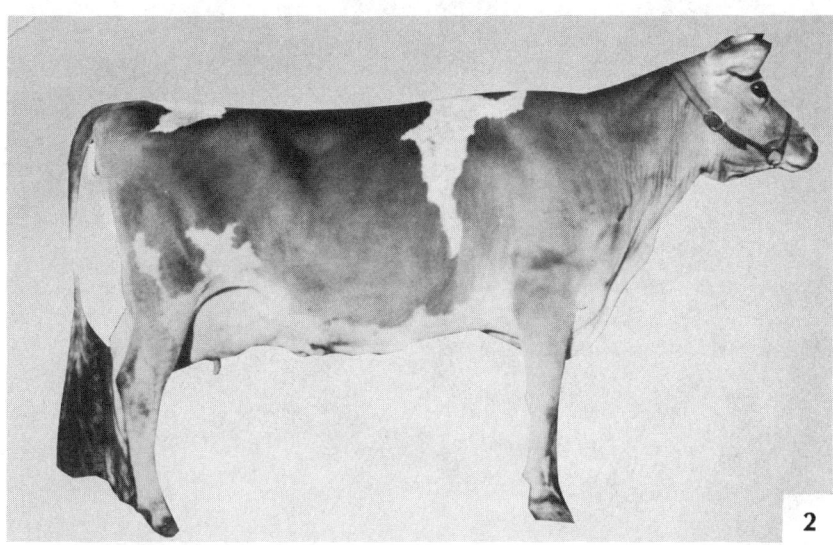

Fig. 3-47. Jersey cow No. 2. (Courtesy, The American Jersey Cattle Club; photo by Horton Studio)

Fig. 3-48. Jersey cow No. 3. (Courtesy, The American Jersey Cattle Club; photo by Strohmeyer & Carpenter)

Fig. 3-49. Jersey cow No. 4. (Courtesy, The American Jersey Cattle Club; photo by Strohmeyer & Carpenter)

DAIRY CATTLE JUDGING TERMINOLOGY

The ability to describe a dairy animal accurately aids greatly in formulating a word picture of the animal in your mind and in giving a correct set of reasons without repetition. This requires a broad knowledge of terms or expressions relating to dairy cattle. You should not attempt to memorize the list of expressions; instead, you should become familiar with the terms and their exact meaning in order to formulate expressions that adequately and accurately describe the animal in a precise manner. The use of inaccurate terminology is a serious error.

Breed Type and General Appearance

1. Milky (Figs. 3-2, 3-3, 3-4, 3-5, 3-6, 3-7, 3-8, 3-9 and 3-10).
2. Dairy-like (Figs. 3-2, 3-5, 3-6, 3-8 and 3-10).
3. Clean-cut (Fig. 3-4).
4. Outstanding breed character (Fig. 3-6).
5. Very feminine head.
6. Vigorous, or revealing vigor.
7. Attractive individual (Figs. 3-2, 3-6 and 3-8).
8. Harmonious blending and correlation of parts.
9. Extremely masculine.

10. Typical size and type for the breed.
11. Resembling the true-type model.
12. Symmetrical and pleasing.
13. More stylish (Fig. 3-2).
14. True type and evenly balanced.
15. Greater finish and refinement.
16. Evenly balanced.
17. More quality and correctness of fleshing.
18. More bloom.
19. More snap and alertness.
20. Smoother throughout.
21. Plenty of scale (Fig. 3-8).
22. Handicapped by age.
23. Tremendous substance (Fig. 3-8).
24. Greater scale.
25. More style in walking.
26. Best-balanced animal in the class.
27. A strong, straight topline (Fig. 3-2).
28. Broad, level rump.
29. Wider and more level at the pins (Figs. 3-3 and 3-8).
30. Rough over the rump.
31. Sloping in the rump.
32. Low at the pins.
33. High at the tailhead.
34. Lacking in general appearance.
35. Lacking balance.
36. Lacking in breed type.
37. Lacking quality and refinement.
38. Lacking style.
39. Lacking snap and alertness.
40. Too refined.
41. Does not blend well.
42. Weak loin.
43. Easy in the back.
44. Short-coupled.
45. Short, close-knit kind.
46. Too narrow at pins.
47. Too small.
48. Does not carry out as smoothly over the rump.
49. Out of condition.
50. Does not have fitting and bloom.
51. Outstanding in the class.
52. Easy top of the class.

53. Easy bottom of the class.
54. Two-pair class.
55. Small, but typey heifer.
56. Longer, more open type.
57. Too coarse.
58. Plain in the head.
59. Long, narrow head.
60. Small pig eye.
61. Wry face.
62. Roman nose.
63. Large, coarse horns.
64. Bright, snappy eye.
65. Plain throughout.
66. Lacking character.
67. Standing squarely on her hind legs.
68. Not as good in the hind legs.
69. Crooked in the hind legs.
70. Sickle-hocked (too much "set" to the rear legs).
71. More head character.
72. More pleasing head and neck.
73. Especially good in the rear.
74. Easy in the loin.
75. Stronger in the back.
76. Lacking strength of top.
77. Lacking smoothness at the tailhead.
78. Possessing greater size and scale.
79. More stretch.
80. A larger, growthier cow.
81. A bigger, more rugged, stouter, more vigorous-looking cow.
82. Impressive style and carriage.
83. Longer from the hooks to the pins (longer-rumped) (Fig. 3-2).
84. Longer and more leveled out over the rump.
85. Broad and moderately dished forehead.
86. Straight, strong, level topline (back).
87. Thurls set high and wide apart.
88. Clean-cut and free from patchiness.
89. Hard, clean, flat, high-quality bone (Fig. 3-29).
90. Short, strong pasterns.
91. Neat, clean hocks (free from puffiness).
92. Short, compact, well-rounded feet with deep heel and level sole.
93. More correct set of feet and legs.
94. Legs straight, wide apart and squarely placed.
95. Cow-hocked (too close at the hocks).

96. Toed-out, or splay-footed (Fig. 3-31).
97. Toed-in, or pigeon-toed.
98. A rough, coarse individual that lacks quality and refinement throughout.
99. A short, steep-rumped cow.
100. A draggy, listless-looking heifer.
101. Wry tail.
102. Weak pasterns and shallow heel (Fig. 3-30).

Dairy Character

1. More angular.
2. More milky.
3. Clean-cut head and neck.
4. Smoother over the shoulder.
5. A smooth blending of neck and shoulders.
6. Wedge-shaped withers.
7. Silky hair and hide.
8. More open-ribbed.
9. Free of excess flesh.
10. Short, thick neck.
11. Open in the shoulders.
12. Weak behind the shoulders.
13. Heavy and rounding thighs.
14. Thick and meaty throughout.
15. More rounding over the rump.
16. Too thick and meaty.
17. Thick, meaty thighs.
18. A great deal of smoothness and quality.
19. Sharp over the shoulders.
20. Heavy over the shoulder.
21. Slightly patchy at the pins.
22. Blending well at the shoulders.
23. Long, lean neck that blends smoothly into the shoulders.
24. Clean-cut throat, dewlap and brisket.
25. Deep and refined about the flanks.
26. Incurving thighs.
27. Flat-thighed cow.
28. From the rear view, wide apart thighs, providing ample room for the udder and its rear attachment.
29. Loose, pliable skin.
30. A short, thick, beefy heifer.
31. Extremely sharp over the withers.
32. Winged shoulders (Fig. 3-26).

Body Capacity

1. Pinched at the heart.
2. Narrow-chested.
3. Cut in behind the shoulders.
4. Full in the heart girth.
5. Deeper in the heart. (Figs. 3-2 and 3-4).
6. Filled in nicely behind the shoulders.
7. Greater spring of rib.
8. Well-sprung ribs.
9. Flat-ribbed.
10. Narrow.
11. Deeper body.
12. Greater capacity.
13. Tremendous spring of rib and depth of body.
14. Stronger constitution.
15. Deeper through the heart.
16. Long, deep, strongly supported barrel (Fig. 3-2).
17. Significant increase in depth and width of the barrel toward the rear.
18. Full in the crops.
19. Full at the elbows.
20. Wider in the chest floor.

Mammary System

1. Larger udder (Figs. 3-2, 3-4, 3-5 and 3-10).
2. More symmetrical udder.
3. More evenly balanced udder (Figs. 3-2 and 3-5).
4. Stronger-attached udder.
5. Greater udder quality.
6. Stronger-attached front udder (Fig. 3-2).
7. Stronger-attached rear udder (Fig. 3-8).
8. Wider and higher-attached rear udder.
9. Teats more evenly spaced.
10. Beautifully uddered cow (Fig. 3-2).
11. More typical udder.
12. A balanced udder.
13. More uniformly placed teats.
14. A soft, pliable udder.
15. More level floor of udder.
16. Larger and more numerous veins (Fig. 3-7).
17. More veins on the udder (Fig. 3-10).
18. Longer, larger veins.

19. More firmly attached udder.
20. Milky kind of udder.
21. Carries up and out behind better.
22. Firmer in the front attachment.
23. Tilted udder.
24. Weak attachment in front.
25. Weak attachment in rear udder.
26. Broken udder attachment.
27. Easy attachment.
28. Indifferent attachment.
29. Short front udder.
30. Unbalanced udder.
31. Weak quarter.
32. Slightly easy half.
33. Cut up between quarters.
34. All teats too close together.
35. Hard, meaty udder.
36. Pendulous udder.
37. Short veins.
38. Underdeveloped rear quarters.
39. Strutting teats.
40. Congested udder.
41. Hard lumps in the udder.
42. Loose in front attachment.
43. Funnel-shaped teats.
44. Rear teats too close together.
45. Ill-shaped udder.
46. Badly halved and quartered udder.
47. A strongly attached, well-balanced, capacious, fine-textured udder (Figs. 3-4 and 3-5).
48. A moderately long udder showing average cleavage between halves.
49. More strongly attached foreudder (rear udder).
50. Possessing an udder with more uniform width from front to rear and top to floor than any other cow in the class (Fig. 3-2).
51. Possessing teats of uniform size (Fig. 3-10).
52. Possessing teats of medium length and diameter.
53. Cylindrical teats that are squarely placed under each quarter.
54. Well-spaced teats from both the side and the rear views.
55. Large, long, tortuous branching mammary veins.
56. A heifer with an overdeveloped udder.
57. A fatty udder, undesirable in heifer calves or yearlings.

SELECTING DAIRY CATTLE

Selecting dairy cattle includes not only judging cattle on type and conformation but also evaluating pedigrees and production performance.

Some judging contests, including the FFA National Contest, include classes to be placed on type alone and classes to be placed on type, production and pedigree.

The purpose of this kind of contest is to train contestants in evaluating not only type but also records of production and factors indicating ability of animals to transmit desirable type and high production to their offspring.

The criticism of this kind of judging contest is the difficulty of securing cows whose records have been made under similar conditions of feed and care. Other factors that must be adjusted for are age, number of times milked daily and length of the milking period.

Records should be adjusted to 2X – 305 day – mature age basis for comparison, but adjustment factors do not equalize environmental factors of feed and management and may not be accurate when they are applied to the performance of an individual cow. Adjustment factors are the most accurate when they are applied to groups and shown as averages.

Much of the inaccuracy could be eliminated if all the cows in the class were from the same herd and tested under the same environmental conditions.

Additional information on this kind of contest may be secured from Bulletin No. 4, *Future Farmers of America National Contest,* published by the Future Farmers of America and the Future Farmers of America Foundation, Inc., in cooperation with the Agricultural Education Service of the Department of Education, Federal Security Agency, Washington, D.C.

In evaluating a pedigree, give the most emphasis to the close-up ancestors of the individual concerned. Ancestors past the third generation have little influence. The inheritance from an animal in the fourth or previous generations is diluted so greatly that its contribution is very limited.

One great bull or cow in the fourth generation contributes about 6 percent of the hereditary material of the animal in question, while the sire and dam each contributes 50 percent to the animal's inheritance.

Due to the great variation in the amount of factual information listed in pedigrees, it is difficult to give a definite figure or percentage that will apply to the majority of pedigrees. However, the following rule of thumb may be used as a guide in the evaluation of a two- or three-generation pedigree.

Assuming that the information in the first and second generations of a pedigree is comparable, about 75 percent of the pedigree score on the cow should be determined from the first generation — the parents — and about 25 percent from the second generation — the grandparents.

In your evaluation of a three-generation pedigree, the percentage figures should be about 70 percent for the first generation, 20 percent for the second and 10 percent for the third.

However, if factual information is lacking in a generation, the total score of the pedigree should be reduced accordingly. For example, if the first generation of a two- or three-generation pedigree is completely blank of factual information, the total score of this particular pedigree should not be more than 25 to 30 percent of the maximum pedigree score.

In evaluating the four following pedigrees, note that Cow No. 1 greatly excels the others. She has a type rating of Very Good and two records that average 508 pounds butterfat. She is an animal of high-producing capacity combined with proper type. The average of two records is somewhat more reliable as an indication of her producing ability than only one record. The more records a cow has, the more reliable the information becomes.

All the direct close-up ancestors show high-producing ability or proven transmitting ability for high production and desirable type.

Both the sire and maternal grandsire are proven with 10 or more daughter-dam pairs with relatively high records and a satisfactory increase in production of daughters over dams. The type of the daughters is also satisfactory.

The dam of Cow No. 1 has not only three records with an acceptable average but also two tested daughters with satisfactory records.

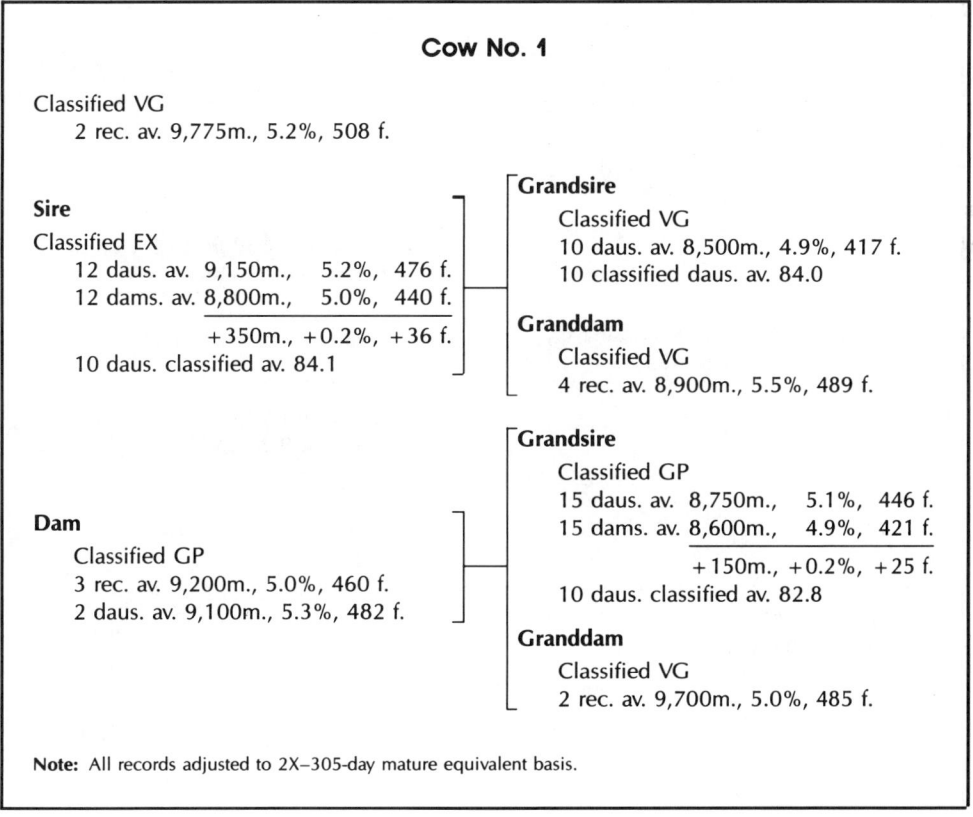

Cow No. 1

Classified VG
    2 rec. av. 9,775m., 5.2%, 508 f.

Sire
Classified EX
    12 daus. av. 9,150m.,   5.2%, 476 f.
    12 dams. av. 8,800m.,   5.0%, 440 f.
               +350m., +0.2%, +36 f.
    10 daus. classified av. 84.1

  Grandsire
    Classified VG
    10 daus. av. 8,500m., 4.9%, 417 f.
    10 classified daus. av. 84.0

  Granddam
    Classified VG
    4 rec. av. 8,900m., 5.5%, 489 f.

Dam
  Classified GP
  3 rec. av. 9,200m., 5.0%, 460 f.
  2 daus. av. 9,100m., 5.3%, 482 f.

  Grandsire
    Classified GP
    15 daus. av. 8,750m.,   5.1%, 446 f.
    15 dams. av. 8,600m.,   4.9%, 421 f.
              +150m., +0.2%, +25 f.
    10 daus. classified av. 82.8

  Granddam
    Classified VG
    2 rec. av. 9,700m., 5.0%, 485 f.

Note: All records adjusted to 2X–305-day mature equivalent basis.

The paternal grandsire has 10 tested and classified daughters, and the two grand-dams each have two or more acceptable records and all with high-classification ratings.

Cow No. 2 has one high record and a satisfactory rating. Her sire is not proven, but the three tested daughters and eight classified daughters give an indication of his transmitting ability, especially when backed by a proven sire. Both grandsires are proven with five or more daughter-dam pairs.

The dam and both granddams each have one satisfactory record, and each has a classification rating.

This pedigree is not as complete as the pedigree of No. 1, but the available information is very promising.

Cow No. 3 has a high-type classification rating and one good production record, and both sire and dam are classified "Excellent," with indications of satisfactory type in the second generation.

However, the production in the first generation is completely blank. This type of pedigree is called a "padded" pedigree. Careful reading will show that the production data listed under the sire and dam are only "filler," for they refer to their respective sires.

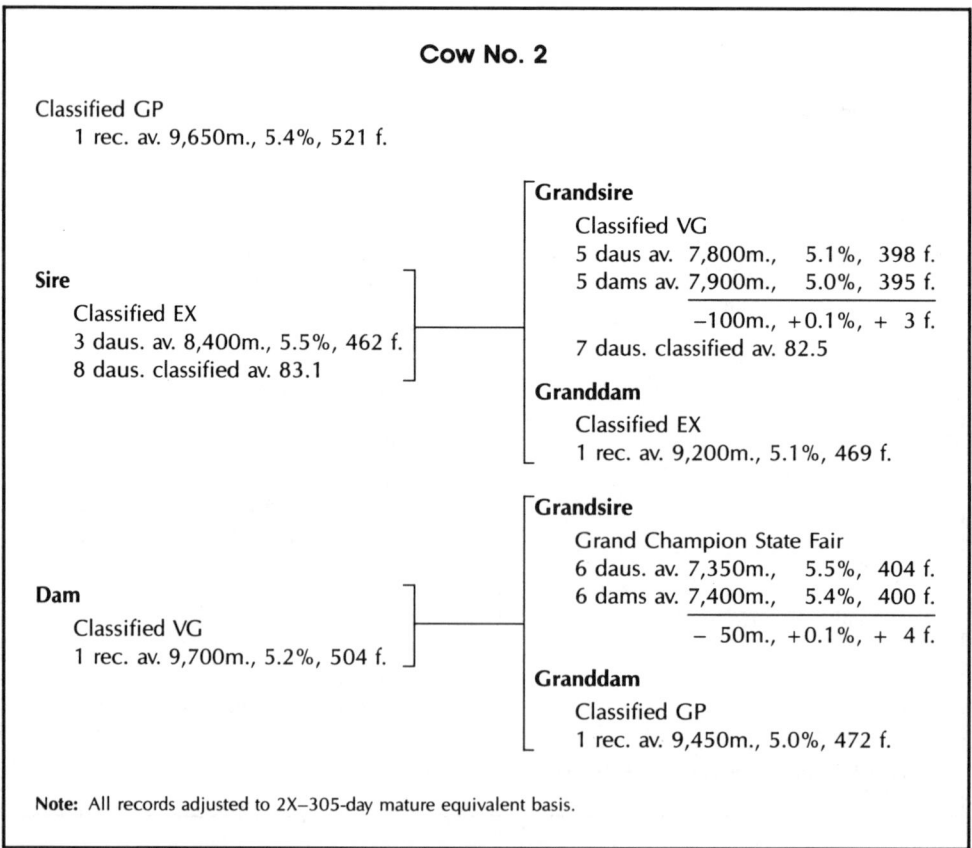

Cow No. 2

Classified GP
   1 rec. av. 9,650m., 5.4%, 521 f.

Grandsire
   Classified VG
   5 daus av. 7,800m.,   5.1%,   398 f.
   5 dams av. 7,900m.,   5.0%,   395 f.
                              −100m., +0.1%, + 3 f.
   7 daus. classified av. 82.5

Sire
   Classified EX
   3 daus. av. 8,400m., 5.5%, 462 f.
   8 daus. classified av. 83.1

Granddam
   Classified EX
   1 rec. av. 9,200m., 5.1%, 469 f.

Grandsire
   Grand Champion State Fair
   6 daus. av. 7,350m.,   5.5%,   404 f.
   6 dams av. 7,400m.,   5.4%,   400 f.
                              − 50m., +0.1%, + 4 f.

Dam
   Classified VG
   1 rec. av. 9,700m., 5.2%, 504 f.

Granddam
   Classified GP
   1 rec. av. 9,450m., 5.0%, 472 f.

Note: All records adjusted to 2X–305-day mature equivalent basis.

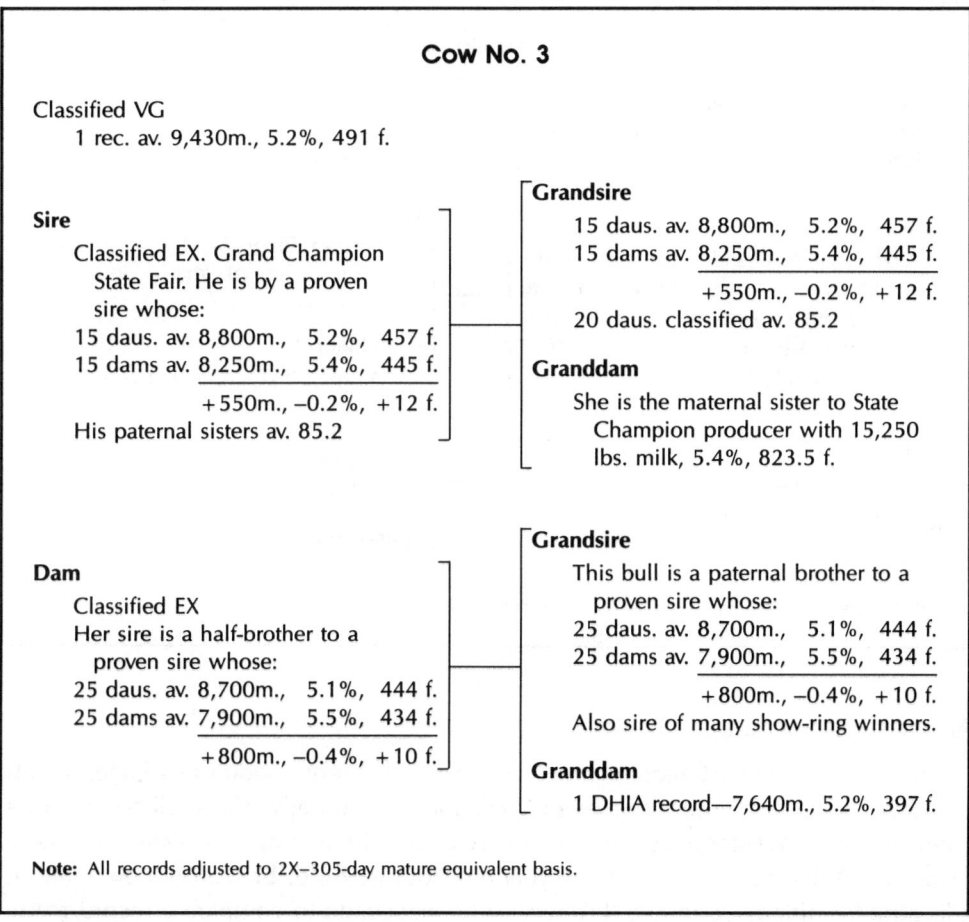

Cow No. 3

Classified VG
  1 rec. av. 9,430m., 5.2%, 491 f.

Sire
  Classified EX. Grand Champion
    State Fair. He is by a proven
    sire whose:
  15 daus. av. 8,800m., 5.2%, 457 f.
  15 dams av. 8,250m., 5.4%, 445 f.
            +550m., −0.2%, +12 f.
  His paternal sisters av. 85.2

Grandsire
  15 daus. av. 8,800m., 5.2%, 457 f.
  15 dams av. 8,250m., 5.4%, 445 f.
            +550m., −0.2%, +12 f.
  20 daus. classified av. 85.2

Granddam
  She is the maternal sister to State
    Champion producer with 15,250
    lbs. milk, 5.4%, 823.5 f.

Dam
  Classified EX
  Her sire is a half-brother to a
    proven sire whose:
  25 daus. av. 8,700m., 5.1%, 444 f.
  25 dams av. 7,900m., 5.5%, 434 f.
            +800m., −0.4%, +10 f.

Grandsire
  This bull is a paternal brother to a
    proven sire whose:
  25 daus. av. 8,700m., 5.1%, 444 f.
  25 dams av. 7,900m., 5.5%, 434 f.
            +800m., −0.4%, +10 f.
  Also sire of many show-ring winners.

Granddam
  1 DHIA record—7,640m., 5.2%, 397 f.

Note: All records adjusted to 2X–305-day mature equivalent basis.

The two grandsires have acceptable transmitting ability, while the paternal grand-dam is a blank, as the record listed under paternal granddam is not her record but refers to her half-sister.

This pedigree shows "Excellent" type for the sire and the dam, but no production or transmitting ability whatever in the first generation, which is the most important.

The pedigree of Cow No. 4 is even weaker than No. 3. Her record is not large and a championship won at a county fair even for three consecutive years does not mean very much, because fair winnings depend upon the competition. In a small county fair, with only a few entries of mediocre cattle, a championship is not significant.

The sire and paternal grandsire show indications of acceptable type and ability to transmit acceptable type, but no indication of production whatever. The dam has a satisfactory production record and a classification rating, but there is no information about either granddam and not very much about the maternal grandsire.

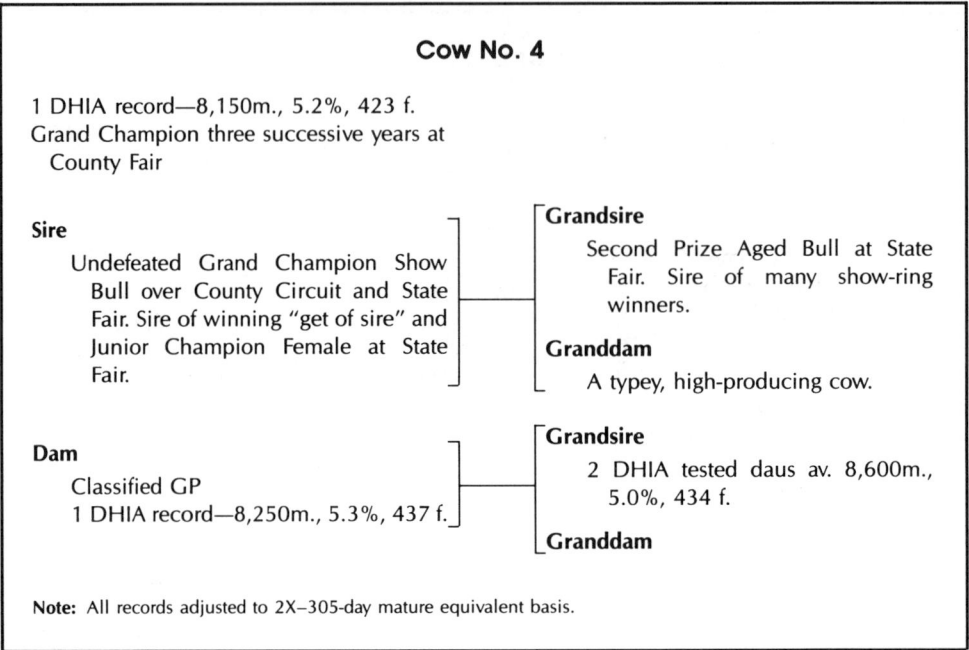

Cow No. 4

1 DHIA record—8,150m., 5.2%, 423 f.
Grand Champion three successive years at
  County Fair

Sire
  Undefeated Grand Champion Show
  Bull over County Circuit and State
  Fair. Sire of winning "get of sire" and
  Junior Champion Female at State
  Fair.

Grandsire
  Second Prize Aged Bull at State
  Fair. Sire of many show-ring
  winners.

Granddam
  A typey, high-producing cow.

Dam
  Classified GP
  1 DHIA record—8,250m., 5.3%, 437 f.

Grandsire
  2 DHIA tested daus av. 8,600m.,
  5.0%, 434 f.

Granddam

Note: All records adjusted to 2X–305-day mature equivalent basis.

Adjustment Factors

In order to compare records made under the different conditions of age, length of test period and number of times the cow was milked daily, adjust all records to a common basis. It is not accurate to compare a record made by a two-year-old with a record made by her dam as a three-year-old, four-year-old, or mature cow without adjusting for difference in age. Likewise, it is inaccurate to compare a record made when the cow was milked four times or three times daily with one when the cow was milked two times daily or to compare a record of 365 days with one of 305 days.

Factors called "age conversion factors" have been developed to adjust all records to mature equivalent (ME) basis. Also, other factors convert a record made under four times or three times daily milking to two times daily milking and also a 365-day record to a 305-day record. The application of these factors adjusts the records to a two times daily milking, 305-day mature equivalent basis, usually shown as 2X – 305 day – ME.

Table 3-1 gives the age conversion factors for adjusting records to mature equivalent basis for cows from 1 year, 6 months of age, to 16 years of age at date of freshening. Since different breeds reach maturity at different rates, separate age conversion factors are shown for the following breeds: Brown Swiss, Ayrshire, Guernsey, Jersey, Holstein-Friesian and mixed breeds.

Also listed are factors for reducing records to a twice-a-day milking basis from a three-times-a-day milking and to adjust a 365-day record to a 305-day basis (see Table 3-2).

TABLE 3-1. DHIA Age Conversion Factors[1]

Age at Freshening	Brown Swiss	Ayrshire, Guernsey, Jersey	Holstein-Friesian	Mixed Breeds
1 – 6	1.718	1.343	1.515	1.429
1 – 9	1.628	1.301	1.446	1.373
2 – 0	1.538	1.262	1.377	1.319
2 – 3	1.469	1.228	1.326	1.271
2 – 6	1.400	1.195	1.275	1.232
2 – 9	1.343	1.168	1.239	1.202
3 – 0	1.286	1.141	1.203	1.172
3 – 3	1.241	1.120	1.167	1.142
3 – 6	1.196	1.099	1.131	1.115
3 – 9	1.166	1.081	1.104	1.091
4 – 0	1.136	1.063	1.077	1.070
4 – 3	1.112	1.049	1.056	1.052
4 – 6	1.088	1.037	1.035	1.036
4 – 9	1.070	1.028	1.026	1.027
5 – 0	1.052	1.020	1.017	1.018
5 – 3	1.040	1.014	1.011	1.012
5 – 6	1.028	1.008	1.006	1.006
5 – 9	1.019	1.003	1.003	1.003
6 – 0	1.012	1.000	1.000	1.000
6 – 3	1.009	1.000	1.000	1.000
6 – 6	1.006	1.000	1.000	1.000
6 – 9	1.003	1.000	1.003	1.000
7 – 0	1.000	1.000	1.006	1.000
7 – 3	1.000	1.003	1.009	1.003
7 – 6	1.000	1.006	1.012	1.006
7 – 9	1.000	1.009	1.015	1.009
8 – 0	1.000	1.012	1.018	1.015
8 – 3	1.000	1.015	1.027	1.021
8 – 6	1.000	1.018	1.036	1.027
8 – 9	1.003	1.021	1.045	1.033
9 – 0	1.006	1.024	1.054	1.039
9 – 3	1.009	1.029	1.063	1.045
9 – 6	1.012	1.035	1.072	1.051
9 – 9	1.021	1.041	1.081	1.059
10 – 0	1.030	1.047	1.090	1.068
10 – 3	1.039	1.055	1.102	1.077
10 – 6	1.048	1.064	1.114	1.089
10 – 9	1.060	1.073	1.126	1.098
11 – 0	1.072	1.082	1.138	1.110
11 – 3	1.084	1.091	1.150	1.119
11 – 6	1.096	1.100	1.162	1.131
11 – 9	1.105	1.106	1.177	1.140

(Continued)

TABLE 3-1 (Continued)

Age at Freshening	Brown Swiss	Ayrshire, Guernsey, Jersey	Holstein-Friesian	Mixed Breeds
12 – 0	1.114	1.112	1.192	1.152
12 – 3	1.123	1.118	1.207	1.161
12 – 6	1.132	1.124	1.222	1.173
12 – 9	1.138	1.130	1.237	1.182
13 – 0	1.144	1.136	1.252	1.194
13 – 3	1.150	1.142	1.267	1.206
13 – 6	1.156	1.148	1.282	1.215
13 – 9	1.162	1.154	1.294	1.224
14 – 0	1.168	1.160	1.306	1.233
14 – 3	1.171	1.166	1.318	1.242
14 – 6	1.174	1.172	1.330	1.251
14 – 9	1.177	1.178	1.339	1.260
15 – 0	1.180	1.184	1.348	1.266
15 – 3	1.183	1.190	1.357	1.272
15 – 6	1.186	1.193	1.366	1.278
15 – 9	1.189	1.196	1.372	1.284
16 – 0	1.192	1.199	1.378	1.288

[1]Data compiled by the Bureau of Dairy Industry, USDA, from DHIA records.

TABLE 3-2. Factors for Converting Production to a 305-Day Basis[1]

Days in Lactation	Factor
240 or less	1.15
241–270	1.06
271–309	1.00
310	0.99
315	0.98
320	0.96
325	0.95
330	0.94
335	0.93
340	0.92
345	0.91
350	0.90
355	0.89
360	0.88
365	0.87

[1]Data compiled by the USDA from DHIA records.

To adjust a 365-day record to a 305-day basis, multiply the record by 0.87, and to adjust a three times daily milking record to a two times basis, multiply by 0.8333, and a four times daily milking record to a two times basis, multiply by 0.7407. For example, an 800-pound butterfat 3X – 365-day record becomes 579.9 pounds of butterfat 2X – 305 days after being adjusted.

The national breed associations also have developed adjustment factors applicable to their respective breeds, based upon official records. These factors differ somewhat from the factors listed here.

Some commercial pedigrees contain terms that may require explanation.

Most pedigrees give the actual production records of individual animals and only show adjusted records in reference to daughter-dam comparison or the average production of the daughters of a sire.

Some pedigrees show complete information, including the age when the records were started, the length of the records and the number of times the cows were milked and the average of all records, while others merely show the highest record, with or without explanatory information. The use of adjustment factors is often necessary in order to make such records comparable.

Sometimes records in pedigrees indicate "DHIA" or "AR" or "ROM" or "Herd Test" (HT or HIR) records. DHIA records are usually made under average farm conditions, while AR or ROM records are usually made of selected individual cows and often under conditions of feed and management more favorable for higher production than DHIA records. Herd Test and DHIA records are made under conditions in which the whole herd is tested; thus, they often are somewhat lower than AR or ROM records.

BREED REGISTRY ASSOCIATIONS

Breed	Association	Officer and Address
Ayrshire	Ayrshire Breeders' Association	Judy M. Disorda Operations Officer 2 Union Street Brandon, Vermont 05733
Brown Swiss	Brown Swiss Cattle Breeders' Association of America	George W. Opperman Secretary-Treasurer Box 1038 800 Pleasant Street Beloit, Wisconsin 53511
Guernsey	The American Guernsey Cattle Club	Bernard M. Heisner Executive Secretary-Treasurer 2105-J South Hamilton Road P.O. Box 27410 Columbus, Ohio 43227

(Continued)

BREED REGISTRY ASSOCIATIONS (Continued)

Breed	Association	Officer and Address
Holstein-Friesian	Holstein-Friesian Association of America	Zane Akins Executive Secretary 1 Holstein Place P.O. Box 808 Brattleboro, Vermont 05301
Jersey	The American Jersey Cattle Club	Maurice E. Core Executive Secretary 2105-J South Hamilton Road P.O. Box 27310 Columbus, Ohio 43227

4

SWINE

When judging swine, you must recognize that the ideal market barrow serves as the basis for the appraisal of the merits in all breeds. It is generally accepted that the producer of swine and the consumers of pork products — as they express their preference in the retail market — influence type in the market hog. The producers are directly concerned with type in the breeding hog as well as in the market hog. The consumers are directly interested in the market hog only. The producers must keep in mind those qualities in type that are favorable to economy in production, such as the ability to produce large litters of thrifty, rapid-growing, early-maturing hogs that are efficient feeders as well as clean, meaty, heavy-muscled hogs. The consumers' influence, however, is expressed very directly and forcefully through the buyers of market hogs and the processors of pork products who pay top prices for the type that yields the highest percentage of the kind and quality of pork cuts that the consumers demand. Accordingly, the producers must attempt to comply as much as possible with producing the type that is desired by the pork trade.

The rather uniform demand of the pork trade for a standard type of ideal market barrow no doubt has been an important factor in encouraging breeders to produce hogs that are very similar in type and usefulness. In general, the breeders are well agreed on the characteristics that have the greatest value to the swine industry. Note these by comparing the barrows, gilts, sows and boars in the different breeds. Attempt to point out, without reference to color, the specific differences in the useful qualities of the barrows that are found throughout these pages. If this is done, the similarity in type will become all the more obvious.

SCORING SWINE

A detailed and persistent study of the important points to consider in swine selection and evaluation (Fig. 4-1) and a discussion of these points are important in learning how to select hogs. Too often students are expected to place classes of

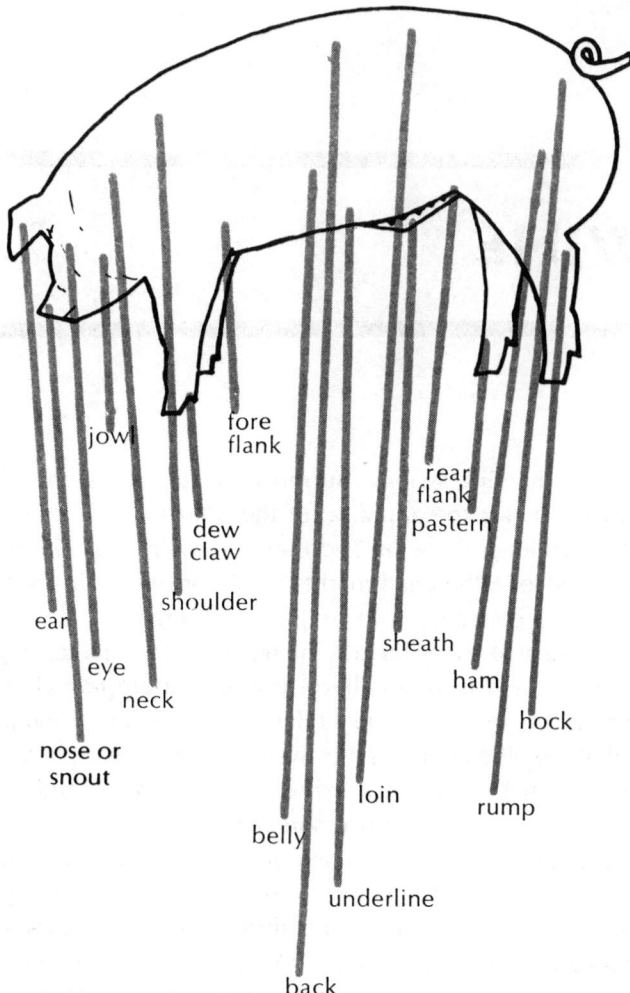

Fig. 4-1. Parts of the market hog. (Courtesy, Purdue University)

swine without previous training in the fundamentals of judging; as a result, they develop bad habits and resort to guessing instead of judging.

To study the points given in the Swine Judging Card, use a placeable class of four hogs. After you have placed the animals on each point, the instructor should discuss the correct placing. You should repeat this exercise several times until you have become thoroughly familiar with the fundamentals of swine selection. When you have mastered this, you may begin using the system of judging animals for final placing.

SWINE JUDGING CARD

Points to Consider   Placing

	1st	2nd	3rd	4th

General Appearance

Moderately wide, thick, meaty, muscular top; long, smooth side; strong, level top; deep-bodied, standing on adequate length of leg; deep, thick, meaty ham with moderate length of shank; smooth shoulder, clean over the top and behind the shoulders; extreme spring to the fore-rib and rear rib; long, deep loin; long rump with a flare to the ham; neat, lean and trim in the head, neck, jowl, along the underline, in the rear flank, at the base of the ham and in the crotch area; and sound, well-placed feet and legs. (See all figures.) . . . .

Form

1. *Head and neck* — Broad, clean-cut and neat; large, prominent eyes; wide forehead; medium-sized ears (should be representative of the breed); neat, trim jowl and cheek and a moderately long, well-developed snout. Neck should be medium length and blend smoothly with the shoulder. (In breeding classes, breed type and character should be considered.) . . . .

2. *Shoulders* — Smooth, clean over the top (free of excess finish), neatly laid in, possessing the same width and depth as the rest of the body. When the animal is walking, it should show evidence of muscling in the shoulder and movement of the shoulder blade . . . .

3. *Back and loin* — Wide, thick, meaty with good length; a strong, firm, extremely heavily muscled, level top . . . .

4. *Sides* — Long, firm, moderately deep, clean and neat in the flanks and free from creases and wrinkles . . . .

5. *Belly* — Trim, clean, firm, straight—not flabby. (In breeding gilts and sows, there should be at least six well-spaced, properly developed teats on each side of the underline.) . . . .

6. *Rump* — Long, thick, muscular, indicating an extreme flare out from the loin and a high tail setting . . . .

7. *Ham* — Deep, thick, meaty, showing outstanding muscle development through the stifle and in the inner and outer portions. Clean and free of excessive fat cover at the base and in the crotch area. Should be firm, clean and heavily muscled, not flabby or wrinkled . . . .

8. *Feet and legs* — Should stand on plenty of hard, clean heavy bone and on moderate (or above) length of leg. Legs should be set straight, wide apart and well-placed on the corners of the body. Joints should be clean, pasterns strong and upright and toes of equal size that do not spread. (Feet and legs are of minor importance in the placing of market hogs.) . . . .

(Continued)

SWINE JUDGING CARD (Continued)

Points to Consider

	Placing			
	1st	2nd	3rd	4th

Finish

A minimum amount of smooth finish, uniformly distributed over the entire body. Overly fat, wasty hogs are undesirable. (Breeding hogs only need to carry enough finish to show the essential features to the best advantage.) . . . .

Quality

Smooth finish and body form; free from wrinkles and flabbiness; clean about the joints and hocks; refined hair, hide, bone, head and ear . . . .

Dressing Percentage

A minimum amount of firm, smooth finish; heavy muscling; high quality and trim middles result in high dressing percentage . . . .

Balance

Harmonious and symmetrical unity of all parts . . . .

Breed and Sex Character

As indicated by strong head, style and breed-type characteristics and ample masculine or feminine sex characters (breeding swine only) . . . .

THE MODERN MEAT-TYPE HOG

During the last 25 years, the swine industry has gone through several type changes. The ideal-type market hog changed from the short, deep, chuffy, excessively fat hog of the early 1950's to the long, lean hog of the mid- to late 1950's and to the moderately long, thick, muscular, trim, heavy-boned hog with minimum amount of backfat of today. The changes have been due to several reasons, but the most frequent reason given is that there was no longer a demand for fat — so the swine industry made an about-face. The result was a long, narrow, rather meatless hog. This hog was light-muscled and did not grow and gain the high degree of efficiency that was so essential to the swine industry. Therefore, some of the unnecessary length was given up for additional thickness, muscling and substance. Swine producers found it necessary to breed, raise and feed the kind that produces a maximum amount of muscle and a minimum amount of fat at 220 pounds in the shortest period of time. This brought about the modern meat-type hog.

The text and illustrations in this section emphasize the distinguishing characteristics of the modern barrow (Animal A in Figs. 4-3, 4-4, 4-5 and 4-6) and the old-fashioned barrows (Animals B, C and D in Figs. 4-3, 4-4, 4-5 and 4-6). These illustrations and explanations pertain to the characteristics of skeletal structure, muscular development and fat deposition of Animals A through D. The points or areas of a hog that can be used as indicators of muscling and finish, either too much or not enough, are illustrated, pointed out and discussed.

First Steps in Evaluating Live Hogs

A basic knowledge of carcass differences is necessary in the evaluation of the live animal. Fig. 4-1 shows the parts of the live market hog, and Fig. 4-2 shows the wholesale cuts of the pork carcass. Confidence and accurate decisions come only after careful observations and considerable experience.

There are five ways that can be observed: front, top, side, rear and walking. The following is a discussion of what you should look for at the various angles during live animal evaluation. This information is based on research findings, practical experience and observations. For the purpose of selecting breeding stock, hog producers must be able to "eyeball" the desirable characteristics in the live animal. The breeding value of an animal is lost when the animal is slaughtered so that it can be

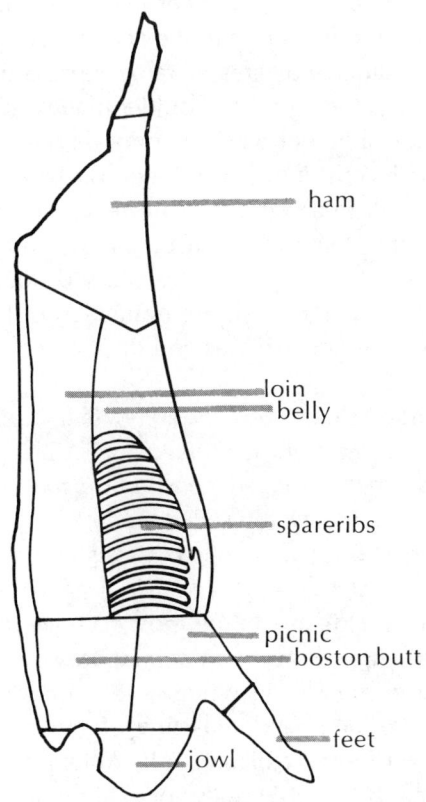

Fig. 4-2. Wholesale cuts of the pork carcass. (Courtesy, Purdue University)

evaluated properly. The discussion that follows will pertain directly to the market hog (barrow); however, the same principles are involved and should be considered when breeding stock are selected to produce the desirable market animal. For breeding animals, those characteristics associated with production, such as femininity, masculinity and breed type, should be studied carefully.

Front View

Modern meat-type Animal A (Fig. 4-3) shows a thick, meaty top (1) that displays a squarer, more desirable turn than either Animal B or C. This could also be described as a more symmetrical turn or a more meat-animal shape than that displayed by either Animal B or C.

Animal A shows considerable muscle development and width of blades in the shoulder area (2). This area can be used as both a muscle and a fat indicator. When a muscular barrow walks, the muscle in the shoulder will bulge and express itself as the muscles contract and relax to allow the barrow to move. If a barrow is carrying too much finish, you will see the fat roll up, or push up, in front of the shoulder blade. There is a certain amount of judgment involved in determining whether the shoulders are coarse and prominent due to the actual skeletal structure of the animal in question, or if the shoulder indicates extreme muscle development. As a meat-type barrow walks, the movement of the shoulder blades should be evident over the top of the shoulders. If the shoulder blades cannot be seen when the animal moves, then the barrow is usually pushing a heavy deposit of fat up over the top of the shoulders. This would result in a very heavy probe at the first rib site.

The jowl (3) is an area that can be used as an indicator of excessive finish. However, not all barrows with heavy jowls are wasty or too fat — so again the judgment factor is of extreme importance in the evaluation of this trait. Animal A has a trim, clean jowl (3) that is free of wrinkles and displays considerable firmness when the animal moves.

Animal A has a wide, clean chest floor (4) that is essential in rugged, fast-gaining, vigorous barrows. The front legs of Animal A are set out on the corners of the body and help provide the rugged constitution that is so necessary in good-doing, fast-gaining, efficient, meaty, muscular, modern hogs. Animal A has very little excessive finish in the chest floor area, which makes for the very trim, clean appearance desired in modern meat-type hogs.

The cannon bone area (5) is one of the most accurate and quickest indicators of bone size, substance and ruggedness of an animal. Research data show that heavier-boned animals have more muscle. In other words, there is a positive relationship between bone size or substance and muscling of swine. Note the very desirable length of leg and the larger foot size displayed by Animal A as viewed from the front.

Old-fashioned Animal B (Fig. 4-3) does not exhibit the uniform symmetrical turn to the top for the meat-animal shape (1) displayed by Animal A. Note the rather flat top that turns at right angles to the ground. This is due to the excessive fat that is

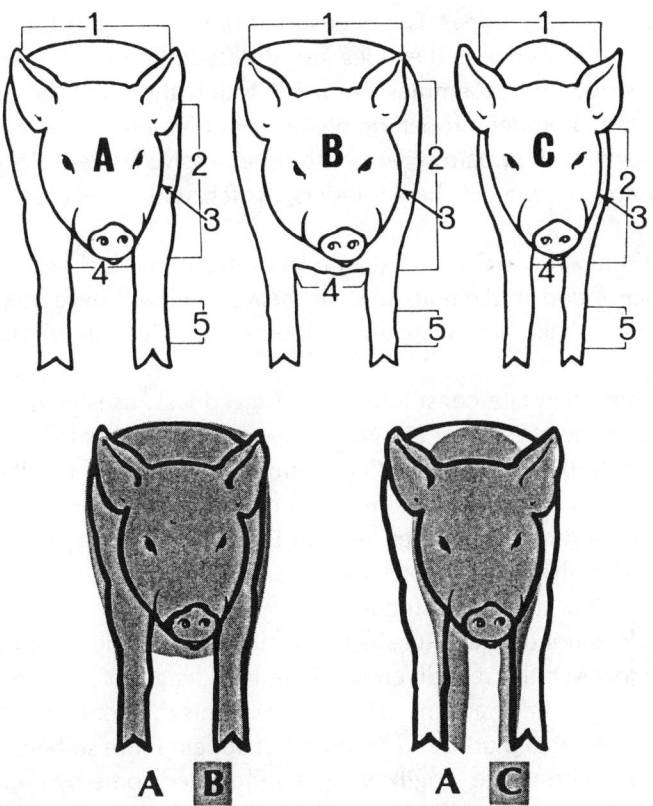

Fig. 4-3. Front view of the modern meat-type (A), old-fashioned (B) and meatless (C) market hogs, with superimposed comparisons. (Courtesy, Purdue University)

deposited along the topline and down over the shoulders on Animal B. Muscles are round, not square. The deposition of fat between and over the body muscle structure is instrumental in giving Animal B the square look.

Animal B does not have the muscular appearance through the shoulder region (2) that is displayed by Animal A. An animal of this kind (B) may be as muscular as Animal A, but excessive finish will often conceal muscle development; however, in most cases, animals of this appearance (Animal B) are usually lighter-muscled and overfinished. Note that Animal B, when compared to Animal A, is extremely smooth. The muscle seams (in between the muscles) are filled with fat to give the animal the smooth appearance that livestock evaluators have desired in the past. The animal is smooth only because it is relatively light-muscled, and the muscle seams have been filled with fat to bring out the smooth effect. Animal B is much wider at the top of the shoulders (2) than at the bottom. This is due to heavy deposits of fat over the top half of the shoulders and down over the top half of the outside of the shoulders. The overfinished barrows will be somewhat wedge-shaped as you view them over the top from either the front or the rear. The widest part of the wedge is the topline and the narrowest part of the wedge is the sides and the underline for the overfinished barrow. When a barrow that resembles Animal B walks, there will be little noticeable movement of the shoulder proper or of the shoulder blades over the top of the shoulder. The overfinished barrow will push up excessive fat deposits over the top of the shoulders and in front of the shoulders, which will be evident from the front view.

The jowl (3) of Animal B is heavy with fat deposits and lacks the trim, clean-cut, firm appearance found in Animal A. A fat, heavy jowl will have several noticeable wrinkles and will shake and wiggle and look extremely sloppy when the animal moves.

Animal B has adequate constitution and ruggedness, as is evident by the width displayed through the chest floor (4). Note, however, that Animal B looks as though it has more depth through the chest than Animal A. Research results have demonstrated that this extra depth is again the result of heavy fat deposits through the chest and along the underline area. When Animal B walks, the loose, sloppy appearance that was noticeable through the jowl will also be particularly obvious back through the throat, chest floor and underline areas.

Animal B does not display quite as much substance in the cannon bone area (5) as Animal A. However, the real difference here is in length of leg. Animal B appears to have much shorter legs than Animal A; however, this difference is rather deceiving. The front legs on Animal B are actually short but appear more so because of the extra depth through the chest. The length of leg is influenced some by depth through the chest and shoulder as well as by waste along the underline. Research has demonstrated that there are certain productive factors, such as feed efficiency and rate of gain, related to length of leg and width and capacity through the chest floor. Today, the trend is to select away from the short-legged barrows and breeding animals because these short-legged individuals are too early-maturing and do not perform as well at the self-feeder or in the farrowing house as the big, stout, heavy-structured, wide-chested hogs that stand up on more length of leg and display more substance of bone and a larger foot size (Animal A).

Meatless Animal C (Fig. 4-3) is too roundly turned over the top (1) and lacks thickness and width over the top (1), which is usually characteristic of a meatless individual. Animal C is very clean (little finish) over the top, and this characteristic, along with the rounded turn to the top, has led many to believe that this type was the right kind. This type of barrow is usually the poor-doing, light-muscled kind that if fed enough will look almost exactly like Animal B.

When compared to the shoulders on Animal A, those (2) on Animal C lack thickness, width, depth and evidence of muscling. Animal C will have, as a rule, very little fat over the top of the shoulders and down over the outside of the shoulders — so there will be very little fat that will push up or be noticeable in either of these areas as the animal moves at the walk.

Note the neat, trim, clean-cut jowl (3) displayed by Animal C. This goes along with the meatless appearance exhibited through the entire make-up of Animal C. Occasionally, a barrow that resembles Animal C will have a loose, sloppy jowl to go along with its meatless appearance; thus, the entire animal, not just part of it, should be evaluated.

Animal C is very narrow and shallow through the chest floor (4) and lacks width between the front legs, which is especially evident in slow-growing, poor-doing barrows that do not have the constitution, ruggedness and vigor of barrows represented by Animal A.

Animal C is light-boned (5) and does not have the substance, durability and ruggedness displayed by Animals A and B. Animal C has good length of leg, which usually accompanies the narrow, meatless-type barrows represented by Animal C.

Top View

Modern meat-type Animal A (Fig. 4-4) expresses muscling in the shoulder area (1) from this view as well as from the front view. Barrows that are meaty and muscular are going to have some thickness and muscling through the shoulders. When you evaluate barrows on the basis of percent muscle, the barrow with more shoulder, width and muscle and less backfat will usually have a higher percent muscle than narrow-shouldered, narrow-chested, lighter-muscled, fatter barrows. We're just fooling ourselves if we turn down barrows that have some muscling, thickness and development through the shoulder region. Animal A is neat and clean over the shoulder, and the shoulder movement and the muscle expression can be seen through this region as the animal walks out.

Animal A is not full and smooth back of the shoulders, in the heart-girth or fore-rib area (2). The skeletal structure and muscle development do not permit an animal to be smooth in its appearance in this area. The shoulder blade is attached to the outside of the rib cage — so unless the animal is extremely narrow-chested, the shoulder is going to be noticeable. If the animal is heavily muscled and thick, the shoulder will be even more noticeable, especially if the individual has the desired spring and squareness of rib that is so essential in modern meat-type hogs. A muscular animal that is properly finished should not be required to be extremely neat-shouldered and full back of the shoulders. The full heart-girth or fore-rib area of animals similar to B is smooth and full because of excessive finish over the top of the shoulders and just back of the shoulders in the fore-rib area. Note the correct shape over the top and the tremendous spring of the fore-ribs and rear ribs (2) of Animal A. Animal A tends to taper a bit from the shoulders back to the loin-ham (or rump) connection. This is because the heaviest fat deposit over the top of a barrow is over the shoulders, and the lightest fat deposit is over the rump. The fat is responsible for some of this tapering effect, plus there are no ribs in the loin area proper to help maintain the spread and thickness uniformly over the top and down through the sides.

Animal A has tremendous length, volume, thickness and muscle dimension through the ham (3). Note how the ham flares out from the loin and how far forward

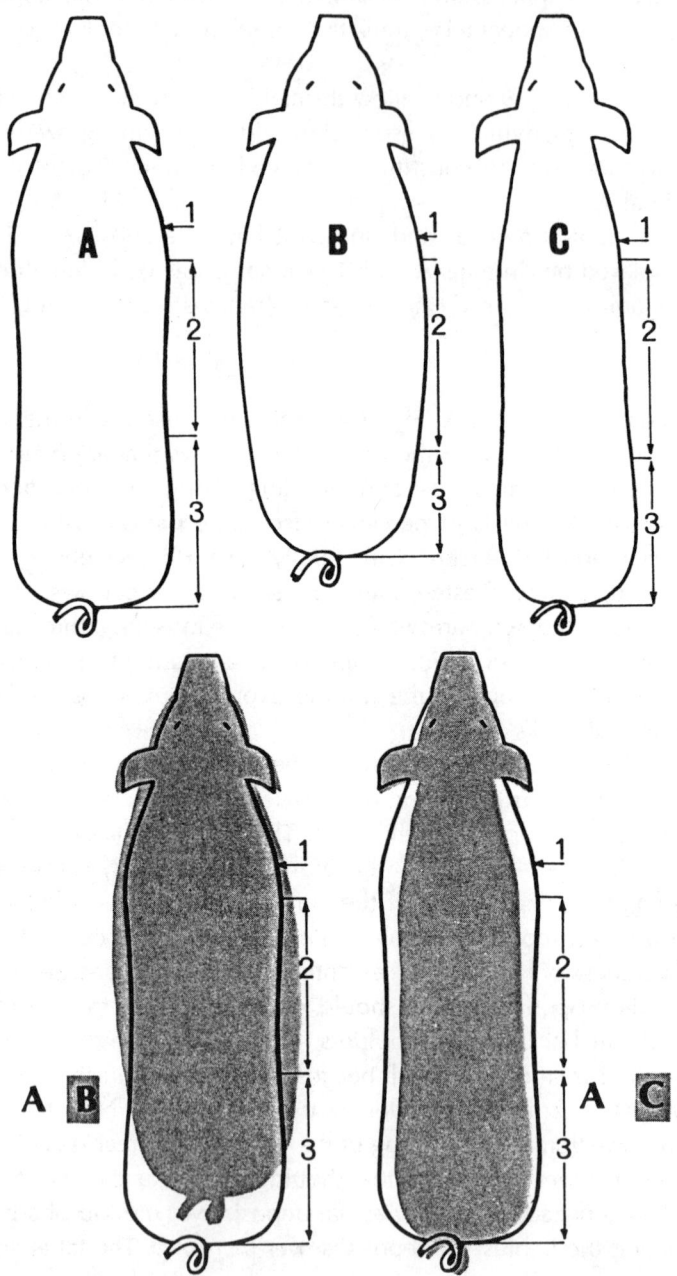

Fig. 4-4. Top view of the modern meat-type (A), old-fashioned (B) and meatless (C) market hogs, with superimposed comparisons. (Courtesy, Purdue University)

the rump or ham ties or hooks up into the loin and side. This represents length of rump. The thickest part of Animal A is through the center of the ham (stifle muscle area). This is the greatest muscle mass in the animal's body. If an animal is meaty and muscular, the ham area should be the thickest portion of the animal when it is viewed from either the top or rear. Observe the extra width and thickness through the shoulders, tapering back slightly into the loin with the thickest, meatiest, heaviest-muscled portion through the center of the ham (stifle muscle area). This is sometimes referred to as the desired "right shape" or the "correct shape" in the modern meat-type barrow.

Old-fashioned Animal B (Fig. 4-4) displays rather uniform thickness from front to rear. The terms commonly used to describe an animal of this type are "smooth" and "uniform." In this case, uniformity doesn't mean a thing except that the animal is carrying excessive amounts of waste fat in several places throughout the general body make-up. Animal B does not show much muscle development through the shoulder area (1) either because it was light-muscled to start with or because the muscling is concealed by too much fat. At any rate, neither animal is the right kind. There will be very little shoulder movement and muscle expression over the top of the shoulders and down over the outside of the shoulders when Animal B moves out. However, there will be excessive amounts of fat pushed up over the top of the shoulders and in front of the shoulders as the animal walks out.

Animal B is very smooth and full behind the shoulders in the heart-girth or fore-rib area (2). Animals that take on this general appearance usually carry extra amounts of fat just behind the shoulders in the fore-rib area, and as a rule they have heavy fat deposits in the elbow pocket (point of the elbow). When Animal B walks, there will be excessive amounts of fat that will roll up just behind the point of elbow (elbow pocket) as the front legs move and the animal shifts its weight forward to facilitate the movement of its body. Note that Animal B is somewhat boat-shaped, with the widest part in the middle of the body. This is the area where the least volume of red meat or muscle is located. Animal B has a rather sharp right-angle turn over the top and in no way exhibits the tapering effect from the shoulders to the loin-ham (or rump) connection displayed by Animal A. In fact, the shape is almost inverted, which is mostly due to fat but which also is influenced by the amount of muscling.

Animal B has a short ham (3) that lacks the thickness, length, dimension and muscling possessed by Animal A. Animal B is short and tapering in the rump. This short, tapering appearance in the rump is quite often associated with animals that are short-bodied, light-muscled and overfinished. Animal B displays no flare to the ham, but it does have a very smooth, uniform appearance throughout, which means too much finish and too little muscle.

Meatless Animal C (Fig. 4-4) is clean (free of excess finish) over the shoulders but shows very little evidence of muscle development through the shoulder area (1). Meatless barrows will be very sharp up over the shoulders, occasionally referred to as being "fishbacked." The narrowness in the chest floor area and the close set of the front legs permit the shoulder to set in rather smoothly on the meatless barrow. For these reasons, Animal C appears very narrow, light-muscled and extremely meatless through the shoulder area (1).

There is very little shoulder development on Animal C; consequently, Animal C appears to be smooth and full just back of the shoulders in the heart-girth or fore-rib area. This is due to the lack of muscling through the shoulders and the general lack of spread, thickness and muscling all down the top of Animal C. Animal C is clean down the top and is more correctly turned over the top than Animal B. Compared to Animal A, Animal C exhibits very little, if any, tapering effect from the shoulder to the loin-ham (or rump) connection. This helps establish the narrow-topped, meatless condition of Animal C.

Animal C displays adequate length of rump (3) but definitely lacks the dimension, thickness and muscling in this area and down through the ham that is so evident in Animal A. Meatless barrows, represented by Animal C, will usually have a reasonably high percent muscle because they have very little waste fat deposited over the carcass. However, animals of this type (meatless) lack acceptable carcass quality and as a rule yield bellies that are too thin to be suitable for bacon production. These reasons, plus the fact that barrows represented by Animal C are slow-gaining, poor-doing barrows that do not have the muscle volume of the modern meat-type barrow, are the best evidence that could be presented as to why Animal C (the meatless barrow) is not the barrow of the future. Animal C would eventually have the same general shape from the top view as Animal B if it were fed long enough and fattened to the degree illustrated by Animal B.

Side View

Modern meat-type Animal A (Fig. 4-5) is clean, firm and trim in the jowl (1), shows extreme muscle development through the shoulder and forearm area (2) and stands up off the ground on correctly set, heavy-boned front legs that possess excellent substance (3). Note that Animal A is free of excessive fat deposits in the elbow pocket (foreflank area) (4) and wrinkles over the shoulder. Animal A is a deep-ribbed, deep-sided barrow that shows excellent length of side (5). At this time, there is no research or evidence available that demonstrates any advantage to extreme length of body or side in the modern meat-type hog. Animal A is deep and loose in the rear flank (6), which is quite acceptable under present-day evaluation standards. In past years, it was practically unheard of for a champion animal not to be tucked up in the rear flank. Research has yielded considerable evidence that this extra depth and fullness in the flank was due to nothing but fat deposited between the skin and the abdominal wall all along the underline, but particularly in the flank region. Thus, you must make sure the rear-rib depth and flank depth and looseness are true body volume that will provide you with a visual measure of internal body dimension in all hogs and pig-carrying capacity in females. Note again the length of leg; the straight, correctly placed hind leg; the heavy, rugged bone; and the moderately long, strong pasterns (7) possessed by Animal A. The strong pasterns and straight hind legs help

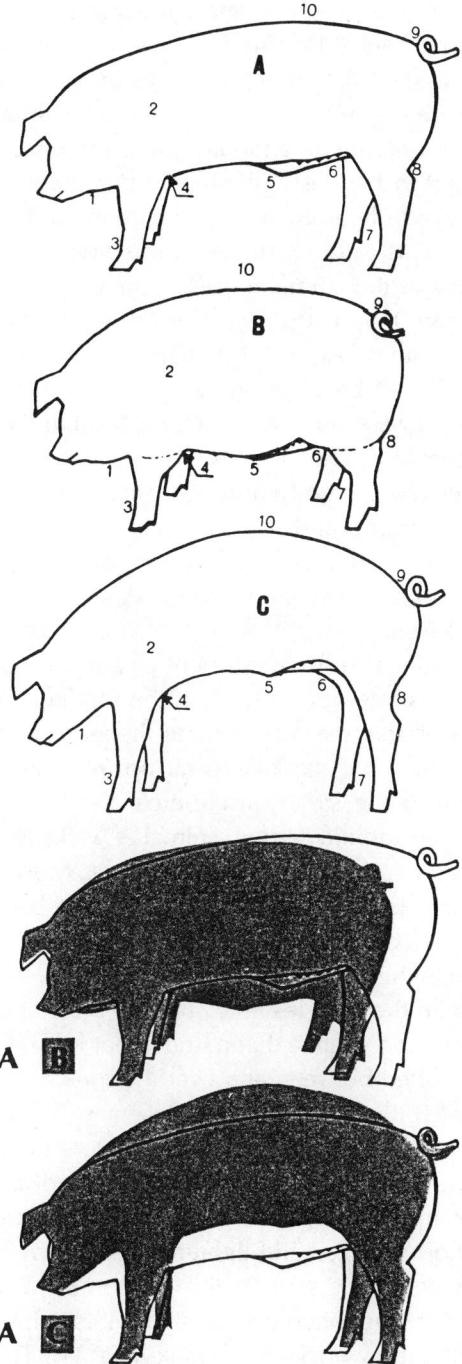

Fig. 4-5. Side view of the modern meat-type (A), old-fashioned (B) and meatless (C) market hogs, with superimposed comparisons. (Courtesy, Purdue University)

an animal express and display its muscling and correctness. Animal A does not have the extremely deep bulging ham (8) desired by livestock (swine) evaluators in the past. The ham ties in above the hock joint and leaves the joint clean and neat when the animal is viewed from either the side or the rear (8). The skeletal structure and muscle development do not permit an animal to have the extreme depth and bulge to the ham that was sought in the past. This highly unnatural condition of extreme depth and bulge of ham is mainly brought about by excessive fat deposits in the area of the cushion of the ham. This causes wrinkles at the base of the ham and a general lack of firmness in this area as the animal moves. Animal A has a high tail setting (9) that makes for a deeper ham than is displayed by Animal B, even with the extra fat deposits at the base of the ham on Animal B. When the tail sets (or fits) higher up on to the back, muscle length will be increased and will result in more ham volume than when the tail setting is low. Animal A has a long, level, uniformly arched topline (10), which is a desirable characteristic of modern meat-type hogs. Usually a strong, high arch to the top is related to tightly muscled, tightly wound, poorly structured hogs that are steep in their rump and that lack the durability and soundness that Animal A exhibits. Note also the long rump that is emphasized by the level top that carries from the tip of the nose to the tail setting of Animal A.

When Animal A walks out, note the firm, trim, clean-cut jowl; the muscle expression through the shoulder and the movement of the shoulder blades; the long, deep, clean underline that is free of excess fat in the foreflank and rear flank areas; the length of leg, the heavy bone, the correct set to the feet and legs and slope to the pasterns; the long, meaty ham that fits or ties far up into the side and the muscle expression and movement in the stifle muscle area; the high tail setting; and the long, level, uniformly arched topline. When Animal A walks, you will note that it is very firm throughout. This is an indication of muscling. Muscle is firm, and fat is loose and sloppy. Watch the jowl, underline, rear flank and base of the ham. When the muscles in these areas contract or tighten, there will be a definite degree of firmness displayed. Then, when these same muscles relax, they will still display firmness in a relaxed state. If the muscles relax in these areas and become very loose and sloppy, then there is a good chance that part of what looked like muscle or body depth was simply fat deposited between and over the muscle tissue.

Old-fashioned Animal B (Fig. 4-5) is wasty, loose and flabby in the jowl (1) and extremely thick and heavy through the shoulders (2). Note the wrinkles in the jowl and over the shoulder. These wrinkles are nothing but rolls of fat. There will be very little, if any, evidence of muscle expression through the shoulder and forearm of Animal B. Animal B is shorter-legged and lighter-boned (3) when compared to Animal A. The less-than-average bone size in the cannon bone area would tend to support the lack of muscling mentioned previously for this animal. There is evidence of heavy fat deposit in the elbow pocket (foreflank area) (4). This will be especially noticeable when Animal B walks, as the point of the elbow will push back the rolls of fat that are present in this area. Animal B has a short, flabby underline and a deep body with numerous wrinkles along the side (5). The extremely deep side of Animal

B is mainly due to excessive waste and flab all along the underline and into the flank area. There has been some thought that additional depth indicated extra capacity and resulted in faster-gaining, better-doing animals. Research studies have demonstrated that animals get extra capacity from body length (up to a certain point) and not depth. Body depth stops at the point of the elbow, as this is directly in line with the sternum. Within a population of similar age and weight, depth of body does not vary much at all. The ribs spring out from the backbone (vertebral column) and connect to the breastbone (sternum). No matter how deep-bodied an animal gets, it is not possible for the body capacity to increase, except for elongation of the rib bones, which is rather slow after hogs reach about 100 days of age on a standard plane of nutrition. There is very little muscle in the lower half of the midsection of swine, so the extra depth is of no great value. When animals fatten, they can only deposit so much fat within their muscles (intramuscular), and the remainder is deposited on the outside or between muscles (intermuscular). This is why the extra depth displayed by Animal B is mainly due to extra flab and waste fat rather than red meat and muscle. Note the deep, full rear flank (6) on Animal B. The flank will appear very loose and sloppy when the animal walks, because this full flank once again is the result of waste fat. The rear legs (7) of Animal B are short, are set under the body too far, and do not have the clean-cut, high-quality appearance demonstrated by Animal A. The lack of substance of bone, weak pasterns and incorrect placement of the rear feet and legs are evident in the illustration depicting the old-fashioned barrow (Animal B). The deep, bulging appearance of the ham (8) possessed by Animal B is mainly due to extremely large deposits of fat on the cushion of the ham. Research results have shown that this extra depth is predominantly fat and really gives a false impression of red meat and muscle development in the lower ham area. The deep ham, illustrated by the drawings representing Animal B, carries way down into the hock joint area, and many times this will result in a coarse, puffy condition in this area. The tail setting (9) on Animal B is low, and consequently, tends to make the ham a bit shallower than it would actually seem when the individual animal is observed. Note that Animal B has a rather short, flat, unevenly arched topline (10). A short, flat, uneven top is often associated with overfinished, wasty, light-muscled hogs. Note also the short, steep rump displayed by Animal B. This characteristic is usually typical of light-muscled, overfinished hogs.

When Animal B walks out, note the loose, sloppy, wasty jowl; the presence of wrinkles on the jowl, shoulders, sides, underline and at the base of the ham; the loose, flabby appearance in the foreflank and rear flank areas, along the underline and at the base of the ham; the light bone, crooked legs and weak pasterns that result in incorrectly placed feet and legs; the short side; the short ham; the low tail setting; the uneven topline; the general lack of muscling throughout; and the evidence of excessive finish over the entire skeletal make-up. This is an old-fashioned barrow, and there are still many hogs produced and marketed today that would fit this description rather closely. Old-fashioned barrows have often been referred to as cob-rollers; pumpkin seeds; peanuts; the short, chuffy kind; the lard type; and so forth.

Meatless Animal C (Fig. 4-5) is extremely neat and clean in the jowl (1) and shows considerable firmness throughout this area when it is walking. There is very little evidence of muscle development through the shoulder and forearm areas (2) of Animal C. Note the length of the front legs (3) displayed by Animal C and also observe the definite lack of substance of bone that is especially evident in the cannon bone area (3). As mentioned previously, less substance and ruggedness will usually accompany the lack of muscling, ruggedness and durability, or vice versa. Animal C is clean and neat in the foreflank (4), along the underline (5) and in the rear flank (6) areas, which is especially noticeable when Animals B and C are compared. Note that the meatless barrow (Animal C) resembles Animal A more closely from the side view than does Animal B (old-fashioned barrow). This is why it is so important for swine judges to be aware of the true differences between Animals A and C. Animal C shows adequate length of side. In most cases, the meatless-type barrows (animals similar to C) are never too short; if anything, they are usually long and snaky-looking. Animal C could be criticized for not having enough depth of body (5) and capacity, and in most cases, the criticism is justified. The meatless barrows (Animal C) will usually be poorer-doing, less efficient animals that do not grow and muscle up as fast because they do not have the capacity and the genetic potential to do so. Observe the cut-up appearance in the rear flank area (6) displayed by Animal C. This is because Animal C is not carrying a lot of waste fat; therefore, it hasn't filled the flank areas (4 and 6) with waste and flab. Animal C stands on too much length of rear legs (7) and definitely lacks substance of bone (7). Note the "cat ham" appearance of the ham (8) on Animal C when it is viewed from the side. The ham (8) is very shallow and narrow and simply shows the evidence that Animal C is light-muscled throughout. The tail setting (9) is low, and the rump is short and steep, both of which contribute to the short, shallow, narrow ham (8). Animal C has a long topline (10), which is strong and highly arched (10). Again, the long, narrow, generally high top, with a slight weakness just behind the top of the shoulders, is a good trademark of the meatless barrow (Animal C). Meatless barrows (Animal C) usually display some degree of sharpness over the shoulders and fore-rib area to go along with their uneven topline. Through close comparison of Animals A and C, the differences between the modern meat-type barrow and the meatless barrow should become quite evident.

Rear View

Modern meat-type Animal A (Fig. 4-6) has a long, level top, is squarely turned over the top and has an extremely desirable meat-animal shape (1). The *extremely* meaty barrows that are carrying a minimum amount of backfat will actually have a groove down their top immediately above the vertebra of the back (dorsal vertebral process). Many of the *extremely* heavily-muscled barrows are too tightly wound and lack production value and confinement adaptability. When a heavy-muscled barrow (Animal A) of this description is observed, it will have an average backfat probe or

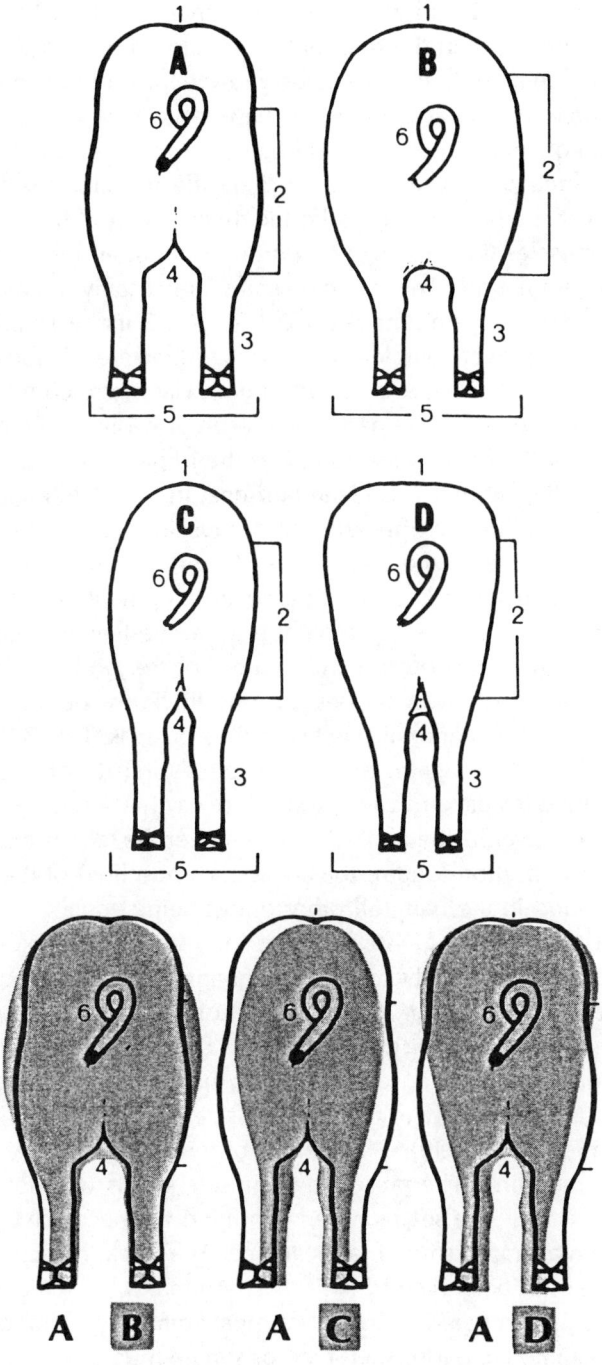

Fig. 4-6. Rear view of the modern meat-type (A), old-fashioned (B), meatless (C) and meatless-fat (D) market hogs, with superimposed comparisons. (Courtesy, Purdue University)

measure from 0.7 to 1.0 inch and will yield high percent muscle. The above description is more the exception than the rule, but more and more market barrows should fit this description. The ham (2) is the greatest muscle mass of the animal's body and should be trim, thick, muscular and should have large total dimensions. Note the *flare* to the ham (2) of Animal A. The muscling causes the ham to flare out or spring out from the loin area somewhat. Because the stifle muscle area is located right through the center of the ham, it should be the thickest part of the ham. A muscular hog should have muscles that are visible upon movement and that flex and express themselves upon stimulation. The muscle seams are usually visible in the meaty barrows carrying a desirable amount of backfat (0.7 – 1.3 inches). Animal A exhibits the heavy bone and correct rear leg structure (3) that are characteristic of most modern meat-type hogs. Animal A is cut up in the twist or crotch (4). The deep, full crotch commonly referred to in the past was a result of waste fat. During the feeding period, fat deposits in the crotch area and gives the impression of additional muscle development. Note that Animal A is clean and neat in the crotch and at the base of the ham and that it exhibits a visible amount of muscle through the inner ham area (4). The weight or volume of the ham is greatly influenced by the thickness and length it possesses, and the additional muscle development of the inner ham on Animal A is an excellent example (4). Note the set to the feet and legs of Animal A. The feet (5) are set wide and out on the corners of the body. Animals that carry above-average muscle must have wide-set legs and thickness between them to carry the vast amounts of red meat and muscle tissue they possess. This is illustrated in the drawing of Animal A. Once again, the higher tail setting (6) of Animal A is mentioned. The tail setting is neat and clean and in no way has a countersunken appearance that is evident in overfinished barrows. The higher the tail setting (6) fits into the rump, the greater the portion (depth, thickness and muscling) of the total pork carcass that will be found in the ham, all other things being equal.

Old-fashioned Animal B (Fig. 4-6) does not show any muscle expression over the top (1), and in general, has a rather square, right-angle look over the top. Note that Animal B is wider at the top than it is at the bottom. This is the result of heavy fat deposits down the top and along the loin edge. The fat deposits give Animal B a smooth, symmetrical, full-looking appearance. Animal B lacks muscle expression through the ham (2), particularly in the stifle muscle area. Note how the ham tapers toward the bottom; this is a definite indication of too much finish and a lack of the desired degree of muscling. The presence of muscle seams will not be seen in the ham of Animal B because the seams have been filled with excessive amounts of fat, which give the ham a round, smooth appearance. The thick, muscular, high-volume dimensions are the ones possessed by Animal A. Animal B has a rather square, right-angle look when it is compared to the more meat-animal appearance of Animal A. This square, right-angle, table-top, shelf-type or square-muscled look can be used as an indication of excessive finish. Note the short legs and average bone size (3) displayed by Animal B, which is usually associated with light-muscled, less-productive hogs. Animal B possesses a deep, full crotch and an inner ham area (4).

This area is loaded with fat, which is highly responsible for the deep, full-looking appearance. The muscle structure in the inner ham region in no way forms and develops to give the deep, full look displayed by Animal B. Also note the sagging appearance in the twist area just below the tail. The heavy fat deposits in this area, and at the cushion or base of the ham, cause wrinkles to form in these areas, and the end result is a loose, sloppy, flabby appearance that is particularly evident when the animal is walking. Note that the feet and legs (5) of Animal B are set too close together and under the center of the body. A close set of the feet and legs usually indicates a lack of muscle in the animal that possesses this characteristic. Animals that have this appearance and lack muscling tend to fatten faster; thus, they become the overfinished, light-muscled hogs that are undesirable according to present evaluation standards. The tail setting (6) on Animal B is rather low and somewhat counter-sunken into the body of the animal. This is the result of excessive fat build-up around the tail. The lower tail setting of Animal B (6) results in less total ham volume when it is compared to the higher tail setting displayed by Animal A. As a rule, a low tail setting will accompany a short, steep rump.

Meatless Animal C (Fig. 4-6) possesses a rounded turn to the top (1) and indicates very little muscle development over the top (1). The ham area (2) is rather flat and expressionless with very little evidence of the desired muscle development through the stifle (center ham). Note the long-legged, fine-boned traits (3) displayed by Animal C. These traits quite often accompany the lack of muscling. Animal C is extremely cut up in the twist or crotch area (4) and shows almost no evidence of muscle development through the inner ham region (4). The hams from Animal C would be lean with a minimum amount of fat on them, but they would have very little dimension and would lack the quantity demanded by the consumer at today's self-service supermarket meat counter. Note the narrow stance (5) of Animal C. The narrow placement of the rear legs is very typical of meatless barrows and is one of the best traits to use in identifying animals of this nature. The tail (6) is placed rather low on Animal C and will be accompanied by a short, steep, narrow rump and a narrow, shallow ham.

Meatless-fat Animal D (Fig. 4-6) is a meatless barrow that was fattened to the same degree as Animal B, the difference being that Animal B was a more muscular, wider-structured individual to start with than Animal D. Note the very square, right-angle, table-top look exhibited over the top (1) on Animal D. Since there was less muscle to start with in Animal D, the finished product is more exaggerated than Animal B. Animal D displays a rather "pear-shaped" appearance from the rear, with the top being considerably wider than the bottom. This is truly an indication of a lack of muscling. Note the flat, expressionless appearance to the ham (2) of Animal D. The fine bone (3) characteristic is another good indicator of a lack of muscling. Animal D has the same deep, full appearance in the twist or crotch area (4) as that displayed by Animal B. This is the result of heavy fat deposits in this area, plus the lack of muscle development through the inner ham region (4). The rear legs (5) are set very close together and indicate that Animal D has a very low muscle volume through the ham region. The additional thickness displayed by Animal D, when compared to Animal C, is essentially all fat. Note the countersunken appearance of the tail (6) on Animal D, as well as the wrinkles at the cushion of the ham and in the upper crotch or twist region. All these characteristics point toward an overfinished, light-muscled animal. The same general characteristics are evident in any type hog when it is fat (compare Animals B and D).

It is extremely important for breeders, feeders, packers, judges, professional hog producers, or anyone else connected with the pork industry, to keep in mind that visual appraisal is not the only tool to use in the selection and improvement of modern meat-type hogs. Information from swine-testing programs can be very helpful in assisting the pork producers of today to achieve the goals of tomorrow. The real differences between Animals A, B, C and D are due to their genetic makeup or breeding. Management will help express the genetic potential (either good or bad) of each hog. Since genetics is primarily responsible for the differences, it is the combination of subjective (visual appraisal) and objective (physical measures) methods that provides the best way to produce modern meat-type hogs.

Walking

The muscling of the pig can be seen while the animal is walking. Distinctive movements of the shoulder blade mean that there is not a heavy deposit of fat over the shoulder. In the well-muscled pig, the muscles flexing or moving in the forward part of the ham can be easily observed.

The firmness of the fleshing is very evident as the pig is moving. A heavy roll of fat, located above the foreflank (see Fig. 4-1) of the pig means too much fat, particularly in the shoulder area.

Both the desirable market animal and the desirable breeding animal should have the characteristics of Animal A in Figs. 4-3, 4-4, 4-5 and 4-6. Plus, they need to have the correct slope to the shoulder, set to the front and rear feet and legs and slope to the pasterns. The toes should be the same length, and the animal should walk out with style and with ease of movement. Hogs should gain 1 pound for every 2 pounds of feed consumed and should weigh 240 pounds in 150 days or less. Such animals need rugged framework, correct structure and above all heavy bone and sound feet and legs that will carry the weight while they are being raised in confinement and on through the breeding cycle if they are breeding animals. The barrows of tomorrow will consistently yield 40 to 45 percent ham and loin or 55 to 60 percent muscle (lean), depending on the degree of muscling and finish. These questions need to be answered: Is the barrow of tomorrow here today? Are the judges and evaluators on the right track in using the present standards, measures and ideals? Are all segments of the industry considered in the show ring and on the rail? Is the marketing structure justly recognizing the advantages of the modern meat-type barrow?

The demand for a carcass that has an increased percentage of lean meat and a relatively small percentage of separable fat has resulted in the modern meat-type. This is a comparatively recent accomplishment; and as the various pictures show, the modern meat type is not limited to one breed. The general pattern of change is relatively uniform in all the breeds, and champions are found among them all. The potential for change in type is inherent in all breeds, and the competition among breeds brings out the best that each breed has to offer in compliance with the public demand for more lean meat and less fat.

The modern meat type encompasses the characteristics that contribute to the highest degree of economic efficiency from the consumer's point of view, which ultimately resolves itself into the most profitable product for the producer as well.

In judging modern meat-type barrows, you should note that there is less variation in the value of the different parts of the hog carcass than there is in the carcass of the market steer or lamb (Table 4-2). This factor emphasizes the importance of stressing balanced development of all parts of the market hog. While it is customary to place the greatest emphasis on the hams and loins, you will notice in Figs. 4-7, 4-8, 4-9 and 4-10 that the shoulders and sides of prize-winning barrows are, in general, as acceptable in development as the parts involving the higher-priced cuts.

Inasmuch as smooth and balanced development is very important in swine, you should attempt to get a good idea of the **general appearance** before undertaking to make a detailed analysis. This can be done by making observations at a distance of about 25 feet from the hog. Note from this point the length, depth, thickness, muscling and balance; the strength and uniformity of arch and muscle development over the back, loin and rump; the turn over the top and the firmness, trimness and amount of finish down the top, in the jowl, along the underline, in the flanks, at the base of the ham and in the crotch region; the muscling and meatiness of the ham and the neatness of the head and neck; the smoothness and general refinement (quality); the substance of bone, the soundness and the quality of the feet and legs.

In good barrows the **back** and **loin** are wide, meaty and heavily muscled, showing good length; a strong, smooth, uniform arch; a correct turn to the top; and a minimum amount of hard, firm finish (Figs. 4-7, 4-8, 4-9 and 4-10). Hogs do not have to be extremely thick to be heavy-muscled. A hog with less thickness and a minimum amount of fat cover will yield a higher percentage of lean meat than a thicker hog that gets the additional thickness from excess fat cover. The topline in most of the modern breeds shows a fairly strong arch, with the highest point midway between the shoulders and the rump (Figs. 4-7, 4-8, 4-9 and 4-10). Hogs that carry their backs up high and display excellent toplines also invariably have a cleaner, neater, straighter-carried bottomline. They do not show a loose, flabby belly as do hogs that have a flat top (back); consequently, they do not have a bottomline that hangs very low and loose. From both the producer's and the consumer's viewpoints, hogs that possess a strongly arched topline are more heavily muscled through the back and loin than those with a flat, poorly supported back and a weak, narrow loin. From the processor's viewpoint, a neat, trim bottomline is much more desirable than a loose, flabby one because less trimming is necessary along the bottom side when the bacon belly is trimmed out. A narrow, weak back and a thin, narrow loin are very objectionable.

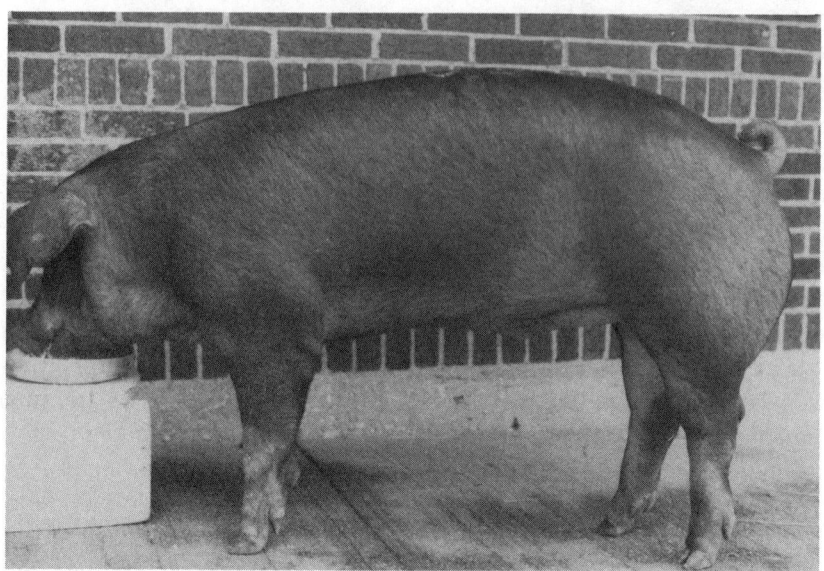

Fig. 4-7. Outstanding Duroc barrow, one of the Top Premier Barrows at the 1984 Illinois State Fair. (Courtesy, United Duroc Swine Registry)

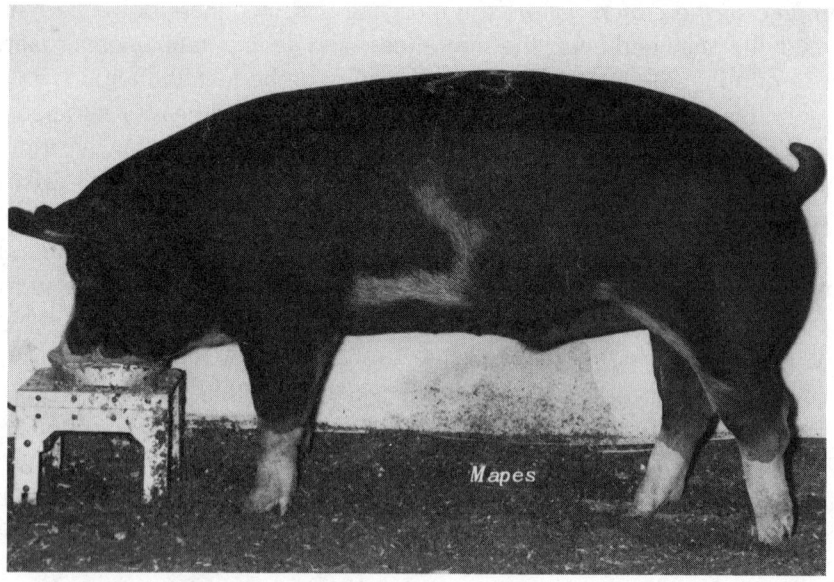

Fig. 4-8. Modern Berkshire barrow. (Courtesy, American Berkshire Association)

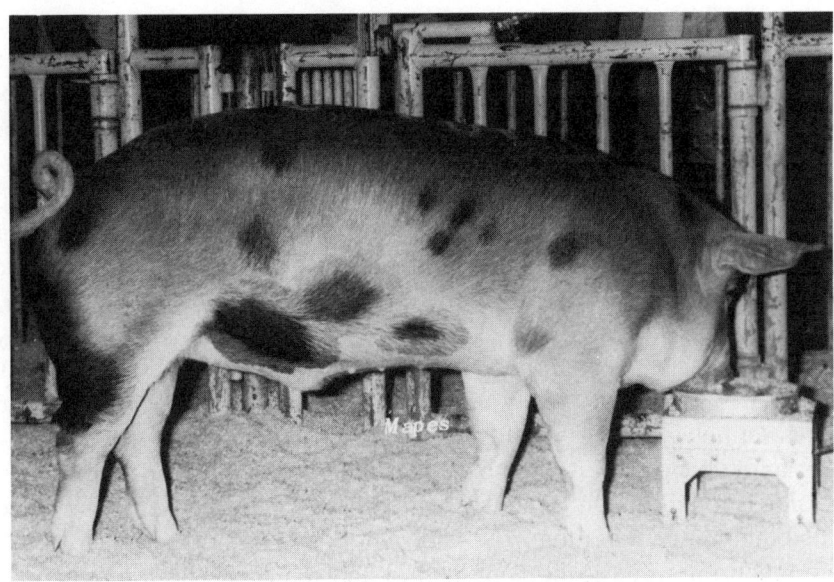

Fig. 4-9. Champion Berkshire crossbred barrow, a result of a Berkshire and Duroc cross. (Courtesy, American Berkshire Association)

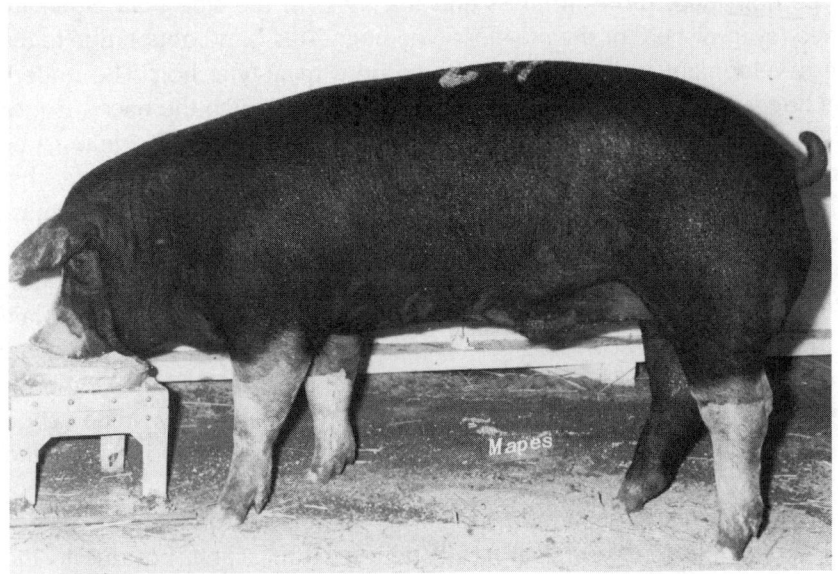

Fig. 4-10. Outstanding Poland China barrow. (Courtesy, The Poland China Record Association, Inc.)

The *rump* is long and sufficiently full to curve smoothly with the arch of the back. The rump should flare out from the loin, indicating outstanding muscle development. The tail setting should be high to allow for maximum depth of the ham. A steep, flat or narrow rump meets with serious objection. A long rump adds to the size and muscle volume of the ham.

While the market **ham** includes the rump, in the live hog the rump is usually considered as a separate part (Fig. 4-1). In all classes of swine judging, the ham receives careful consideration for its shape (dimensions), smoothness and trimness, as well as for its size and muscling. An ideal ham is deep, thick, meaty, muscular and trim (Fig. 4-7). It is thick from front to rear, wide and muscular from side to side (or plump and bulging), smooth, firm and trim at the base and in the crotch area. When a hog walks or moves, the muscle through the stifle should bulge and express itself, and the muscle seams throughout the ham should be clearly evident. While Figs. 4-7, 4-8, 4-9 and 4-10 show very satisfactory ham development in barrows, there is, nevertheless, some variation in the size and/or shape. Hogs with light-muscled narrow hams are discriminated against in the swine industry today.

The *side* is long, straight, smooth and free from excess fat trim in the flanks and along the bottomline (underline). The flank is firm and free of excess fat trim, as the sides are much improved if the flank is firm and requires little trimming of loose, flabby fat. There was some concern as to whether the modern meat-type hog would have enough finish along the side and bottomline to have the desired thickness and firmness of side. Researchers have demonstrated that the heavily muscled, modern meat-type hogs have more firmness and thickness in the side than either the old-fashioned (overfinished) or the meatless-type hogs. This is, no doubt, due to the extra muscle development in the sides of the modern meat-type hog. The underline in the live hog is, in the carcass, a part of the belly from which the bacon is produced (Fig. 4-2). Note, therefore, that inasmuch as a high-quality bacon side must be firm, all flabby tissue must be trimmed off. The more flabby the bottomline, the more trimming that has to be done and the less profitable is the side. Shallow, thin, wasty, overfinished, light-muscled and wrinkled sides are discriminated against.

The *shoulders* in the best barrows show the same general depth and width as the rest of the body. They are wide and muscular from front to rear and neatly laid in to conform with the general width of the body. The shoulders must not be coarse and open or flat on top and should not be extremely thick and heavy or protruding outward (winged or out in the shoulder). Deep seams and wrinkles should be avoided. If hogs are meaty and heavily muscled down the top and through the ham, it only seems reasonable to expect to find the same degree of muscling through the shoulder and down into the arm and forearm region. The extremely meaty hogs will have some shoulder on them, but it will be due to muscling, not due to incorrect skeletal structure. Cattle, sheep and horse producers look to the shoulder, arm and forearm areas in cattle, sheep and horses respectively as indicators of muscling. These areas would seem fully as reliable as indicators of muscling in swine. Again, research data, as well as carcass show results, have stressed the importance of muscling through the shoulder and down into the forearm when the most desirable top-cutting carcasses are selected. Hogs possessing narrow, pointed, light-muscled shoulders are objectionable.

The **belly** influences the dressing percentage favorably when it is neat, trim, firm and smooth. When this condition prevails, the flanks and heart girth are firm, clean, trim and free of any excess finish. A hog should have some depth, width, and thickness between the front legs. This gives the animal constitution and capacity to utilize feed and provide space for the vital organs of the body. A narrow, shallow condition between the forelegs is to be discouraged.

Note the medium-sized, broad, clean-cut, neat **heads** in Figs. 4-7, 4-8, 4-9 and 4-10. Large, prominent eyes; wide forehead; thin, high-quality ears; neat, trim, smooth jowls and cheeks; and a moderately long, well-developed snout are characteristics of a neat head. Heavy and flabby jowls may characterize a general lack of firmness and muscling throughout the body. Wrinkles over the forehead, large and coarse ears and small eyes (pig eyes) are objectionable.

The **neck** is medium in length and should show adequate width and fullness to blend smoothly with the shoulder and head. The top of the neck should merge neatly into the general arch over the topline. Any deviation in the topline detracts from the smoothness of the arch and the overall balance of the animal.

The length of **leg** varies somewhat in the different breeds. Greater length of leg and more substance of bone are two characteristics sought in modern meat-type hogs, provided the other previously mentioned traits and characteristics are also present. Individuals possessing these traits are usually more rugged, heavier-muscled, faster-gaining, more efficient hogs that have a higher level of productivity, which results in more useful animals of greater economic value. Excessive length of leg is undesirable, especially when associated with a narrow, shallow body. Short-legged individuals have more frequent reproductive problems and tend to be slower, less efficient gainers that lack the ruggedness and stamina desired in the modern meat-type hog. The legs must be straight and strong and show ample (above-average) substance of high-quality bone. The joints must be clean-cut, the pasterns should be strong and moderately long and the toes should be equal in size and held well together. Where the inside toe is too small, too much weight is placed on the outside toe and the leg twists inward at the pastern and hock (or knee). If the outside toe is too small, the opposite situation will result. The front legs should be set wide apart as an indicator of strong constitution and adequate chest capacity. The rear legs should be set wide and out on the corners of the body. Animals that are heavily muscled must have wide-set legs (both front and rear, especially rear) and thickness between them to carry the vast amounts of red meat and muscle tissue developed in this area. The rear legs often appear considerably longer than the front legs. The front and rear legs should be of about equal length. Note the comparative effect of these two conditions on the topline. The highest point of the shoulders should be on the same level as the highest point on the rump.

There should be a fairly heavy coat of straight, soft, smooth hair. Curly hair and whorls (swirls) are objectionable, as they tend to indicate a general lack of quality and refinement.

CHARACTERISTICS USED IN SELECTING
AND EVALUATING LIVE HOGS

Quality

Quality in hogs involves smoothness in general body outline and skin; refinement in the hair, head, ears and bone; neatness about the head and jowls; trimness in the underline and middle, in the flanks, at the base of the ham and in the crotch region; and firmness of finish, all which contribute directly or indirectly to the value of hogs for the production of pork products.

Constitution

While constitution is not of much concern to the buyer of market hogs, nevertheless, it has tremendous significance in that it is characterized by a wide, full chest, wide-set front legs and adequate width and spread through the shoulder and fore-rib regions, all which contribute to the pork cuts they involve. Besides, market hogs with good constitutions are usually rugged, thrifty, rapid-gaining individuals. In feeder and breeding hogs, constitution is important because it is a good indication that they are good feeders and that the latter have the necessary activity and ruggedness to wear well and to put on a fast, efficient gain.

Hogs should have a pleasant temperament; they should not be restless and irritable. The narrow, shallow, extremely streamlined hogs are often a disappointment, as they are usually poor feeders and annoying rovers.

Style

Hogs that move about smoothly and alertly with their backs neatly arched and that hold themselves together well have style. A hog with style commands the attention of the judge or the buyer much sooner than does one with sluggish habits. A stylish barrow shows his merits to a better advantage than one that lacks style. Style is also a good indicator of quality.

Substance

Substance refers to the amount or size of bone. In a barrow the quality and clean-cut features of the bone and joints are more important than the amount of bone, as long as the bone is adequate to carry the barrow handily. Breeding hogs should be heavy-boned, which means they should have at least above-average bone size for the breed they represent. Research data show that heavier-boned animals have more muscle. In other words, there is a positive relationship between bone size or substance and muscling in hogs. Bone size also indicates durability and ruggedness. The best place on the live animal to determine bone size or substance is the area from the knee down to the pastern joint. There is very little flesh on the bone in this area, and you can readily determine and compare substance of bone on several hogs by looking at the bone in this region.

Trimness

Trimness in hogs is indicated by a smooth, firm, neat head, jowl, foreflank, underline, rear flank, ham and crotch area, in contrast to a flabby, soft, wasty condition. The top should be clean, and there should not be any evidence of excess fat deposition (countersunken) around the tail setting. Trimness is an indispensable characteristic in a quality hog.

Balance

Balance in swine involves a consistent and efficient development of all parts. Uniformity in depth, width, thickness, muscling, meatiness, finish, correctness and smoothness of all parts contributes to a harmonious and symmetrical unity of parts that form a highly functional unit.

Heritability Percentage

The economic traits involved in swine production are influenced by both heritability and environmental factors. These genetic factors are expressed as heritability percentages (Table 4-1).

Most of the desirable carcass qualities are highly heritable. This means that

TABLE 4-1. Heritability Percentages

Trait	Heritability	Environment	
	......(percent)......		
Carcass length	65	35	
Backfat thickness	50	50	
Loin-eye area	50	50	Highly
Percent lean cuts	40	60	Heritable
Feed efficiency	35	65	
Growth rate	30	70	
Weaning weight	15	85	Highly
Number farrowed	10	90	Responsive
Number weaned	10	90	to
Birthweight	5	95	Environment

producers can improve the conformation of their breeding herds quite rapidly by using carefully selected boars.

Table 4-1 indicates the heritability-environment ratio for each trait as a percentage.

Thus, with two or more farrowings per year, the traits with 30 percent or more heritability can be improved quite rapidly.

JUDGING IMMATURE BREEDING HOGS

The ideal market type serves as a standard for the evaluation of breeding hogs also. It should be noted that the breeder's standard for an ideal gilt or an ideal boar that weighs around 180 to 225 pounds, which is an acceptable weight for a market barrow, is basically the same as the ideal for the market barrow. Those gilts and boars that are developed for breeding purposes should have a smooth body, a strong back, sound feet and legs, outstanding muscling and excellent size and scale. Surplus fat on young breeding stock is produced at an unnecessary expense, and it has an undesirable effect on the usefulness of breeding animals. Many hog producers let their young boars and gilts run to the self-feeder with their market hogs (barrow) until they are five months of age or weigh between 180 and 220 pounds. At this time the producers can select the fastest-gaining, most efficient gilts and boars for breeding stock. The producers should probe (Figs. 4-41 and 4-42) the animals they have selected for replacement or breeding stock so that they can be sure they are selecting animals that will improve their herds and the breed. Some breeders even go so far as to Sonoray every animal (boar or gilt) selected for breeding or herd replacement stock. The Sonoray instrument will give the producer an ultrasonic measure of backfat thickness and loin-eye area, both of which are used in selecting meatier, heavier-muscled hogs with a minimum amount of backfat. To make maximum use of backfat probes and Sonoray information in a selection program, you need to consider the following carefully.

For probes or Sonoray data to be comparable, the animals should have been fed as nearly alike as possible. Self-feeding a high-energy ration to hogs until they are five months old or 180 to 220 pounds is necessary to remove those animals that produce a high proportion of fat to lean. This is essentially the purpose served by the many boar-testing stations located throughout the United States. After the breeder has selected replacement boars and gilts when they are five months of age or 180 to 220 pounds, then they are "grown out" to a moderate degree of finish.

Note carefully the splendid type that prevails in the mature sows and boars, regardless of breed, in Part 4. The depth, width, muscling, smoothness, trimness, substance, correctness and quality that prevail so uniformly in them are very suggestive of those same qualities in the ideal market barrow. The majority of the mature sows and boars were grown out in the same manner as were the gilts and boars just mentioned in the previous paragraph. Moreover, this type of mature breeding hog is the type that produces the ideal carcass contest barrow that serves as the standard for market barrows.

Breeders have spent many years of close study becoming familiar with the acceptable qualities that they now have in the standard that has been set up for young breeding swine. In judging barrows, you appraise the finished product on the basis of its value to the pork-processing and pork-consuming trade. In selecting breeding stock, and particularly immature animals, you face the problem of appraising them for a long, useful, pig-producing life, which involves characteristics that very directly influence many important economic phases in general pork production.

Immature breeding animals occasionally appear rather long and somewhat upstanding when they are compared to finished barrows. However, these characteristics, among many others, have been uniformly adopted by breeders as indicators of long and useful pork-producing performance. You must learn to recognize that slight variations in characteristics such as depth, width, fullness of ham, etc., are probably more significant in thinner-fleshed breeding animals than they are in market barrows. The reason for this is that in the rather thin-fleshed growing animals, the body depth and form in general are influenced more by the muscle and bone structure than are slight variations in these same characteristics in the finished barrows. Therefore, slight variations in these qualities in thin hogs are magnified when the hogs reach maturity to a point in excess of similar slight variations as they are found in more fleshy hogs. For this reason, it is very important to be critical of deficiencies that may occur in growing swine that are in comparatively thin flesh. Young breeding animals, particularly swine, have a greater proportionate length of leg to depth of body than have mature animals. Therefore, they are more difficult to judge accurately than market barrows or mature hogs because these and many other parts do not all develop into the mature form in the same proportions in which they occur in the growing pigs.

Size in purebred swine receives considerable attention. The prevailing practice of attempting to get hogs to market at an early age calls for the use of breeding stock that transmit the inherent tendency to grow rapidly and take on an adequate finish long before they are mature. Market hogs should weigh 200 pounds in five months or less. A hog of this type is much more efficient in the conversion of feed to pork than a hog that requires seven to eight months to reach 200 pounds. This is a very important factor in the pork trade, which stresses a large amount of lean in pork cuts. The tendency to develop good weight at an early age is important. An active growth factor is a heritable quality that is helpful in producing market hogs at an early age during which the quantity of feed required per unit of gain is at the minimum during the life of the hog. Boars and gilts selected for breeding and replacement stock should weigh well over 225 pounds at six months of age. It is not uncommon to find boars and gilts that weigh 300 pounds or more at six months of age. Immature boars or gilts that have exhibited this type of growth potential will develop into 500- to 800-pound mature boars and sows with little or no effort, and many will approach 1,000 pounds under excellent care and management and still maintain a high level of reproductive efficiency.

Length of leg, depth, width, thickness, muscling, meatiness, constitution, substance, balance, quality and feeding capacity are important factors to consider in profitable pork production. In no case should lack of muscling, shallow rib development and general lack of constitution be tolerated, regardless of what you think constitutes the correct length of leg in growing pigs.

Sow Pigs — Gilts

Sow pigs that are developed for breeding purposes should have good *size* (Figs. 4-11, 4-12, 4-13 and 4-14). Well-grown gilts should weigh well over 225 pounds at six months of age. If these gilts have been highly fitted, they probably weigh over 300 pounds. Remember that when there is adequate depth of body in growing gilts that they "body down" upon maturity, with the proportion of body depth to length of legs generally being quite acceptable. This is another reason for trying to select for some length of leg in young gilts because after they mature and body down, the short-legged gilts may have too much weight for their legs to carry, or their underline may be too near the ground for maximum reproductive efficiency. As the gilts mature, the body increases more in depth than the legs do in length.

There must be ample evidence of good *feeding qualities,* as shown in Figs. 4-11, 4-12, 4-13 and 4-14. Gilts must have adequate depth, excellent thickness and muscling and good length. They must be roomy sows with good middles and strong constitutions. The head should be moderately long. Gilts should have a moderately long, strong, well-developed snout; strong but neat jaws; good eyes; and a moderately long, neat neck. These qualities are also characteristic of reasonably early-maturing hogs. Shallow, narrow, light-muscled hogs are invariably impossible feeders and poor doers.

Sows that prove to be good producers are generally kept in the herd for a number of years. Their ability to wear well is determined by a vigorous *constitution,* as is indicated by a wide and deep chest, adequate heart-girth development, deep and well-sprung ribs, a substantial middle and excellent substance and ruggedness combined with ample quality throughout (Figs. 4-11, 4-12, 4-13 and 4-14). There must also be evidence of thrift, moderate activity and a pleasant temperament.

Quality gilts are smooth in general outline and in the skin and refined in the hair (Figs. 4-11, 4-12, 4-13 and 4-14). They are neat and clean-cut about the head, jowls and neck and trim along the underline. Flabbiness in a gilt is good evidence of coarseness and a general lack of muscling. The bone should be sound, hard, smooth and heavy in size and substance. Quality is also influenced by freeness in activity and movement.

The *underpinning* is very important in brood sows. A bad set of legs and feet often makes a sow rather awkward and helpless. The legs must be well-placed and easily controlled. They must be of excellent quality — heavy-boned, straight, strong. The pasterns must be strong and moderate in length, and the toes should be held well together and should be equal in size. The knees and hocks should be set wide and correctly placed. Too much set to the legs (sickle-hocked), too straight hind legs (post-legged) or too close at the hocks is very objectionable.

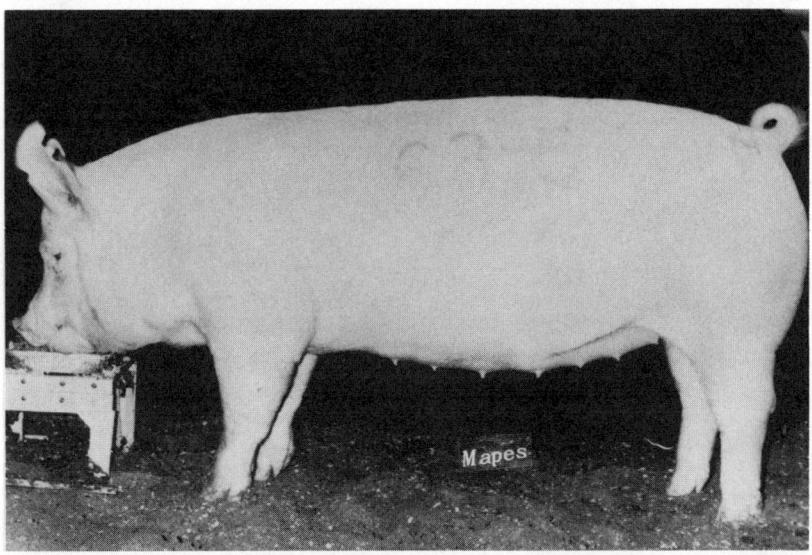

Fig. 4-11. This Yorkshire gilt was a class winner at the 1984 National Barrow Show. The gilt was sold to a breeder in Taiwan for $1,550. (Courtesy, American Yorkshire Club, Inc.)

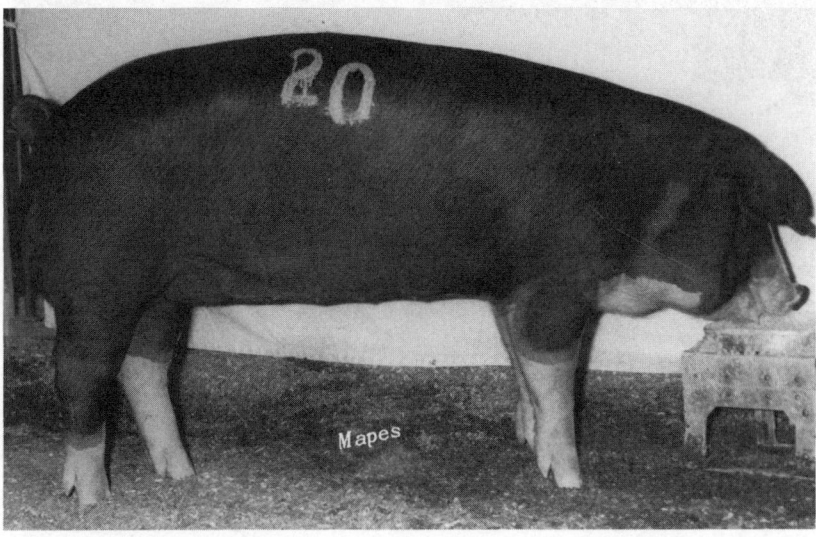

Fig. 4-12. Modern Poland China gilt. (Courtesy, The Poland China Record Association, Inc.)

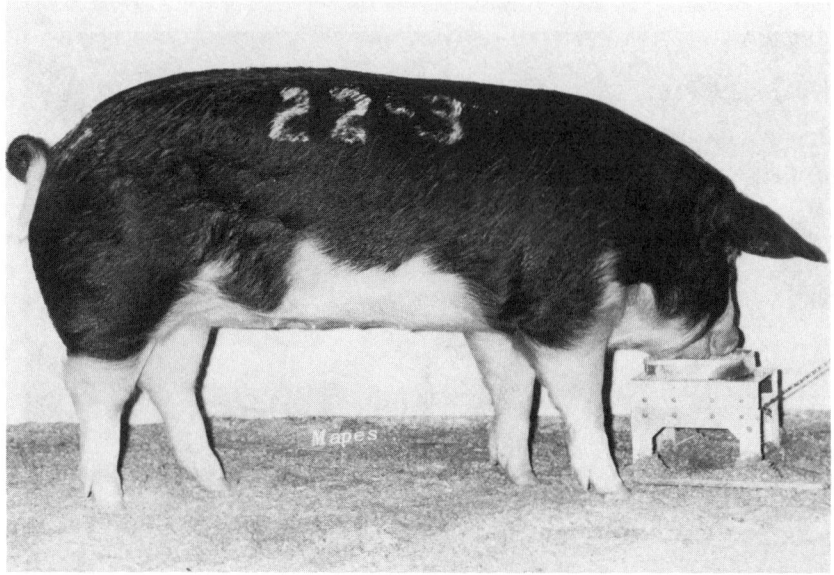

Fig. 4-13. Model Berkshire gilt. (Courtesy, American Berkshire Association)

Fig. 4-14. Reserve Grand Champion Yorkshire Gilt at the 1984 National Barrow Show, Austin, Minnesota. (Courtesy, American Yorkshire Club, Inc.)

While general refinement contributes to matronly qualities, in addition, gilts should have a mild temperament. Also, there should be six well-developed and uniformly placed teats on each side of the underline. Blind or inverted teats are very undesirable because they do not function and because they are inherited.

Superior and distinctive development in all the useful qualities always contributes to breed character. In addition to this, purebred gilts should show well-developed breed qualities, such as desirable color and all the characteristics about the head, ears, etc., that are peculiar to the particular breed.

A desirable balance in all the body parts should be sought. The same efficiency should prevail in all the parts, which should be smoothly joined. The top of the shoulders should be on the same level as the highest point on the rump. It is a common fault for gilts to be too low in front. This interferes with the smoothness in the arch in that the highest point will be at the loin-ham connection instead of at the middle of the back.

Young Boars

Outstanding young boars must be more than merely "male" pigs. They must show definite promise of developing into big, stout, rugged boars. This is indicated by good length, thickness, muscling, strength of arch and substance of bone (Figs. 4-15 and 4-16). They must show strong evidence of a vigorous constitution in a thick, roomy chest and in long, well-sprung ribs. Feeding qualities are indicated by an adequate middle development, as contrasted to extreme shallowness, and by a rugged, thrifty condition throughout the body. Ample depth, width and general vigor reflect early-maturing qualities.

Extremely masculine traits are important. In young boars these are apparent in general ruggedness, size of bone and vigor. Masculinity should not be confused with viciousness, which is equally as undesirable as it is unnecessary. Two well-developed testicles should be plainly visible in a reasonably prominent scrotum (Figs. 4-15 and 4-16).

Quality in boar pigs should be expressed in hard, clean-cut, smooth bone; in a smooth coat of heavy hair; in general smoothness of outline, particularly in the shoulders; and in general neatness about the head and neck. Extremely heavy or thick shoulders in young boars are an indication of coarseness.

Young boars should show unmistakable evidence of soundness. A high, smooth, well-developed top; straight, strong, well-developed legs; moderate length; strong pasterns; toes of equal size and held together well; and a straight, true, active walk are very necessary attributes for soundness (Figs. 4-15 and 4-16). Excessive activity is often due to an irritable temperament. Both these traits characterize "ranting" and "hard-feeding" boars. A practical nervous stability is always evident in well-behaved, vigorous, good-doing boars.

Boars should be distinctive in breed character. The term "growthy" is often used to describe boars that are big, rugged and well-developed for their age.

Fig. 4-15. Painting of a model Poland China boar. (Courtesy, The Poland China Record Association, Inc.)

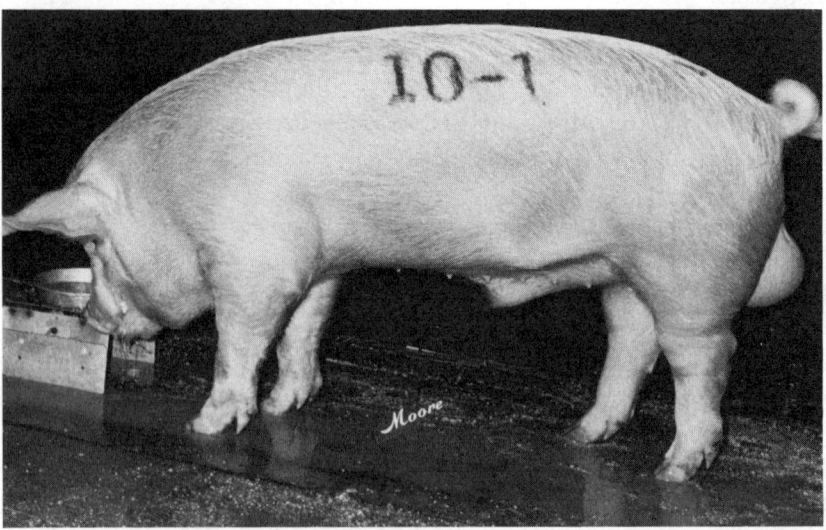

Fig. 4-16. Grand Champion and Top-Selling Yorkshire Boar at the 1984 CHY Ark City Conference. (Courtesy, American Yorkshire Club, Inc.)

DISTINGUISHING BREED-TYPE CHARACTERISTICS

A genuine interest in swine and an unprejudiced appreciation for beneficial pork-producing qualities have been responsible for the uniform excellence in the useful type of the modern breeds. There are, however, a number of breed-type characteristics that make it possible to distinguish one breed from another. These may be considered as the breed "trademarks." Color, set of ears, variations in dish of face, etc., are among these breed differences.

You should begin early in your livestock selection and evaluation training to develop an appreciation for breed-type qualities because they have been, and still are, a very definite inspiration to the breeder who is trying to produce uniformity in all body characteristics. Also, some breed-type characteristics are difficult, if not impossible, to obtain, except as they appear combined in the purebred animal. Hence, they serve as some assurance of good breeding. A high degree of excellence in breed-type characteristics is always a source of pride to a breeder. The breeder should recognize that this pride must be based on excellence and high performance in the productive characteristics. Accordingly, breed type is closely associated with useful type.

The useful characteristics in the various breeds should conform to those qualities in the market hogs, sows and boars that have previously been described. The breed-type characteristics and those points that receive major consideration in the judging of different breeds, which have a direct bearing upon the quality and general usefulness of each breed, are discussed as follows.

MODERN BREEDS OF SWINE

Berkshire

The Berkshire is black with six white points that occur rather uniformly on the face, the four feet and the tip of the tail (Figs. 4-17 and 4-18). White may occur on the jowl, ears or foreflanks and frequently on the sides and shoulders. The white should not extend beyond the six white points.

Erect ears are characteristic of this breed. They should be medium-sized and neatly carried. The face has a pronounced dish and is rather short and broad. A fairly heavy coat of fine, smooth hair generally prevails. Coarse hair and whorls are discriminated against very severely.

This breed is distinctive for its style and quality, and judges place considerable emphasis on these points. The best specimens are long and display excellent balance. They are medium in length of leg. In general, judges of modern Berkshires select larger, growthier hogs that are stronger-topped, longer-sided, longer-legged and cleaner about the head and jowl. Judges look favorably upon a strong, moderately arched back; a long, wide, muscular rump; thick, meaty hams; a trim middle; a neat underline; and short, strong pasterns (Figs. 4-17 and 4-18).

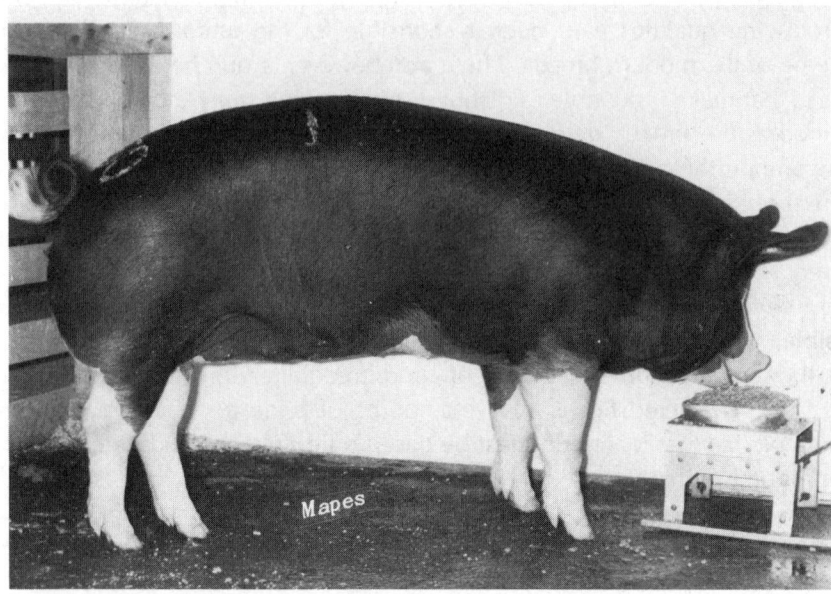

Fig. 4-17. Ideal Berkshire gilt. (Courtesy, American Berkshire Association)

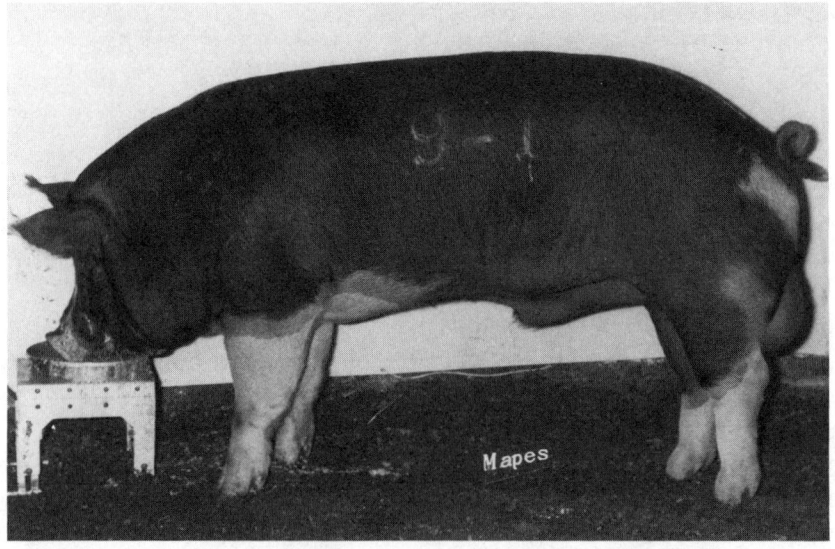

Fig. 4-18. Outstanding Berkshire boar. (Courtesy, American Berkshire Association)

Chester White

The Chester White has white hair and white skin (Figs. 4-19 and 4-20). The ears droop and break down over the eyes. They should be medium-sized and thin-textured. Black hair disqualifies in the show ring, and blue spots in the skin meet with disfavor. A good coat of smooth hair is preferred. A curly coat of coarse hair is vigorously criticized. Whorls are discriminated against.

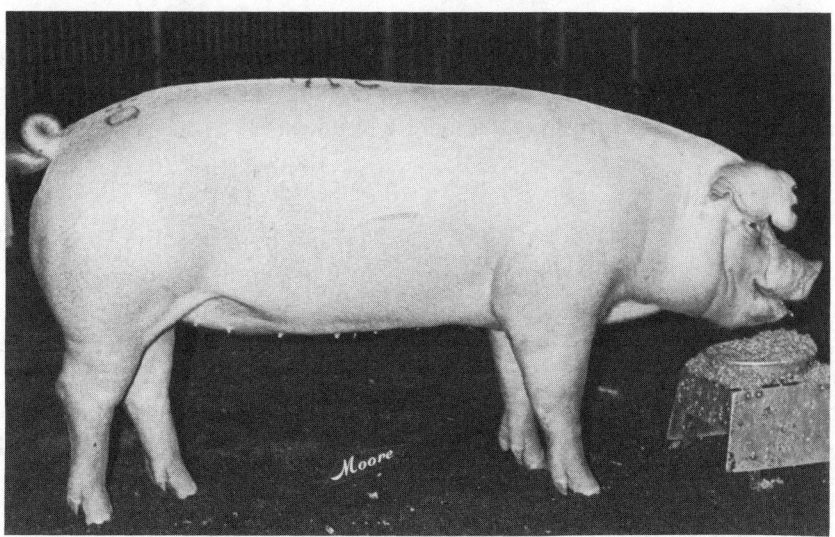

Fig. 4-19. Modern Chester White gilt. (Courtesy, Chester White Swine Record Association)

Fig. 4-20. Typical Chester White boar. (Courtesy, Chester White Swine Record Association)

Modern show prospects exhibit a rather strong and smoothly balanced arch, with the highest point at the middle of the back. Chester Whites are long and moderately wide with good depth. Emphasis is placed upon plump, smooth, firm hams that show excellent muscle development. Smoothness in the shoulder, neatness along the underline and ample, clean-cut, strong bone are stressed. The legs should be straight, and the toes should be held together well. A narrow chest floor, too small a space between the knees and hind feet that are placed too close together are serious objections.

A strong, vigorous constitution, as indicated by well-sprung fore-ribs and a wide chest floor, characterizes the best specimens. Neatness about the head, ears and jowl and only a moderate dish in the face meet with favor.

Duroc

The Duroc is red (Figs. 4-21 and 4-22). While there is some variation in the shade of red, medium red is the most common. Black spots sometimes occur in the hair. These meet with objection. White spots, very occasional in their occurrence, disqualify an animal from registry. Whorls in the hair likewise disqualify an animal from registry and in the show ring.

The back is carried in a pronounced arch, with emphasis placed upon the arch being balanced from the neck to the tailhead, with the highest point in the middle of the back. Inasmuch as this is a large breed, some length of leg must be shown, especially in immature hogs. However, there must be evidence of adequate depth of body and good feeding qualities. Judges stress good depth and width in the region of the chest; trimness through the jowl, along the underline and in the crotch region; thick, meaty hams; strong loins; and strong, muscular tops (Figs. 4-21 and 4-22).

Fig. 4-21. This impressive Duroc gilt was one of the Champion Pair of Gilts at the 1984 Southeastern Duroc Congress. The pair sold for $4,400. (Courtesy, United Duroc Swine Registry)

Fig. 4-22. Top-selling Duroc boar at the 1984 Southeastern Duroc Congress. This outstanding boar sold for $6,200. (Courtesy, United Duroc Swine Registry)

The head is, as a rule, rather neat with only a moderate dish in the face. The ears should be drooped and medium-sized and carried well enough apart in front so as not to obscure vision.

Well-laced legs and a straight and free walk are important. In general, while the Duroc is good on its feet and pasterns, it should be well-balanced. In some specimens, the inside toes are too small, a condition that gives rise to too much weight on the outside toe. The invariable result is weakened pasterns and crooked legs. The opposite condition is also found in the Duroc.

Hampshire

The Hampshire is black with a white belt around the body at the shoulders (Figs. 4-23 and 4-24). The white extends down over the forelegs. While the width of the belt varies, it preferably should be confined to the shoulder. The Hampshire Swine Registry lists the following color patterns as objectionable:

> . . . White high on the hind legs, white belt too wide (more than one-third the length of the body); white from belt running back to the underline to meet white on hindquarters; spots in belt; mixture of white hairs with the black; or roan color. Disqualifications: solid black, incomplete belt; more than two-thirds white; any white on head; one or both front legs black; white higher than ham on hind legs.

Whorls in the hair disqualify an animal from registry and in the show ring.

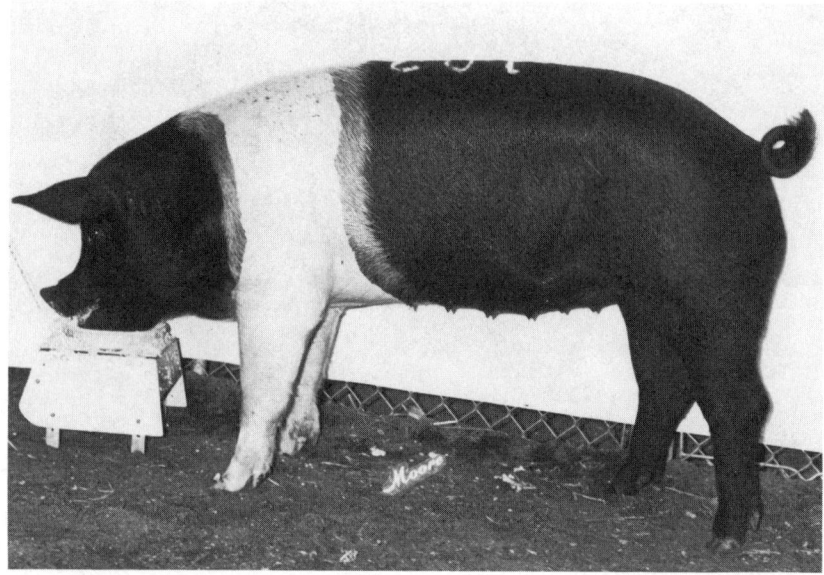

Fig. 4-23. Impressive purebred Hampshire gilt. (Courtesy, Hampshire Swine Registry)

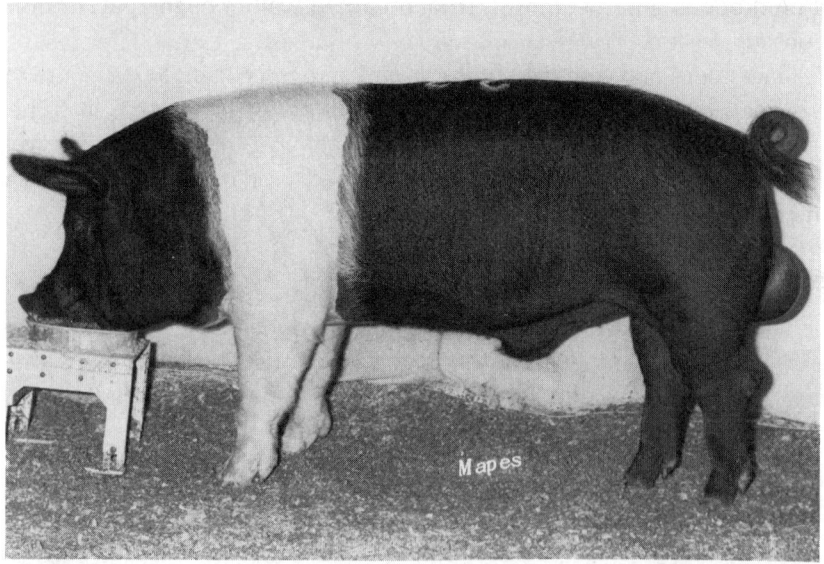

Fig. 4-24. Modern Hampshire boar. (Courtesy, Hampshire Swine Registry)

The Hampshire is far above average in size. In quality, the breed is distinctive. As a rule, the body is smooth, the underline trim and the jowl neat. A minimum amount of smooth, firm finish is typical of the breed. A strong, well-balanced arch and a firm, well-developed topline win favor.

Emphasis is placed upon uniform width, neatly laid shoulders, ample width at the rump and deep, thick, meaty hams (Figs. 4-23 and 4-24). This breed is noted for its muscling, and as a result, Hampshires or Hampshire crossbreds have been consistent winners of the on-foot and carcass championships at many of the leading shows. This breed possesses excellent substance of bone that is neat, clean-cut and high in quality. It is important that the pasterns show strength.

While the head is somewhat long and not very wide, extremes should be avoided. The ears are carried erectly but incline forward and outward. Coarseness in any part is severely criticized.

Landrace

The Landrace is white and has large, drooping ears that shield the eyes from the side (Figs. 4-25, 4-26 and 4-27). However, the head has adequate width to permit excellent forward vision. The body of the Landrace should be long and uniform in depth. Pronounced smoothness in a slightly arched back with uniform width throughout the body; deep, well-laid, smooth shoulders; good length and depth in the sides; deep, well-rounded hams; and smoothness throughout are prominent characteristics of the breed (Figs. 4-25 and 4-26). Powerful, straight, well-set legs with

Fig. 4-25. Purebred Landrace gilt elected to the Hog College at the 1984 National Barrow Show, Austin, Minnesota. (Courtesy, American Landrace Association)

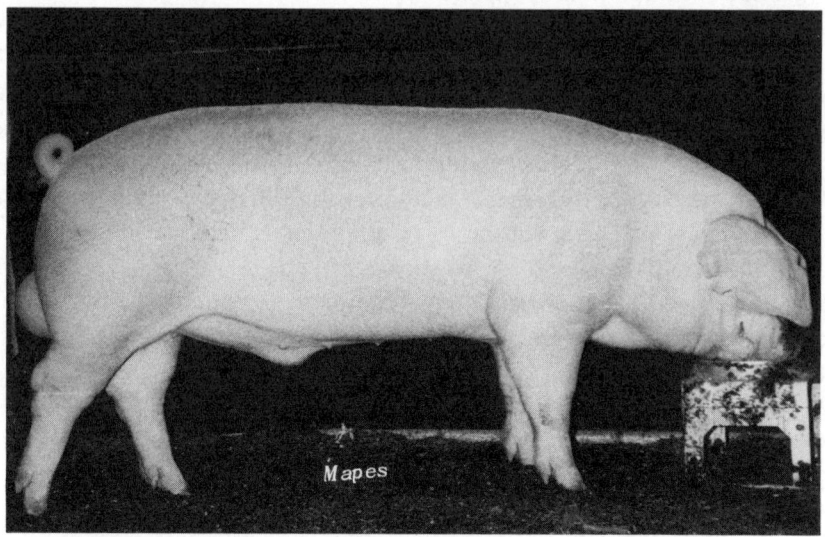

Fig. 4-26. Outstanding purebred Landrace boar elected to the Hog College at a National Barrow Show, Austin, Minnesota. (Courtesy, American Landrace Association)

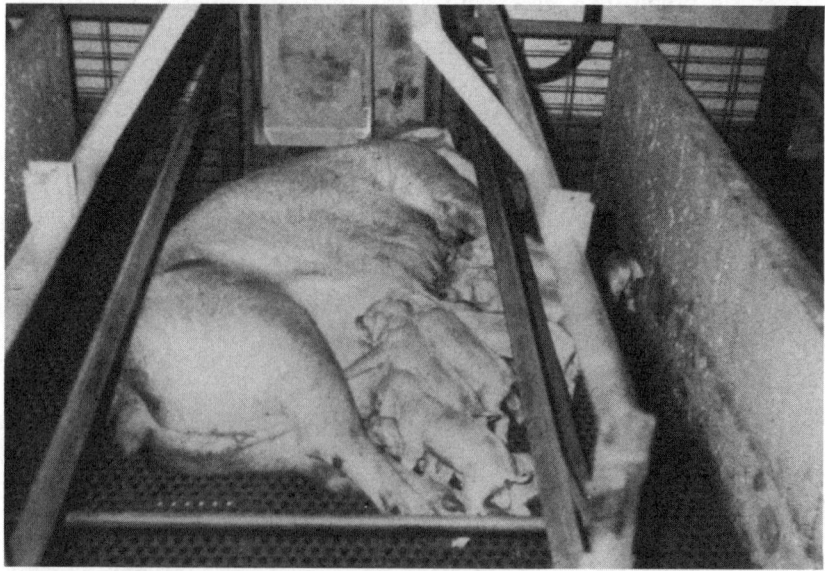

Fig. 4-27. Landrace sow and litter portray the large litter size and high productivity typical of the Landrace breed. (Courtesy, American Landrace Association)

strong pasterns and uniformly developed toes are desirable for a free and easy walk. Six or seven well-developed and evenly spaced teats on each side are desired. Fewer than six teats on either side, black in the hair coat and erect ears are all disqualifications.

Poland China

The Poland China is black with white on the forehead, on the four feet and on the tip of the tail (Figs. 4-28 and 4-29). Preferably the white should be confined below the knees and hocks. Too broad a strip in the face is not favored, and while some white has been accepted on the jowl and along the lower part of the shoulders and sides in recent years, it is not popular. The tendency is to favor less white on the face, the legs and the tail to add to the distinctive appearance of the breed. The drooped, fine-textured ears are of excellent quality.

The Poland China is above average in size. Good depth and width usually prevail, and ample length is seldom lacking in the best specimens. It is important to emphasize trimness in the underline; firmness in the flanks; width, thickness, muscling and meatiness down the top and through the hams; and neatness in the head, ears, jowls and shoulder (Figs. 4-28 and 4-29). This breed is distinctive for its uniformly large ham and excellent quality.

The strong arch this breed carries should be balanced. Low shoulders and a high rump meet with objection, as this invariably shifts the highest point of the topline from the middle of the back where it belongs to the highest point of the rump. This

Fig. 4-28. Model Poland China gilt. (Courtesy, Poland China Record Association, Inc.)

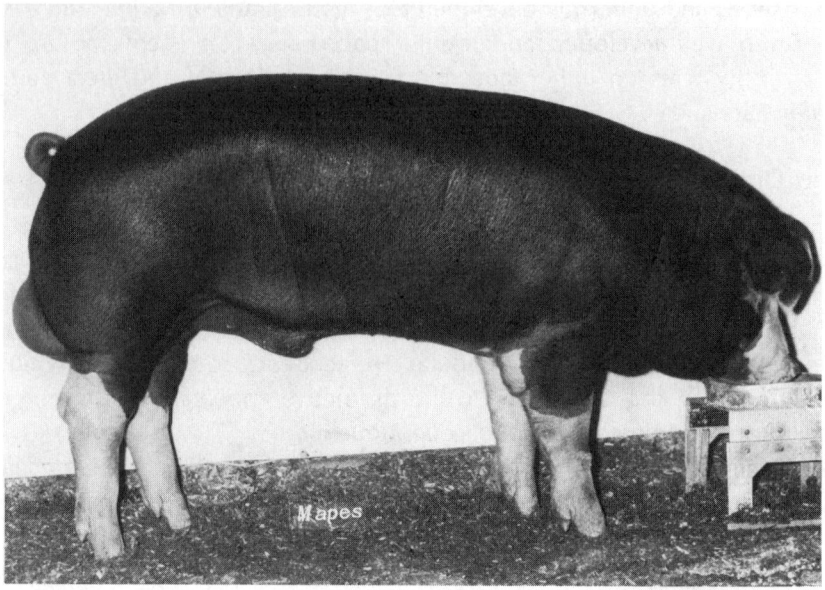

Fig. 4-29. Ideal Poland China boar. (Courtesy, Poland China Record Association, Inc.)

condition also makes the rump seem too steep and pushes the rear legs forward. The rear legs often appear long when the rump is too high.

An ample amount of bone is characteristic of this breed. It must be quality bone. The toes should be of equal size and well held together, and the pasterns should be short and strong. In the standard of excellence for this breed is found the following rule regarding whorls: "Animals with whorls should not be exhibited or recorded."

Spotted

The ideal Spotted is 50 percent black and 50 percent white, distributed in clearly defined spots over the entire body (Figs. 4-30 and 4-31). There must appear not less than 20 percent nor more than 80 percent white on the body, or disqualification rules apply, and the animal will be ineligible for entry in the breed association herd books. Sandy or brown shades also disqualify. "Boars with whorls are disqualified from registry. Sows with whorls may be recorded, but not shown or sold in public sale."

Neatness about the head, ears and jowls is encouraged in the Spotted. A trim middle that is free from flabbiness and firmness at the base of the hams is emphasized. Large, well-balanced, sound, smooth, active individuals with good feet and legs win favor (Figs. 4-30 and 4-31). This breed is above average in muscling and shows good depth, width and length. The ears are drooped and neatly carried.

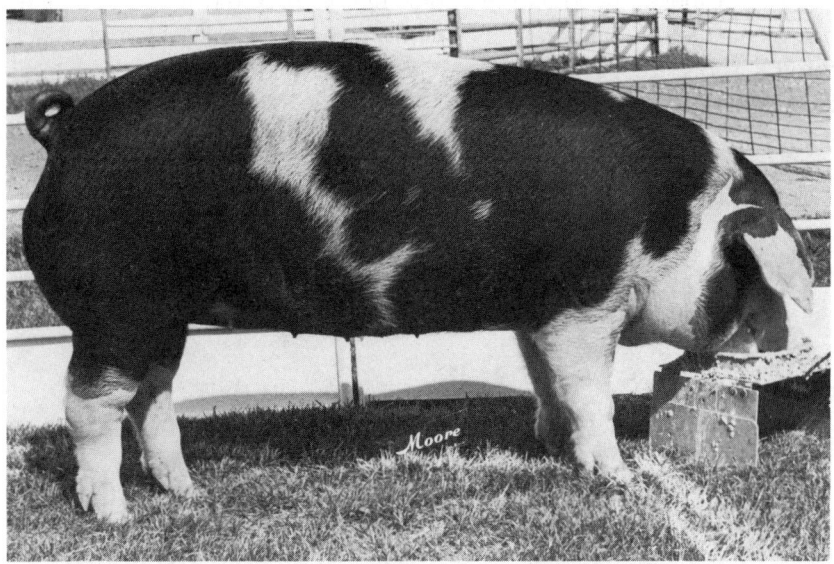

Fig. 4-30. Ideal Spotted gilt. (Courtesy, National Spotted Swine Record, Inc.)

Fig. 4-31. Outstanding, free-striding Spotted boar. (Courtesy, National Spotted Swine Record, Inc.)

Tamworth

The color of the Tamworth varies from light to dark red (Figs. 4-32 and 4-33). Golden red hair that is comparatively free from black hair is preferred. Black spots and light gray spots meet with disfavor. Coarse, curly hair and whorls are discriminated against.

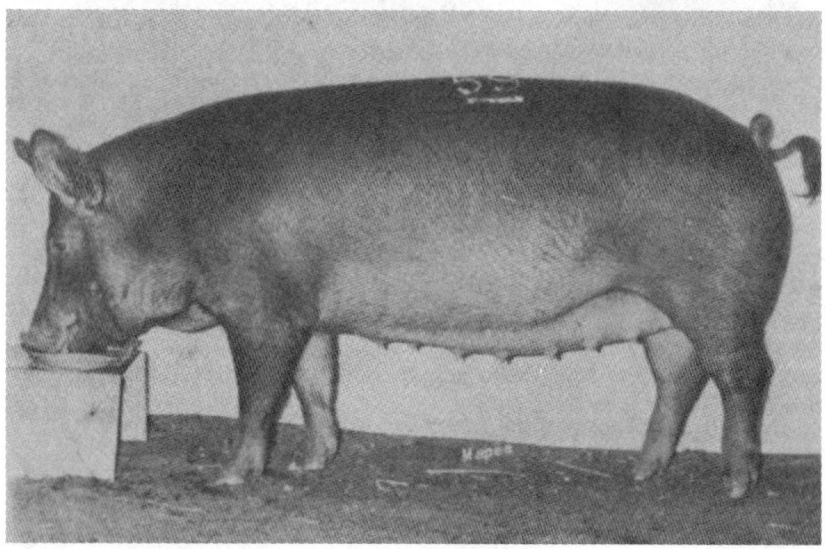

Fig. 4-32. Champion Tamworth bred gilt. (Courtesy, Tamworth Swine Association)

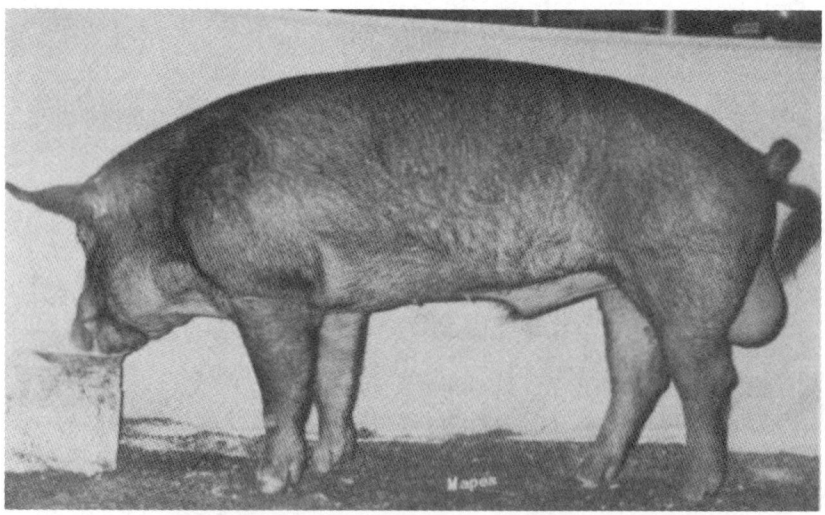

Fig. 4-33. Modern Tamworth boar. (Courtesy, Tamworth Swine Association)

A smooth topline, trim underline and balanced muscle development are characteristic of the model animals. Any tendency to shortness of body, compactness and thickness is severely criticized. Soundness in the feet and legs, trimness and muscling are required in the ideal specimens.

Yorkshire

The Yorkshire is white (Figs. 4-34 and 4-35). Black hair in the coat is undesirable. Erect ears and a pronounced dish in the face are typical of Yorkshires. There

Fig. 4-34. Modern Yorkshire gilt. (Courtesy, American Yorkshire Club, Inc.)

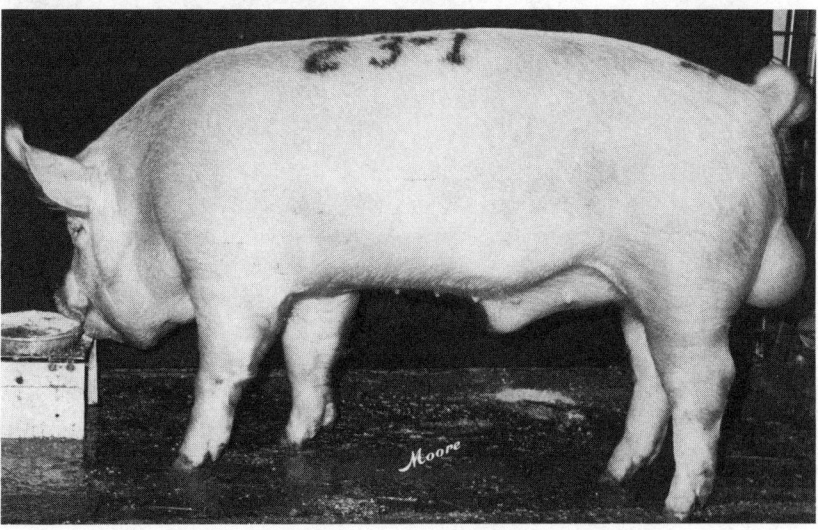

Fig. 4-35. Modern Yorkshire boar. (Courtesy, American Yorkshire Club, Inc.)

must be ample evidence of smoothness, firmness and quality in this breed. It is highly important that the width and depth be uniform and that trimness prevail along the entire underline.

The best specimens exhibit good style and a high degree of muscle development. The rump should carry well back with more width than is displayed over the back and loin, and any tendency toward flatness over the rump does not win favor. Shapely hams that are firm and heavily muscled are preferred (Figs. 4-34 and 4-35).

The hair coat should be high quality, straight and well-groomed. Whorls are subject to disqualification.

Hereford

The Hereford is the youngest breed of hogs to be recognized by the National Association of Swine Records. During the early part of the twentieth century, a strain of hogs whose color markings resembled the red-and-white markings of Hereford cattle was developed.

All Hereford hogs offered for entry in the National Hereford Hog Record must have some white in the face, must be at least two-thirds red and must have at least two white feet and no white belt extending over the shoulders, back or rump (Figs. 4-36 and 4-37). The ideal markings are white face, including ears, white feet, underline and switch of tail; all other parts red, either light or dark, the latter being preferred.

Herefords are known for their docile dispositions. Herefords mature at an early age; with proper care they will reach 200 to 250 pounds at 5 or 6 months of age.

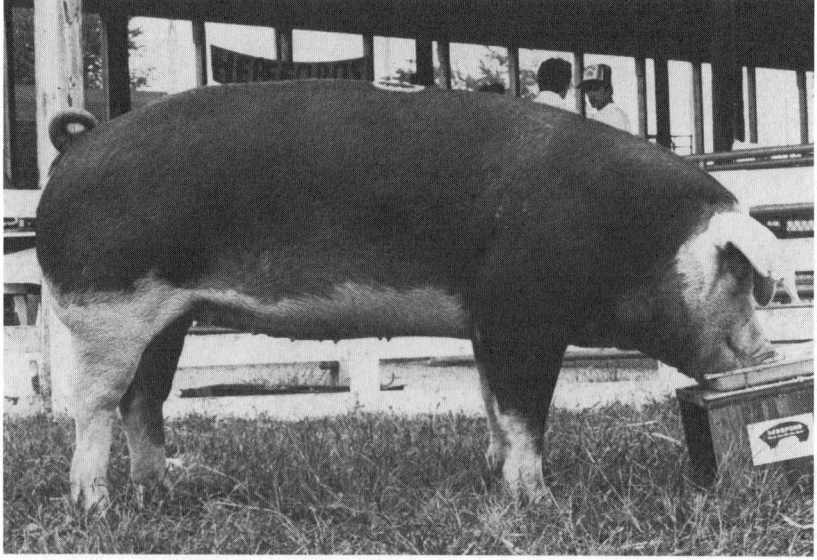

Fig. 4-36. Top-selling Hereford gilt at a National Hereford Hog Show and Sale, Fairfield, Iowa. (Courtesy, National Hereford Hog Record Association)

Fig. 4-37. This modern Hereford hog was the Grand Champion and the top-selling boar at a National Hereford Hog Show and Sale, Fairfield, Iowa. (Courtesy, National Hereford Hog Record Association)

SOME INHERITED DEFECTS IN SWINE

Inherited defects that are occasionally found in swine are the results of matings that carry the factor for the defect. You should know the significance of these as well as the breed association regulations, if any, regarding such defective animals. Although you need to be aware of what these defects are, you should not become too "detailish" in emphasizing and discussing the extremely minor details.

Whorls are swirls of hair that vary considerably in their pattern (Fig. 4-38). They are often found in the flanks, jowls, forehead and snout, where they are not as conspicuous or as objectionable as those along the topline — on top of the shoulder, over the loin-rump coupling and at the tailhead. All breed associations discriminate against whorls on the topline in breeding swine. However, all of them do not disqualify pigs with these whorls from registry. Each breed association's position on this matter appears in the respective discussions under breed type. Experimental data indicate that sows and boars with whorls on the rump do not produce pigs with whorls in other locations unless they also carry the factor for location in these other places. Thus, the possibility that sows and boars with whorls on the flanks, jowls and forehead will produce offspring with whorls on the topline is rather remote.

Official judges generally discriminate heavily against whorls and, for the most part, will disqualify pigs with whorls. To discriminate against an animal for whorls means that it may still be placed, but obviously down the line. Disqualifying the animal means that it definitely goes to the bottom, or in official judging, that it is barred from showing. Technically, therefore, an official judge should be careful not to disqualify an animal unless the breed association has definitely committed itself on this point. However, inasmuch as all breed associations discriminate against whorls, the judge may determine how severe the penalty should be, except to disqualify, unless in the latter case the breed association rules support the disqualification.

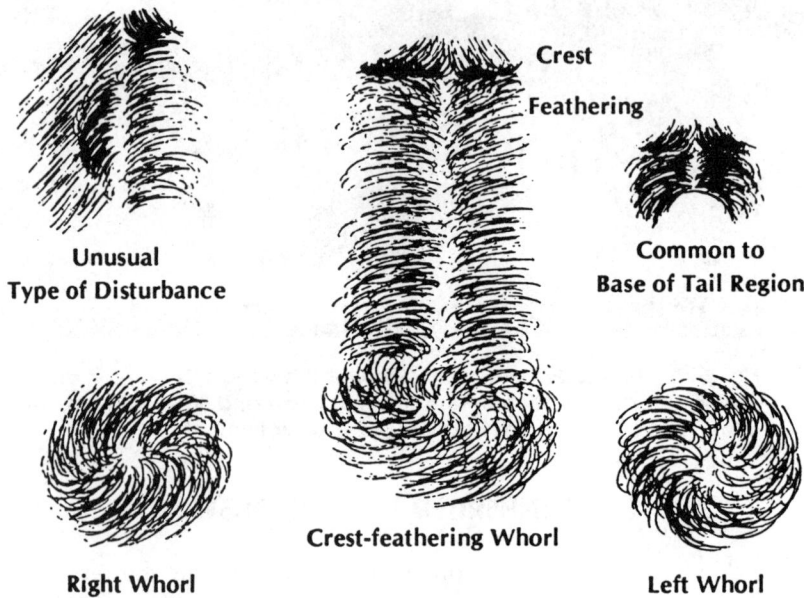

Fig. 4-38. Pen sketch of different types of whorls or hair disturbances along the backline of swine.

Small Inside Toes

Small inside toes, especially on the rear feet, but also on the front feet, may cause some concern inasmuch as they tend to distort the alignment of the feet, weaken the pasterns and place an undue amount of weight on the outside toe. This also is an inherited defect.

When the inside toe is too small, the pastern is weakened; and inasmuch as the pressure is unequal on the two toes, the leg is forced out of alignment toward the improperly supported side, which forces the pastern inward and the hocks together.

There is no specific breed association ruling on this point except that all breed associations discriminate against weak pasterns and crooked legs.

Inverted or Blind Teats

Inverted or blind teats are teats in which the end of each has failed to emerge.

These have been observed in a number of breeds of swine. This condition cannot be corrected by any practical means.

However, some teats that appear inverted in young gilts emerge when the gilts approach a more mature age and especially when the teats begin to develop preparatory to the gilts' farrowing.

Normal sows should have at least six well-placed normal teats on each side of the underline. Some sows have seven pairs. The seriousness of this inherited defect is in proportion to the number and location of defective teats. This defect has been discussed in the literature on swine, and breed associations make mention of it. They require well-developed and well-placed teats; hence, a gilt or a sow with one or more of the permanently inverted teats meets with considerable disfavor. She should not be used as a breeder.

The Poland China, Duroc, Landrace and Yorkshire breed associations require each gilt to have at least six functional teats on each side of the underline to be eligible for registration.

Underdeveloped Ears

Several of the major breeds of swine have a defect that involves an underdevelopment of one or both of the ears. While this may seem to be a minor matter, this external ear defect affects also the inner ear and the skull. In all the cases that have been examined, there has been no opening in the flesh or bone into the inner ear, and the functioning part of the inner ear has been so completely developed that the pig cannot hear on the affected side. Pigs with both ears of this type cannot hear at all. In no case should a pig of this kind be used for reproductive purposes because this defect, too, is inherited.

Unequal Number of Toes, Kinky Tails and Skin Tumors

An unequal number of toes, kinky or "screw" tails and skin tumors commonly called inherited warts, even though they may be an inch or more in diameter, are disturbances that you should recognize. The kinky tail is often encountered in the show ring, and while little emphasis is placed upon it, you should understand that it is brought about by imperfections in the inheritance of swine that cause two or more of the tail vertebrae to fuse and form a rigid angle. The defect may occur near the end of the tail or near the base. Extreme cases observed near the base of the tail may be so badly kinked that the tail is typically screw-shaped.

Defective Color

The inheritance of defective color should be considered in connection with each breed in which such defects are known to exist. The appearance of colors that are foreign to a breed, such as white in red breeds, is more serious than unusual variations in the amount of white in breeds in which white is typical.

Cryptorchidism

Some boars have only one testicle in the scrotum and the other one up in the body cavity. In some cases, both testicles are up in the body. These pigs are known as *cryptorchids.* This defect is inherited. Boars with this condition should not be used as breeders and should be turned down in any class. Boars with both testicles up in the body cannot reproduce because the temperature in the body cavity is too high to permit normal development and formation of the sperm. Boars with only one testicle up in the body and the other one in the scrotum can reproduce. However, when the defective boar pigs which they sire are castrated, the testicle is usually left up in the body, since it is very difficult for the amateur to remove. The result is that the pigs in many cases develop the offensive odor of boars, and the pork produced from them is inedible. This is especially true if they are allowed to become six or more months old before they are slaughtered.

Scrotal and Umbilical Hernias

In scrotal hernia, the scrotum is enlarged as the result of loops of intestines having descended into it. Umbilical or navel hernia is the result of loops of intestines passing through the abdominal wall at the navel, which forms a filling immediately underneath the skin. Both these inherited defects are subject to disqualification, and not in any case should animals with these defects be used as breeders.

THE SKELETON OF THE SWINE — ITS INFLUENCE ON FORM AND FUNCTION

While very little direct reference is made to the skeleton in the judging of swine, there are some rather basic details that are necessary for a comprehensive appreciation of the principles involved. You should aim to become familiar with these and to understand their bearing upon judging (Fig. 4-39).

There is a great deal of variation in the form of swine. This variation is influenced primarily by the amount and distribution of fat, by the size of the muscles and by the length, shape and size of the bones. The influence of muscle and fat upon form has previously been discussed under "The Modern Meat-Type Hog."

The characteristics in body form that give rise to variations in function, such as early maturity, thrift, activity and feeding qualities, are depth, width, length, strength of topline and soundness of feet and legs. There are, of course, many qualities in swine that influence function but are not apparent in form. These have previously been discussed.

The bones of the back are divided into a number of groups. There are 7 cervical (neck) vertebrae, 14 or 15 dorsal (back), 6 lumbar (loin), 5 sacral (in the rump) and 18 to 20 coccygeal (tail). The variations that occur in the total length of the body are influenced by the variations in the length of the vertebrae within the neck, back, loin and rump, the latter three being of major importance.

The bones vary in length and size. The variations in the length of the bones have probably the greatest influence on form. The length of the body cannot be increased as much by adding finish to a hog as can the depth and the width. Therefore, the variations that are noted in the length of the hogs are more directly influenced by the skeleton than are the variations in width and depth.

A number of hogs are very short in the rump. In some hogs the neck is very long, while in others that have the same total body length, the neck is rather short. Therefore, you should observe that the vertebrae within the different groups do not vary in length to the same degree. This, of course, is very significant, since it is possible, by selection, to produce pigs with considerable length when that is required and shortness when that is desired, as in the neck.

The number of rib pairs in pigs vary, usually 14 or 15, although some 16 or 17 pairs have also been found. There are 15 pairs in the skeleton in Fig. 4-39. The

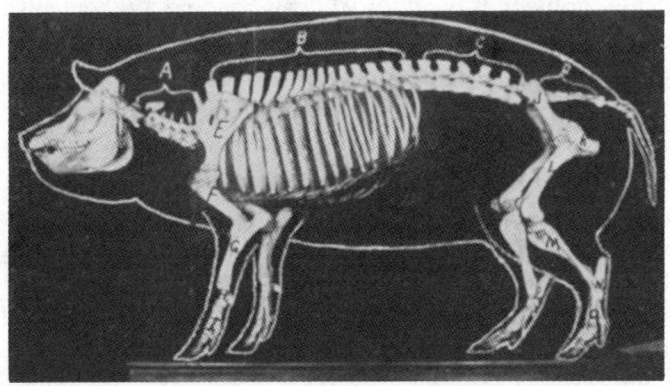

Fig. 4-39. The skeleton of a hog and the names of the bones.

A. 7 cervical vertebrae
B. 15 thoracic vertebrae
C. 6 lumbar vertebrae
D. 5 sacral vertebrae
E. Scapula
F. Humerus
G. Ulna and radius
H. Carpal bones of knee
I. Metacarpal bones of foot
J. Ilium or hip bone
K. Ischium
L. Femur
M. Tibia and fibula
N. Tarsal
O. Metatarsal

manner in which the ribs spring from the backbone has a direct influence on the width and the capacity of the live hog. Likewise, the length of the ribs influences the total depth. Within the ribs lie the heart, lungs, other important vital organs and, in part, the digestive organs. The latter extend back into the pelvic cavity. A strong constitution and good feeding qualities depend fundamentally on the capacity in the body cavity.

The variations in the height of swine are very largely due to differences in the length of the bones of the forelegs and the rear legs. The bones of the forelegs are the scapula (shoulder blade), humerus, ulna and radius, carpals or knee bones and metacarpals below the knee. The bones of the rear legs are the ilium or hip bone, ischium or bone of the buttocks, femur, tibia and fibula, tarsal or hock bones and metatarsals below the hocks.

The bones of the legs and feet seem to vary somewhat independently of each other. Many tall hogs are acceptably short in the pasterns, while others that are medium in height have long pasterns.

WHOLESALE CUTS OF THE PORK CARCASS

A knowledge of the comparative value of the major wholesale cuts of the pork carcass (Fig. 4-2) will help you understand the problems involved in judging swine in general and market hogs in particular. Indeed, this information is basic to successful swine judging. No attempt, however, should be made to apply a specific mathematical analysis to the problem at hand, or you can become too "detailish" and invariably confused in your appraisal. In arriving at comparative values, use the values in Table 4-2 as general information. Desirable development is sought in all the cuts. No matter how well-developed a barrow may be in the ham and loin, it would be difficult to place him very high in a class of well-balanced barrows if he has inferior shoulders, even though they are somewhat less valuable.

The hams, which comprise nearly 30 percent of the total weight of these cuts, represent over 30 percent of their total value. In your judging of breeding animals, do not specifically apply these values because the shoulder region, which involves the chest and, consequently, determines the constitution, is fundamentally important. This is the chief reason why the breeders of all the pure breeds of swine have so uniformly developed good shoulders as well as good hams, resulting in a balanced hog.

TABLE 4-2. Weight and Price Relationship of the Five Primal Cuts of the Pork Carcass[1]

Pork Cut	Price/Lb. Fresh	Weight on Carcass Basis	Total Weight of 5 Cuts	Value in 200-Lb. Hog	Total Value of 5 Cuts
	($)	(%)	(%)	($)	(%)
Ham	1.50	20.5	29.3	43.95	30.24
Loin	1.55	17.0	24.3	37.66	25.91
Belly	1.42	14.5	20.7	29.39	20.22
Boston butt	1.42	10.0	14.3	20.31	13.98
Picnic	1.23	8.0	11.4	14.02	9.65
Total	7.12	70.0	100.0	145.33	100.00

[1]Based on May 1, 1985, price quotation from *The National Provisioner.* Prices vary with change in supply and demand for different cuts.

Details concerning the comparative value of the cuts of lesser importance are not given in Table 4-2, since in general, they are of minor concern in judging swine, except as they show variations in rounding out the general form and in their effect on quality, which applies particularly to the jowls and the fat cuts along the topline.

EVALUATING PORK CARCASSES

The quantity and the quality of edible pork that can be obtained from a hog carcass determine the real value of the carcass and in turn of the live animal. Observations of live hogs have been correlated with the type of carcass on the rail. The direct or objective carcass measurements and the subjective or observed conformation and quality specifications set forth by the USDA Grading Service for grading pork carcasses will be discussed as follows.

Objective Carcass Evaluation

The objective evaluation of swine carcasses includes measuring the backfat thickness, the loin-eye area, the carcass length, the percent of lean cuts or percent of ham and loin, the percent of carcass muscle and the dressing percentage or yield.

Backfat Thickness

One of the main goals in swine improvement is to reduce backfat thickness. Hogs with less backfat and more muscling have higher-valued carcasses.

Backfat is quite easily measured on swine carcasses. With a steel ruler calibrated in tenths of inches, thickness of fat is measured at three places: (1) opposite the first rib, (2) opposite the last rib and (3) opposite the last lumbar vertebrae. Figs. 4-40 and 4-41 show the location of these three measurements.

The three backfat measurements on a carcass might be 1.9, 1.2 and 1.1 inches, and the backfat of the carcass is expressed as an average of the three measurements. For example, 1.9 + 1.2 + 1.1 = 4.2 ÷ 3 = 1.4 inches.

Backfat is highly associated with the percent lean in the carcass. The lower the backfat, the higher the amount of lean in the carcass and vice versa.

The U.S. Department of Agriculture has established grades for swine carcasses based upon the backfat thickness and weight of the hog. Table 4-3 shows the different specifications.

Most of the market barrows and gilts are sold weighing from 190 to 240 pounds. Not all animals in this weight range, having less than 1.1 inches of backfat, would actually be called Utilities. Hogs that have less than 1.1 and are quite heavily muscled will be called No. 1's. Hogs that are thin and poorly muscled, and appear to have grown poorly, will be graded Utility.

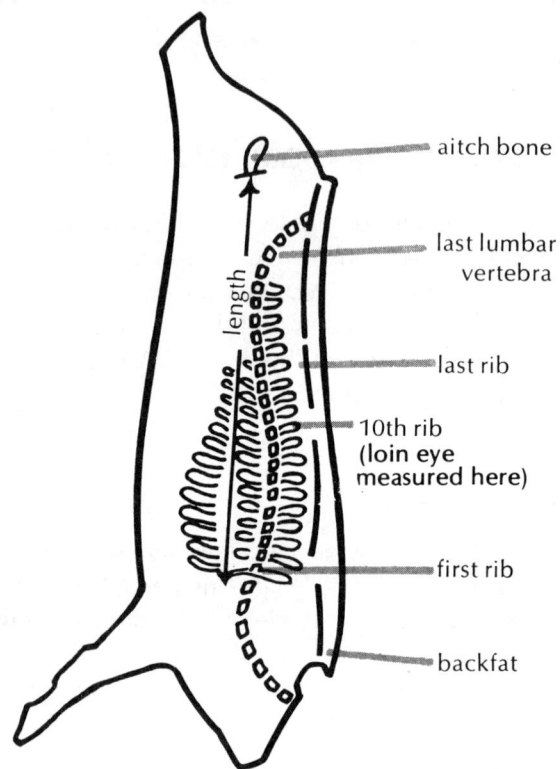

Fig. 4-40. Bones in the swine carcass on which points for measuring backfat, length and loin-eye area are located. (Courtesy, Purdue University)

TABLE 4-3. USDA Grades for Barrows and Gilts and Specifications for Carcasses[1]

Carcass Weight or Length	Average Backfat Thickness (Inches) by Grade				
	U.S. No. 1	U.S. No. 2	U.S. No. 3	U.S. No. 4	Utility
Under 120 lb. or under 27 in.	1.0 – 1.3	1.3 – 1.6	1.6 – 1.9	1.9 – 2.2	Less than 1.0
120 – 164 lb. or 27 in. – 29.9 in.	1.1 – 1.4	1.4 – 1.7	1.7 – 2.0	2.0 – 2.3	Less than 1.1
165 – 204 lb. or 30 in. – 32.9 in.	1.2 – 1.5	1.5 – 1.8	1.8 – 2.1	2.1 – 2.4	Less than 1.2
205 – 255 lb. or 33 in. – 36 in.	1.3 – 1.6	1.6 – 1.9	1.9 – 2.2	2.2 – 2.5	Less than 1.3
Minimum degree of muscling required	Thick	Moderately thick	Slightly thick	Thin	Very thin

[1]Source: *USDA Yield Grades for Barrows and Gilts.*

Probing for backfat thickness. — Swine producers are fortunate to have some accurate ways to measure carcass desirability in the live animal. One of these is measuring the amount of backfat on the live hog, using any one of a number of methods developed for measuring backfat thickness. The most simple and useful one is the steel ruler. The ruler, calibrated in tenths of inches, is pushed through the fat until it rests firmly against the muscle tissue. Fig. 4-41 shows where the three probe sites are located. The first probe is taken behind the shoulder, straight above the elbow; the second probe is at the last rib; and the third probe is straight above the stifle joint (see Fig. 4-1 for location on live animal). All measurements are made approximately 2 inches from the midline of the pig's back.

Fig. 4-42 shows how the backfat probe appears when it penetrates the back of the pig at approximately the last rib. This is a cross-section, reduced in size, from an actual carcass. Note that this pig has considerable thickness of fat, compared to the amount of lean.

Meatiness is difficult to evaluate on the live hog by visual appraisal alone. The backfat probe enables the producer to make more accurate selections for meatiness. After the average backfat thickness has been accurately measured on the live hog, the producer can appraise meatiness in the measured animals by comparing their relative body thickness and the amount of backfat thickness.

To make maximum use of backfat probes in a selection program, you need to carefully consider several factors. For probes to be comparable, the animals should be fed and managed as much alike as possible. Self-feeding a high-energy ration is necessary to remove those animals that produce a high proportion of fat to lean.

The amount of fat produced by the pig varies according to its weight. As the pig becomes heavier, more fat tissue is laid down. Up to 150 pounds, most of the tissue produced is muscle and bone. After 150 pounds, the rate of fat deposition is greatly increased. To properly evaluate and compare backfat measurements on hogs of different weights, you need to adjust the measurements to a standard weight. The standard weight is 200 pounds. Thus, you predict what the backfat of these hogs would be at 200 pounds. Example: If an animal probed 1.6 inches at 220 pounds, then the adjusted probe would be 1.47 inches for 200 pounds. To use Table 4-4, find the weight of the hog and read down, then find the sum of the three probes and read across. Where the two lines intersect is the adjusted backfat thickness. For the most accurate estimate of backfat thickness, the probes should be made when the hog weighs from 180 to 220 pounds.

Loin-Eye Area

Fig. 4-42 shows the location of the loin-eye muscles in the hog carcass. These two muscles are long, running lengthwise along the pig's top, one on one side of the vertebrae and the other on the other side.

In the slaughtering process, the pork carcass is split at the backbone into two halves. Each half contains a loin-eye muscle. To measure the loin eye, cut one of the

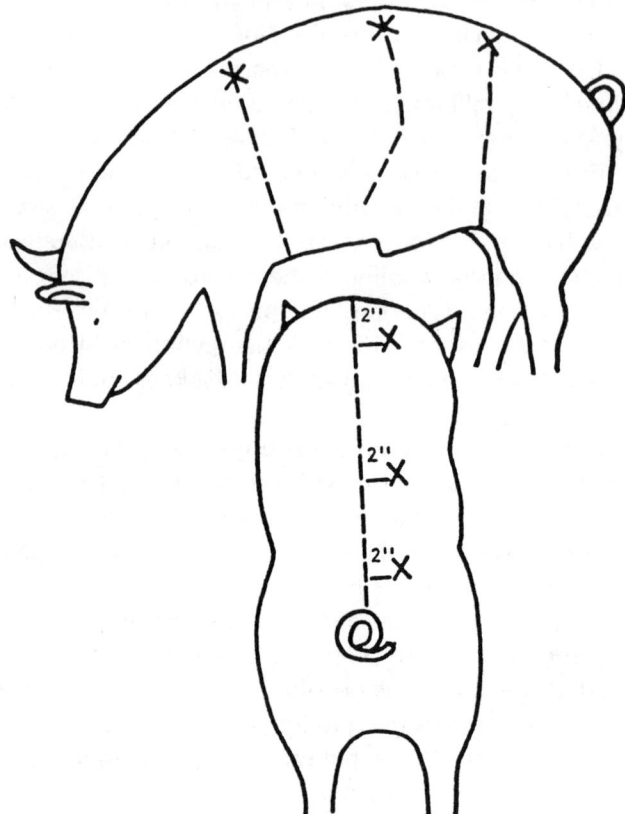

Fig. 4-41. Location of probe sites. (Courtesy, Purdue University)

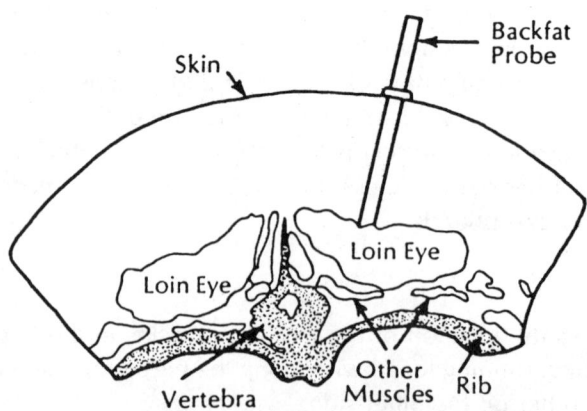

Fig. 4-42. Cross-section through the top part of the hog carcass at approximately the last rib. (Courtesy, Purdue University)

carcass halves between the tenth and the eleventh ribs, counting from the front or anterior end of the carcass. The loin-eye muscle is the cross-sectional surface of the longissimus dorsi muscle (see Fig. 4-42). Fig. 4-40 shows this location.

After the loin eye has been exposed, place a piece of transparent paper over the meat, and then make a tracing of the loin-eye muscle. Measurement of the area of the muscle is made with a *planimeter.* A planimeter is a device that will measure the area, in square inches, by going over the circumference of the loin-eye tracing. You can also measure the loin-eye area by lying a plastic grid over the muscle and then counting the dots that are encompassed by the eye-muscle area. This very simple procedure is becoming a more widely used method than the planimeter for measuring loin-eye areas in carcass evaluation.

The area of the loin-eye muscle is a measure of muscling in the hog. Observe the different sizes of loin eyes in Fig. 4-43. The loin eye is the size of the lean in the pork chop. Which do you think would be most acceptable to the consumer? Loin-eye area is influenced considerably by the genetic make-up of the animal. This means that if animals with large loin eyes are selected, then large loin eyes can be expected to be a characteristic of their offspring.

The area of the loin eye is rather difficult to predict by visual appraisal. However, by using the backfat probe to observe the natural thickness over the loin, you can make a fair appraisal. In the selection of breeding animals, it is most desirable to have littermates slaughtered for carcass information. Having knowledge of littermate cutouts makes the visual appraisal of the live animal much more accurate.

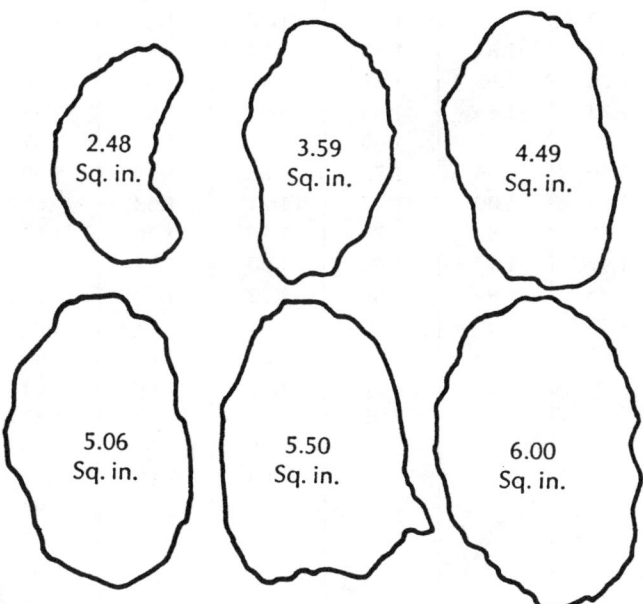

Fig. 4-43. Actual loin-eye tracings from hogs of similar weight and age.
(Courtesy, Purdue University)

TABLE 4-4. Backfat, Adjusted to

Sum of 3 Probes	Actual Weight of Pig						
	175	180	185	190	195	200	205
2.0	0.75	0.73	0.72	0.70	0.68	0.67	0.65
2.1	0.79	0.77	0.75	0.73	0.72	0.70	0.68
2.2	0.83	0.81	0.79	0.77	0.75	0.73	0.72
2.3	0.86	0.84	0.82	0.80	0.78	0.77	0.75
2.4	0.90	0.88	0.86	0.84	0.82	0.80	0.78
2.5	0.94	0.92	0.89	0.87	0.85	0.83	0.81
2.6	0.98	0.95	0.93	0.91	0.89	0.87	0.85
2.7	1.02	0.99	0.97	0.94	0.92	0.90	0.88
2.8	1.05	1.03	1.00	0.98	0.95	0.93	0.91
2.9	1.09	1.06	1.04	1.01	0.99	0.97	0.94
3.0	1.13	1.10	1.07	1.05	1.02	1.00	0.98
3.1	1.17	1.14	1.11	1.08	1.06	1.03	1.01
3.2	1.20	1.17	1.14	1.12	1.09	1.07	1.04
3.3	1.24	1.21	1.18	1.15	1.13	1.10	1.07
3.4	1.28	1.25	1.22	1.19	1.16	1.13	1.11
3.5	1.32	1.28	1.25	1.22	1.19	1.17	1.14
3.6	1.35	1.32	1.29	1.26	1.23	1.20	1.17
3.7	1.39	1.36	1.32	1.29	1.26	1.23	1.20
3.8	1.43	1.39	1.36	1.33	1.29	1.27	1.24
3.9	1.47	1.43	1.39	1.36	1.33	1.30	1.27
4.0	1.50	1.47	1.43	1.40	1.36	1.33	1.30
4.1	1.54	1.50	1.47	1.43	1.40	1.37	1.33
4.2	1.58	1.54	1.50	1.47	1.43	1.40	1.37
4.3	1.62	1.58	1.54	1.50	1.47	1.43	1.40
4.4	1.65	1.61	1.57	1.54	1.50	1.47	1.43
4.5	1.69	1.65	1.61	1.57	1.53	1.50	1.47
4.6	1.73	1.69	1.64	1.61	1.57	1.53	1.50
4.7	1.77	1.72	1.68	1.64	1.60	1.57	1.53
4.8	1.80	1.76	1.72	1.68	1.64	1.60	1.56
4.9	1.84	1.80	1.75	1.71	1.67	1.63	1.60
5.0	1.88	1.84	1.79	1.75	1.70	1.67	1.63
5.1	1.92	1.87	1.82	1.78	1.74	1.70	1.66
5.2	1.96	1.91	1.86	1.82	1.77	1.73	1.69
5.3	1.99	1.95	1.90	1.85	1.81	1.77	1.73
5.4	2.03	1.98	1.93	1.89	1.84	1.80	1.76
5.5		2.02	1.97	1.92	1.88	1.83	1.79
5.6			2.00	1.96	1.91	1.87	1.82
5.7				1.99	1.94	1.90	1.86
5.8				2.03	1.98	1.93	1.89
5.9					2.01	1.97	1.92
6.0						2.00	1.95
6.1							1.99
6.2							2.02
6.3							
6.4							
6.5							

[1]Courtesy, National Association of Swine Records.

200 Pounds Live Weight[1]

When Probed (in Pounds)

210	215	220	225	230	235	240	245	250
0.64	0.62	0.61	0.60	0.59				
0.67	0.66	0.64	0.63	0.62	0.60			
0.70	0.69	0.67	0.66	0.65	0.63	0.62		
0.73	0.72	0.70	0.69	0.67	0.66	0.65	0.64	
0.76	0.75	0.73	0.72	0.70	0.69	0.68	0.67	0.66
0.80	0.78	0.76	0.75	0.73	0.72	0.70	0.69	0.68
0.83	0.81	0.79	0.78	0.76	0.75	0.73	0.72	0.71
0.86	0.84	0.82	0.81	0.79	0.78	0.76	0.75	0.74
0.89	0.87	0.85	0.84	0.82	0.81	0.79	0.78	0.76
0.92	0.90	0.88	0.87	0.85	0.83	0.82	0.80	0.79
0.96	0.94	0.92	0.90	0.88	0.86	0.85	0.83	0.82
0.99	0.97	0.95	0.93	0.91	0.89	0.87	0.86	0.85
1.02	1.00	0.98	0.96	0.94	0.92	0.90	0.89	0.87
1.05	1.03	1.01	0.99	0.97	0.95	0.93	0.92	0.90
1.08	1.06	1.04	1.02	1.00	0.98	0.96	0.94	0.93
1.12	1.09	1.07	1.05	1.03	1.01	0.99	0.97	0.96
1.15	1.12	1.10	1.08	1.06	1.04	1.02	1.00	0.98
1.18	1.15	1.13	1.11	1.09	1.06	1.04	1.03	1.01
1.21	1.19	1.16	1.14	1.11	1.09	1.07	1.05	1.04
1.24	1.22	1.19	1.17	1.14	1.12	1.10	1.08	1.06
1.27	1.25	1.22	1.20	1.17	1.15	1.13	1.11	1.09
1.31	1.28	1.25	1.23	1.20	1.18	1.16	1.14	1.12
1.34	1.31	1.28	1.26	1.23	1.21	1.18	1.16	1.15
1.37	1.34	1.31	1.29	1.26	1.24	1.21	1.19	1.17
1.40	1.37	1.34	1.32	1.29	1.27	1.24	1.22	1.20
1.43	1.40	1.37	1.34	1.32	1.29	1.27	1.25	1.23
1.47	1.44	1.40	1.37	1.35	1.32	1.30	1.28	1.26
1.50	1.47	1.44	1.40	1.38	1.35	1.33	1.30	1.28
1.53	1.50	1.47	1.43	1.41	1.38	1.35	1.33	1.31
1.56	1.53	1.50	1.46	1.44	1.41	1.38	1.36	1.34
1.59	1.56	1.53	1.49	1.47	1.44	1.41	1.39	1.36
1.62	1.59	1.56	1.52	1.50	1.47	1.44	1.41	1.39
1.66	1.62	1.59	1.55	1.53	1.50	1.47	1.44	1.42
1.69	1.65	1.62	1.58	1.55	1.52	1.49	1.47	1.45
1.72	1.68	1.65	1.61	1.58	1.55	1.52	1.50	1.47
1.75	1.72	1.68	1.64	1.61	1.58	1.55	1.53	1.50
1.78	1.75	1.71	1.67	1.64	1.61	1.58	1.55	1.53
1.82	1.78	1.74	1.70	1.67	1.64	1.61	1.58	1.56
1.85	1.81	1.77	1.73	1.70	1.67	1.64	1.61	1.58
1.88	1.84	1.80	1.76	1.73	1.70	1.66	1.64	1.61
1.91	1.87	1.83	1.79	1.76	1.73	1.69	1.66	1.64
1.94	1.90	1.86	1.82	1.79	1.75	1.72	1.68	1.67
1.98	1.93	1.89	1.85	1.82	1.78	1.75	1.72	1.69
2.01	1.96	1.92	1.88	1.85	1.81	1.78	1.75	1.72
	2.00	1.95	1.91	1.88	1.84	1.80	1.77	1.75
		1.98	1.94	1.91	1.87	1.83	1.80	1.77

Carcass Length

Carcass length in swine is measured from the forward edge of the first rib to the forward edge of the aitch bone. The points of these measurements are shown in Fig. 4-40.

Of all the standard carcass measurements, length is least associated with carcass value. Some of the longer breeds, however, are superior in litter size and mothering ability. This may be one reason why length has been included as a desirable characteristic of the meat-type hog.

Percent Lean Cuts of Percent Ham and Loin

The critical measure of the desirability of a carcass is the total value of the trimmed cuts that it contains. In the evaluation of large numbers of carcasses, cutting out, weighing and evaluating various parts of the carcass is not economically feasible. At the present time, we determine what part or percentage of the carcass is made up of some of the high-priced cuts. Note Table 4-5 for the expected yields of the four lean cuts by USDA grade.

TABLE 4-5. Expected Yields of the Four Lean Cuts by Grade[1, 2]

Grade	Minimum USDA Yields by Grade	Realistic USDA Yields by Grade
U.S. No. 1	53% and over	58% and over
U.S. No. 2	50 – 52.9%	52.0 – 57.9%
U.S. No. 3	47 – 49.9%	49.0 – 51.9%
U.S. No. 4	47.0% and less	48.9% and less

[1]Courtesy, USDA Consumer Marketing Service.
[2]Percent lean cuts are based on chilled carcass weight.

The lean cuts of swine are the trimmed ham, loin, Boston butt and picnic shoulder (see Fig. 4-2). If the weight of these cuts in relation to the carcass weight is known, the value of that carcass can quite readily be estimated. To speed up carcass evaluation even further, the percent of ham and loin can be used. The percent of these highest-priced cuts (ham and loin) reflects carcass value quite well compared to percent of lean cuts, and in some cases, it is superior. The lean cuts and ham and loin are trimmed to a small, uniform thickness of fat before these values are obtained. With the exception of the ham, loin, picnic and Boston butt, the remainder of the wholesale cuts are classified as fat cuts (Fig. 4-2).

The most limiting factor associated with the use of percent lean cuts or percent ham and loin is the variability due to cutting and trimming. Because of this variability, these percentages are most accurate when used within a given pork-cutting operation. Consequently, these percentages differ between plants and with regard to seasonal demands for the respective wholesale cuts. Thus, a percent carcass muscle equation based entirely on objective measurements was developed.

Percent Carcass Muscle

The most practical method of evaluating carcass leanness now appears to be through the use of a regression equation, similar to that which has been used for several years to establish beef and sheep yield grades. Hot carcass weight, loin-eye area at the 10th rib and fat depth over the loin eye at the 10th rib can be accurately measured. Place a plastic grid over the loin eye and count the dots or squares that fall within the boundaries of the l. dorsi (loin-eye) muscle. Do not include the small muscles surrounding the longissimus. Convert to square inches by dividing the number of dots or squares by 10. Using a fat probe or another instrument graduated in $^1/_{10}$-inch increments, measure the fat depth at the $^3/_4$ point over the loin eye (Fig. 4-44). Divide the longest axis of the loin-eye muscle into quarters. Measure the fat depth of the $^3/_4$ point from the outer edge of the muscle to a point at the inner edge of and perpendicular to the skin.

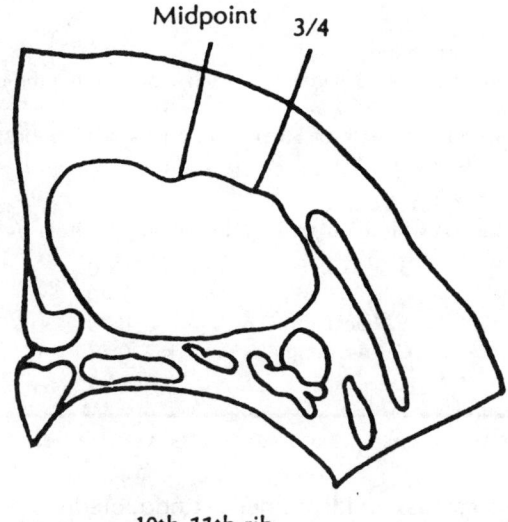

Fig. 4-44. Three-quarter fat measurement for estimating percent carcass lean. (Courtesy, Purdue University)

Use the standard formula to calculate percent lean containing approximately 10 percent fat or use the recommendations and procedures found in Tables 4-6 and 4-7 to estimate it.

Calculation of percent carcass muscle is a more accurate estimate of carcass cutability than calculation of percent lean cuts and percent ham and loin, because it is based entirely on objective measurements and is not influenced by cutting and trimming variation, as is the case with cutout percentages. Thus, the use of percent

TABLE 4-6. Estimating Percent Carcass Lean[1]

I. Standard formula for predicting pounds of muscle containing 10 percent fat.

2.1 + .45 (hot carcass weight in pounds)
  +   5 (loin-eye area in square inches)
  − 11 (³/₄ backfat thickness in tenths of inches at the 10th rib)
  = pounds muscle (containing about 10 percent fat)

$$\text{Percent lean} = \frac{\text{Pounds muscle}}{\text{Hot carcass weight}} \times 100$$

II. Short formula. Use a 55 percent lean for a base and adjust for the following factors.

A. Loin-eye area
  1. Use a base of 5.0 sq. in.
  2. Add 0.3 to the base of 55 percent for each 0.1 sq. in. that the loin eye is over 5.0 sq. in.
  3. Subtract 0.3 from the base of 55 percent for each 0.1 sq. in. that the loin eye is under 5 sq. in.

B. Three-fourths backfat thickness
  1. Use a base of 1 in.
  2. Add 0.7 to the base of 55 percent for each 0.1 in. fat that the fat is less than 1 in.
  3. Subtract 0.7 from the base of 55 percent for each 0.1 in. fat that the fat is more than 1 in.

C. Carcass weight
  1. Use a base of 160 pounds.
  2. Add 0.06 to the base of 55 percent for each 1 pound that the carcass weight is less than 160 pounds.
  3. Subtract 0.06 from the base of 55 percent for each 1 pound that the carcass weight is greater than 160 pounds.

D. Example
  1. 172-pound carcass with a 5.8 sq. in. loin eye and 0.9 inch ³/₄ backfat thickness.

Base =	55.00
Loin-eye area =	+ 2.40
³/₄ backfat thickness =	+ 0.70
Carcass weight =	− 0.72
	57.38 percent lean

[1]Adapted from Roger E. Hunsley's AS-301 and AS-401 bulletins. Animal Sciences Department, Purdue University.

muscle to determine carcass and live price is undoubtedly a better indicator of value than cutout percentages. (Note later section for calculating the live price for a group of hogs.)

The following standards represent the present minimum carcass standards for describing the meat-type hog.

Length — 29.5 inches or more

Backfat — 1.6 inches or less (maximum for No. 1 hog)

Loin-eye area — 4.5 square inches or more

Percent muscle — 55 percent or more (carcass weight)

Fig. 4-45 shows the relationship between average thickness of backfat, carcass length, hot carcass degree of weight and degree of muscling for pork carcasses.

TABLE 4-7. Examples of ³/₄ Backfat Thickness and Percent Lean on Different Groups of Pigs[1,2]

I. Group No. 1 — Pigs with loin-eye area of under 4.5 sq. in.

	Average	Range	Most Pigs
Percent lean	48.62	43.6 –54.7	46.0 –52.5
³/₄ backfat	1.51	0.8 – 2.2	1.0 – 1.8
Average backfat	1.48	0.97– 1.97	0.97– 1.97

³/₄ backfat thickness vs. average backfat thickness.

50 percent of pigs: ³/₄ backfat thickness is from 0.1 above to 0.1 below the average backfat thickness.

73 percent of pigs: ³/₄ backfat thickness is from 0.2 above to 0.2 below the average backfat thickness.

The distribution above and below the average backfat is essentially equal. Mostly depends on the degree of muscling.

II. Group No. 2 — Pigs with average backfat greater than 1.4 in. and loin-eye area greater than 5.4 sq. in.

	Average	Range	Most Pigs
Percent lean	53.88	50.8–56.5	53.2–56.5
³/₄ backfat	1.41	1.1– 1.8	1.1– 1.8
Average backfat	1.54	1.3– 1.83	1.3– 1.83

³/₄ backfat thickness vs. average backfat thickness.

60 percent of pigs: ³/₄ backfat thickness is from 0.2 below to 0.4 below average backfat thickness.

90 percent of pigs: ³/₄ backfat thickness is from equal to 0.4 below average backfat thickness.

III. Group No. 3 — Pigs with average backfat greater than 1.4 in. and loin-eye area between 4.5 and 5.4 sq. in.

	Average	Range	Most Pigs
Percent lean	50.57	48.7–54.7	48.7–51.7
³/₄ backfat	1.53	1.0– 1.9	1.3– 1.9
Average backfat	1.54	1.3– 1.83	1.3– 1.83

³/₄ backfat thickness vs. average backfat thickness.

75 percent of pigs: ³/₄ backfat thickness is from 0.1 above to 0.2 below average backfat thickness.

83 percent of pigs: ³/₄ backfat thickness is from 0.2 above to 0.2 below average backfat thickness.

The distribution above and below the average backfat is equal. Depends on degree of muscling.

IV. Group No. 4 — Pigs with average backfat less than 1.0 in. and loin-eye area greater than 5.9 sq. in.

	Average	Range	Most Pigs
Percent lean	59.35	58.0–60.2	58.0–60.2
³/₄ backfat	0.75	0.6– 0.9	0.6– 0.9
Average backfat	1.08	0.9– 1.2	0.9– 1.2

³/₄ backfat thickness vs. average backfat thickness.

100 percent of pigs: ³/₄ backfat thickness is from 0.3 below to 0.4 below average backfat thickness.

(Continued)

TABLE 4-7 (Continued)

V. Group No. 5 — Average pigs.

	Average	Range	Most Pigs
Percent lean	55.32	51.5 –60.92	53.0 –58.0
3/4 backfat	0.97	0.6 – 1.3	0.8 – 1.2
Average backfat	1.18	0.87– 1.37	0.87– 1.37

3/4 backfat thickness vs. average backfat thickness.

65 percent of pigs: 3/4 backfat thickness is from equal to 0.3 below average backfat thickness.

85 percent of pigs: 3/4 backfat thickness is from equal to 0.4 below average backfat thickness.

95 percent of pigs: 3/4 backfat thickness is from 0.1 above to 0.4 below average backfat thickness.

[1]Adapted from Roger E. Hunsley's AS-301 and AS-401 bulletins, Animal Sciences Department, Purdue University.

[2]Note: Heavy-muscled pigs will usually have less fat at the 3/4 measurement than light-muscled pigs.

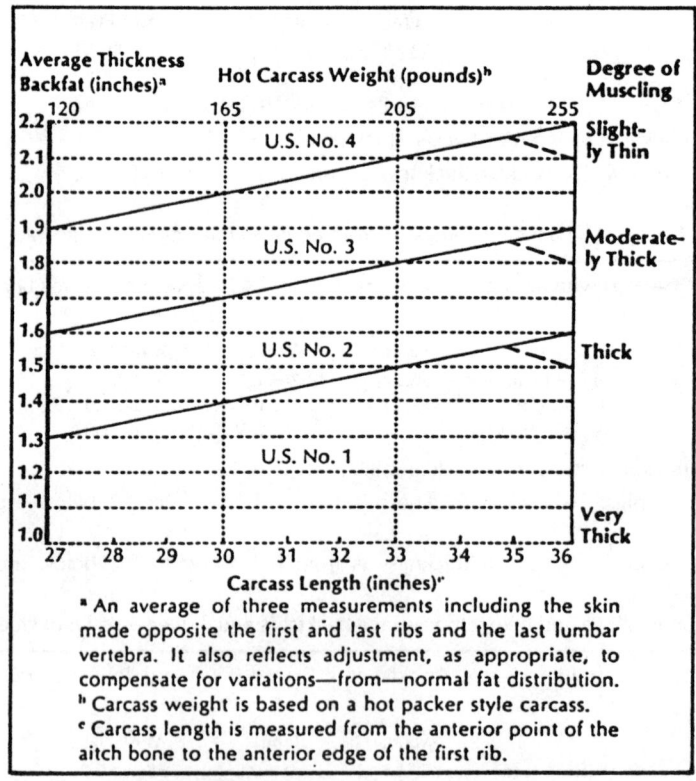

Fig. 4-45. Relationship between average thickness of backfat, carcass length, hot carcass weight and degree of muscling for pork carcasses. (From *USDA Official Pork Carcass Grading Standards*)

Dressing Percentage or Yield

The majority of hogs are purchased on the basis of their live weight. The packer is most interested in how much of the live weight will be represented by carcass weight. In other words, differences in dressing percentage cause differences in the live values of hogs.

$$\text{Dressing percentage} = \frac{\text{Weight of carcass}}{\text{Live weight}} \times 100$$

For example, a 200-pound hog that yielded a 140-pound carcass would have a dressing percentage of 70.

$$\frac{140}{200} \times 100 = 70$$

The two most important factors causing differences in dressing percentage in swine are **fill** and **finish.**

Fill. — Fill is the amount of contents of the digestive tract. The amount of feed and water eaten results in differences in fill. Transporting animals causes loss of digestive contents so that hogs shipped to market would have a higher dressing percentage than hogs at the farm.

Buyers can accurately estimate the dressing percentage of live hogs. A trim, straight middle on a hog is one of the best indicators of a high dressing percentage.

Finish. — Usually the more finish or fat that the hog carries, the higher will be the dressing percentage. This is because more fat is laid down on the carcass with little weight added to the parts removed during the slaughtering process. Table 4-8 shows how the dressing percentage is influenced by backfat thickness.

TABLE 4-8. Average Dressing Percentage for Hogs of Different Slaughter Grades[1]

Grade	Average Backfat (in.)	Dressing Percentage	
		Range	Average
U.S. No. 1	1.3	67–71	69
U.S. No. 2	1.6	68–72	70
U.S. No. 3	1.9	69–73	71
U.S. No. 4	2.2	70–74	72
Utility	1.0	63–69	66

[1]Courtesy, USDA Consumer and Marketing Service.

The average dressing percentage for all hogs is approximately 70 percent.

A higher dressing percentage is not always an advantage in hogs; it might mean selection of fatter animals. However, there are considerable differences in dressing percentage within a slaughter grade. This means that these are high and low dressing percentages represented in hogs that would grade U.S. No. 1. Selecting meaty hogs with somewhat trimmer middles means selecting those with higher dressing percentages as well.

Both the desirable market animal and the desirable breeding animal should have the characteristics of Animal A. Plus, they need to have the correct slope to the shoulder, set to the front and rear feet and legs and slope to the pasterns. The toes should be the same length, and the animal should walk out with style and with ease of movement. Hogs should gain 1 pound for every 2 pounds of feed consumed and weigh 240 pounds in 150 days or less. Such animals need rugged framework, correct structure and above all, heavy bone and sound feet and legs that will carry the weight while they are being raised in confinement and on through the breeding cycle, if they are breeding animals.

Subjective Carcass Evaluation

Conformation, Quality and Finish

Subjective carcass evaluation is based on three factors: **conformation, quality** and **finish.**

Conformation is defined as the relative amount of edible and inedible tissue. The shape of the carcass or of a particular cut is indicative of the lean meat yield in proportion to the yield of fat or waste. In general, conformation determines the carcass that will yield the highest percent muscle, as indicated by an abundance of muscling and thickness with a minimum of external fat (as indicated by smoothness and trimness from end to end of the carcass).

Specifically, the ham should be reasonably short-shanked, fleshed well down on the shank and thick and bulging through the cushion. The loin should indicate meatiness by length, width and fullness, as well as trimness of backfat. Uniformity of depth along the chine, in addition to length and width through the rump, also enhances loin conformation. A desirable side is indicated by length, uniform thickness and freedom from wrinkles. The shoulder should exhibit a muscular trim appearance (revealing meatiness and minimum fat trim). The jowl should be trim and small (as light as possible).

Oftentimes, a pork carcass must be evaluated with the carcass intact. In this case, the ham and loin wholesale cuts are not made available for inspection by the evaluator. While this may leave something to be desired, a number of factors influencing pork **quality** can be determined by visual appraisal of the carcass.

A firm flank and backfat are indications of firmness in the muscle.

The desirable greyish-pink color can be observed in the flank and rib region. The exposed fat should appear white.

Indications of marbling or intramuscular fat streaking, which enhances juiciness, can be observed on the exposed surface of the ham above the aitch bone and on the lumbar lean. **Feathering,** or the amount of fine fat streaking between the ribs, also is correlated with the amount of marbling. A thin covering of fat over the inside of the thoracic cavity, called *overflow fat,* indicates a well-finished carcass (excessive amounts are discriminated against).

A really thorough job of pork carcass evaluation involves observations of fresh-cut muscle surfaces to evaluate **quality.**

Pork quality evaluation is primarily based on four factors: **color, firmness, structure** and **marbling.**

Table 4-9 contains "quality scores" which indicate the various degrees of pork quality. Color, firmness and structure are often combined into one composite score.

TABLE 4-9. Scores Used to Evaluate Pork Carcasses for Quality[1]

Score	Color	Firmness	Structure	Marbling
1	Pale	Soft	Exudative	Devoid
2	Slightly pale	Moderately soft	Moderately exudative	
3	Greyish-pink	Moderately firm	Moderately dry	Moderate
4	Moderately dark	Firm	Dry	
5	Dark	Very firm	Very dry	Abundant

[1]Source: Roger E. Hunsley's AS-301 and AS-401 bulletins, and J. C. Forrest's AS-351 bulletin, Animal Sciences Department, Purdue University.

The third and final factor in subjective carcass evaluation is **finish.** It is important in determining cutout value, which is measured by the ratio of lean to fat.

Indications of excess fat include: (1) excessive backfat (exceeding 1 inch), (2) exceptionally wide back and tapering to a narrow underline, (3) excessive fat along the back of the ham (known as a heavier fat collar) and (4) excessive fat in the flank, along the midline and over the breastbone.

A pork carcass also can contain an insufficient amount of finish, which is indicated by lack of firmness of the cuts.

Grading Pork Carcasses

On April 1, 1968, the grades for swine carcasses based upon backfat thickness and carcass length or hot carcass weight were revised. Table 4-3 and Fig. 4-45 provide objective guides for determining the five USDA hog carcass grades. Medium grade is called Utility. This change was made to improve the coordination of the grade names for all three species of livestock and meat. Utility has been used as a grade name in the other species for many years. Carcasses that have characteristics indicating that the lean in the lean cuts (ham, loin, picnic shoulder and Boston butt) will not have an acceptable quality or that have bellies too thin to be suitable for bacon production are graded U.S. Utility.

U.S. No. 1, U.S. No. 2, U.S. No. 3 and U.S. No. 4 are the four grades for carcasses that have indications of acceptable lean quality and belly thickness. These grades are based entirely on the expected carcass yields of the four lean cuts, and no consideration is given to the development of quality superior to that described as *minimum* for these grades. A more realistic yield for each grade is presented in Table 4-5. The minimum expected yields of the four lean cuts for each of these four grades are shown in Table 4-10.

TABLE 4-10. Minimum Expected Yields of the Four Lean Cuts Based on Chilled Carcass Weight, by Grade[1, 2]

Grade	Percent Yield
U.S. No. 1	53 and over
U.S. No. 2	50–52.9
U.S. No. 3	47–49.9
U.S. No. 4	Less than 47

[1]Courtesy, USDA Consumer and Marketing Service.

[2]These yields will be approximately 1 percent lower if based on hot carcass weight.

The following are the USDA pork grades and their specifications.

U.S. No. 1

1. Acceptable quality lean.
2. High yield of lean cuts.
3. Low yield of fat cuts.
4. Belly at least slightly thick.
5. Slightly firm lean.
6. Slightly firm fat.
7. Slight amount of marbling.
8. Slight amount of feathering.
9. Greyish-pink to moderately dark red color.
10. Thickly muscled ham, loin and shoulder.
11. Lower portion of the ham toward the hock covered with a thin layer of fat.
12. Well-rounded back.
13. Area at the junction of the lower part of the shoulder and the belly depressed in relation to the shoulder and the belly.
14. Area directly anterior to the hip bone depressed in relation to the ham and the loin.

U.S. No. 2

1. Acceptable quality lean.
2. Slightly high yield of lean cuts.
3. Slightly low yield of fat cuts.
4. Belly at least slightly thick.
5. Slightly firm lean.
6. Slightly firm fat.
7. Slight amount of marbling.
8. Slight amount of feathering.
9. Greyish-pink to moderately dark red color.
10. Moderately thick-muscled ham, loin and shoulder.
11. Lower portion of the ham toward the hock covered with a slightly thin layer of fat.
12. Slightly well-rounded back.
13. Area at the junction of the lower part of the shoulder and belly slightly depressed in relation to the shoulder and the belly.
14. Area directly anterior to the hip bone slightly depressed in relation to the ham and the loin.

U.S. No. 3

1. Acceptable quality lean.
2. Slightly low yield of lean cuts.
3. Slightly high yield of fat cuts.
4. Belly at least slightly thick.
5. Lean at least slightly firm.
6. Slightly firm fat.
7. Slight amount of marbling.
8. Slight amount of feathering.
9. Greyish-pink to moderately dark red color.
10. Slightly thin-muscled ham, loin and shoulder.
11. Lower portion of the ham toward the hock covered with a slightly thick layer of fat.
12. Back slightly flat and edge of loin slightly full, resulting in a slight break from the back to the side.
13. Area at the junction of the lower part of the shoulder and the belly only slightly depressed in relation to the shoulder and the belly.
14. Area directly anterior to the hip bone only very slightly depressed in relation to the ham and the loin.

U.S. No. 4

1. Acceptable quality lean.
2. Low yield of lean cuts.
3. High yield of fat cuts.

4. Belly at least slightly thick.
5. Lean at least slightly firm.
6. Slightly firm fat.
7. Slight amount of marbling.
8. Slight amount of feathering.
9. Greyish-pink to moderately dark red color.
10. Thinly muscled ham, loin and shoulder.
11. Lower portion of the ham toward the hock covered with a thick layer of fat.
12. Back flat and edge of loin full, resulting in a definite break from the back to the side.
13. Area at the junction of the lower part of the shoulder and the belly full and smooth in relation to the shoulder and the belly.
14. Area directly anterior to the hip bone full and smooth in relation to the ham and the loin.

U.S. Utility

1. Carcasses that have characteristics indicating they will have a lesser development of lean quality than described as minimum for the other four grades.

TABLE 4-11. Swine—Live Animal

Animal Identi- fication	Live Weight		Dressing Percentage		Length		Loin-Eye Area	
	Estimate	Actual	Estimate	Actual	Estimate	Actual	Estimate	Actual

2. Carcasses that do not have acceptable belly thickness.
3. Carcasses that are soft and oily.

Table 4-11 is a sample of a form for evaluating swine. Such a form should be used when live animals are evaluated and the actual carcass measurements are made.

EXAMPLE FOR CALCULATING LIVE PRICE
FOR HOGS

The following is an example for calculating the live price for a group of hogs.

1. Estimate percent lean (muscle) on a hot carcass weight basis.
2. Furnish current carcass price per hundredweight, the base percent lean (muscle) and its effect on price change and standard dressing percentage.
3. Add or subtract $0.75 to or from the current carcass price per hundredweight for every 1 percent difference between estimated and base carcass percent lean (muscle).
4. Multiply standard dressing percentage (70 percent) by the estimated carcass price.

and Carcass Evaluation Sheet

Average Backfat Thickness		³/₄ Backfat Thickness		USDA Quality Carcass Grade		Percent Muscle	
Estimate	Actual	Estimate	Actual	Estimate	Actual	Estimate	Actual

5. Example: If the current price for carcasses yielding 50 percent lean (muscle) (base yield) equals $70 per hundredweight, and it is estimated that the group of live hogs will yield 52 percent lean (muscle) (carcass weight basis) and will dress 70 percent (standard dressing percentage), then:
   a. Estimated carcass value/cwt. = (52.00 − 50.0) 0.75 + $70.00 = $71.50.
   b. Estimated live value/cwt. + ($71.50) (.70) = $50.05 (rounded to nearest 1.10) = $50.00.
   c. If the actual live value/cwt. is $49.50, then the student's score = 60 − ($50.00 − $49.50)/0.05 (1 point off) = 60 − 10 = 50. (Round prices to nearest $0.10).
6. Assume that the cost of slaughter offsets the value of the by-products.
7. Express your final estimated average price for the group of live hogs on a dollar per hundredweight *live weight basis.*

SWINE JUDGING TERMINOLOGY

The following swine judging terms or expressions can be used to present your reasons for placing swine classes.

General Expressions

1. The meatiest (heaviest-muscled), most correctly finished, highest-quality, smoothest barrow in the class.
2. A typier, more nicely balanced, stretchier gilt than No. 2.
3. More style (class), breed type and sex character than any other gilt in the class.
4. An off-type, light-muscled, overfinished barrow.
5. A plain, short-sided, poorly balanced gilt.
6. A coarse, low-quality, roughly made barrow.
7. A big, rugged, firm, trim, muscular barrow that stands on more timber (heavier bone) than any other pig in the class.
8. A long, clean barrow with more spring of rib, more thickness and spread down his top and more depth, thickness and total dimension to the ham.
9. A big, stout, hard, trim kind of barrow that looks extremely fresh today and possesses a nicer turn over his top and a stronger, more even arch to his back than any other barrow in the class.
10. A heavier-boned, breedier-headed, heavier-muscled gilt that is more even in her teat spacing, more prominent in her underline and straighter and stronger standing on her feet and legs than any other gilt.
11. More desirable composition of gain.
12. More length and elevation to structural make-up.
13. An individual that moves out wider off both ends, indicating more intermuscular dimension.

14. A larger-framed individual.
15. A longer, freer-strided barrow or gilt.
16. A more level-topped, more level-rumped individual.
17. A round-boned, short-muscled, early-maturing individual.
18. A smaller-framed, tighter-wound, earlier-maturing individual.
19. The least potential to develop into a productive brood matron (sow).
20. Least (or most) genetic promise.
21. Longer, looser muscle structure from front to rear.
22. More muscle volume stretched over a larger skeletal framework.
23. An individual that appears to convert that high-priced cereal grain to a more salable, usable product with a greater degree of efficiency.
24. Greater weight per day of age.
25. Taller from the ground up.
26. Squarer-ribbed.
27. Standing on more rugged substance of bone.
28. Lacking in growability, doability and profitability to merit a higher placing in the class today.
29. A broodier, roomier female that appears to have greater capacity to carry that bigger, thriftier litter of pigs.
30. Possessing size, scale, frame and all growth factors.
31. More flexible in his kind as far as going on to higher and/or heavier weights.
32. Outstanding performance, growability.
33. More desirable degree of condition for the breeding pasture.
34. More fit and true in her breeding condition.
35. More desirable composition of gain.
36. Looser, longer muscle structure through the ham.
37. Higher quality.
38. Heavy-fronted.
39. Greater width and muscle definition.
40. Faster-growing.
41. Rough over the loin edge.
42. More uniform fat cover over the rib and loin edge.
43. If the market so demanded.
44. Large frame and skeleton.
45. Bigger when he/she reaches maturity.
46. Higher volume.
47. Longer through stifle and longer from hip to hock and higher tail setting, thus more total muscle dimension to the ham.
48. Thicker through the lower one-third of the body.
49. Overpowering.
50. Heavy-muscled through the biceps and over the top of the rump.
51. Mature-headed.
52. Tremendous profit potential.

53. More ruggedly made.
54. More profitable to the progressive breeder.
55. Big-tailed, strong-jawed, rugged, high-capacity individual.
56. Freer-wheeling.

Head and Neck

1. A cleaner, neater, more alert head.
2. A shorter-necked, heavier-jowled barrow.
3. Cleaner-cut in the jowl than No. 2.
4. A plain-headed, short-necked, wasty-jowled barrow.
5. A fine-headed, small-eared gilt with a soft, flabby, wasty jowl.
6. More desirable (Hampshire, etc.) breed character about the head and ear than any other gilt in the class.
7. A breedy-headed gilt with a correct set to the ears and large, alert, bright eyes.
8. A plain-headed boar that lacks the breed type and sex character to place any higher in this class.
9. Ears carried too erect. Ears too pendulous and close to the head.
10. A head with ample style and excellent breed character.
11. A heavy-headed, cresty-necked gilt that lacks the female quality and refinement of the other individuals in the class.
12. Feminine-headed (gilts); masculine-headed (boars).
13. Short, thick-necked.

Shoulders and Chest

1. Smooth blending of neck into the shoulders.
2. Neater laid in at the shoulder.
3. Too heavy, rough and open in the shoulders.
4. Freer-moving shoulder and cleaner at the point of the shoulder than any other barrow in the class.
5. A barrow wrinkled along the shoulder and pushing a heavy roll of fat over and behind the shoulder — this being especially evident when he moves.
6. A heavier-muscled, cleaner, tighter shoulder than No. 3.
7. Carrying too much grease up over the shoulder.
8. A little heavy in the shoulder.
9. Light-muscled, loose, open shoulder.
10. Sharp-shouldered, narrow-chested gilt.
11. More length, depth and width to her chest chamber and reproductive tract.
12. A wide-chested barrow that stands wider in front, exhibits more muscling through the shoulders and is cleaner over the top and behind the shoulders than any other pig in the class.
13. A gilt with tremendous width through the chest floor, possessing more constitution and capacity to carry a litter of pigs than any other gilt in the class.
14. Flat over the top of the shoulders. Slack in the shoulders. A little out in the shoulders.

15. A coarse-shouldered pig that lacks firmness, particularly at the base. Wrinkled and rough.
16. A bolder-chested, wider-sprung, deeper-ribbed individual.
17. Tighter-shouldered.

Back and Loin

1. A more level-topped barrow with a thicker, wider, heavier-muscled loin than any other pig in the class.
2. More nicely turned over the top with a longer, more uniform arch to the back than any other gilt in the class.
3. A cleaner, more evenly turned top than No. 1.
4. A tighter-framed barrow that exhibits more evidence of muscling (natural width) down the top than No. 3.
5. More spread and thickness over the back and loin with a more level, more even arch to the top than No. 2.
6. A weak-topped, loose-framed, light-muscled gilt.
7. A barrow with an unevenly turned top, lacking evidence of muscling over the back and through the loin and very uneven in the arch to his back.
8. A tight-framed, heavy-muscled barrow that is cleaner than a hound's tooth all down his top.
9. A bit sharp in the back, lacking width and general fullness along the back (fishbacked).
10. A bit flat in the top for his / her breed. Somewhat low in the back. Easy in the topline. Lacking proper support in the topline.
11. Too high-topped. A high-topped pig that is not typical of the breed.
12. Unbalanced arch of top or back. Arch too high over the loin, too low at the shoulders and too flat over the rump.
13. A barrow that is grooved down his top, is cleaner along his top and has more natural thickness and spread over the top than any other in the class.
14. Extreme flare over the back or top, which would indicate an excess of finish in this area.
15. Tied in or narrow or pinched in the loin.
16. An overfinished light-muscled hog with a flat-top, square-top, "shelf" appearance along the top, wider at the top than at the bottom (tapering from top to bottom).
17. A greater arch and squarer rib than No. 2.
18. A longer, more level-topped gilt with a higher, more desirable tail setting.

Rump and Ham

1. A barrow with a wider, more flaring rump and more natural spread and thickness over the rump than No. 2.
2. A gilt possessing a longer ham and rump that ties up higher into the loin and further into the side.

3. Longer and leveler in his rump, higher in his tail setting, with more dimension of ham that is thicker through the center than the No. 2 barrow.
4. Deeper, thicker, wider, meatier, heavier-muscled ham.
5. A short, steep, narrow-rumped gilt that lacks the width, depth, thickness and muscling through the ham to place (to stand) any higher in this class today.
6. A shallow, narrow-hammed pig that is too low in his tail setting and extremely short in his rump.
7. A barrow that is harder, firmer and trimmer about the base of the ham and through the crotch.
8. A firmer, harder, heavier-muscled ham.
9. A thin-skinned ham that is free of wrinkles and very firm at the base (or cushion).
10. A tight-skinned, light-muscled ham that is soft, flabby and wrinkled at the base (or cushion).
11. A barrow that is loose, soft or flabby at the base of the ham and in the crotch.
12. Tapering through the rump and ham, lacking the general fullness and muscling of the other animals in the class.
13. More total dimension to the ham than any other barrow in the class (Fig. 4-46).
14. More flare and dimension to the ham.
15. A heavier-muscled, cleaner-hammed pig, as indicated by the presence of muscle seams and expression (flex or bulge) to the muscles, which is especially evident when the animal moves.
16. Thicker through the center of the ham with stifle muscle area (or stifle) being trimmer and firmer throughout the ham than No. 2.

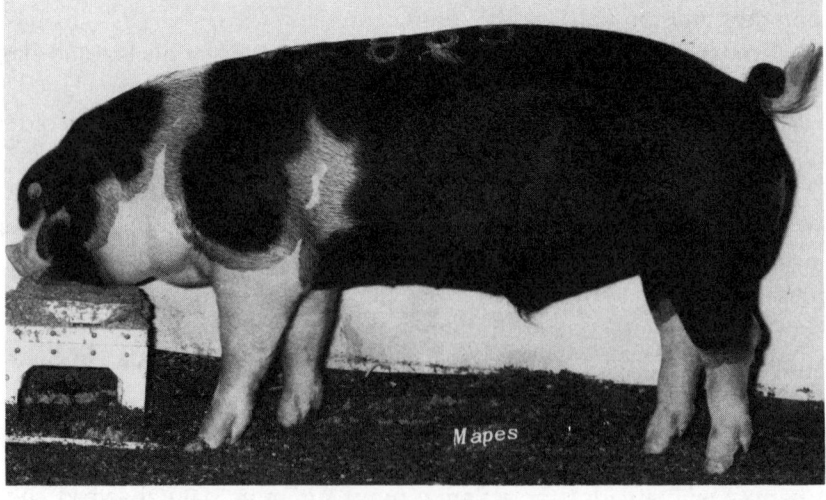

Fig. 4-46. Typical prize-winning Spotted barrow. (Courtesy, National Spotted Swine Record, Inc.)

17. A gilt with a higher tail setting, more flare to her rump and more muscle development and meatiness through the ham area than any other animal in the class.
18. A longer, thicker, deeper, more heavily-muscled ham than No. 4.
19. A narrow-rumped, "cat-hammed," light-muscled gilt.
20. Thicker through the center of the ham than No. 2 but tending to narrow up at the base of the ham and lacking the volume of ham of the other animals in the class.
21. A more shapely, meatier, firmer ham than No. 1.
22. A pig with more flare to the rump and more thickness through the center of the ham; especially trimmer and firmer about the base of the ham and in the crotch than any other barrow in the class.

Side

1. A longer, deeper-sided gilt.
2. A barrow with more length and stretch than any other pig in the class.
3. A long, deep-sided gilt that exhibits more broodiness and roominess than any other animal in the class.
4. A higher-capacity, broodier, roomier gilt with more length and stretch throughout.
5. A long, firm-sided barrow.
6. A short-sided, shallow-bodied gilt.
7. A deep-sided gilt with wrinkles along the side and underline.
8. A stretchy, smooth-sided barrow that is hard, firm and clean along the underline.
9. Flabby and wasty in the flanks.
10. Shallow in the fore-rib and tucked up in the rear flank.
11. A barrow that lacks length of side and will probably not meet the minimum length requirements.
12. Excellent length and very trim all along the underline.

Belly and Flank

1. Firm, trim, smooth belly that is carried well up.
2. Deep in the foreflank and rear flank.
3. Deep, flabby and loose in the rear flank.
4. Neat in the flanks and along the belly.
5. Loose and wasty along the belly.
6. Lets down into a deeper rear flank.

Heart Girth

1. Ample or adequate spring and depth to the fore-rib.
2. Tight in the heart.

3. Narrow and pinched in the heart and flat in the rib (flat-ribbed or lacking spring of rib).
4. Excellent spring to the fore-rib and rear rib, with tremendous depth and squareness to the lower rib.
5. Slack in the heart.
6. Outstanding spring to the fore-rib and well – filled-in heart girth (smooth in the heart girth).

Feet and Legs

1. Standing squarely on the soundest feet and legs in the class.
2. A pig with more rugged, heavier bone, standing straighter and stronger on his feet and legs (could be either front or rear legs or both).
3. Stronger on her pasterns and exhibiting more uniform length of toes when she is compared to the other animals in the class.
4. A heavier-boned, wider-fronted, squarer-standing barrow.
5. A narrow-fronted, light-boned barrow that is very long and weak on his pasterns.
6. Sickle-hocked.
7. Cow-hocked.
8. Toed-out or toed-in — either front or rear legs, usually front.
9. Close at the knees.
10. Back at the knees.
11. Buck knees, or bucked knees.
12. Short inside toe.
13. Standing on more timber.
14. More correct in underpinning (set to feet and legs).
15. Standing on high-quality, clean bone, a gilt that is very correct in the set to her feet and legs.
16. Post-legged — rear legs too straight.
17. Ample bone, but somewhat coarse.
18. Tender in front; walks insecurely.
19. Plenty of clean-cut, hard bone with strong pasterns and a free active walk.
20. Puffy and swollen hocks.
21. Unsound set of feet and legs.
22. A gilt that is not sound on her feet and legs or in her movement as viewed today.
23. A gilt that lacks the substance of bone and the correctness in structure and placement of her feet and legs to carry a litter of pigs and to maintain the reproductive excellence desired in this breed.
24. A barrow that stands too close in front and behind. Weaves when he walks. Walks with feet too close. Not enough space between legs.
25. More give and cushion up front, thus providing her with greater longevity and durability under modern-day confinement systems.

26. Incorrect in underpinning.
27. Gimpy-moving.
28. Stands under at the hocks.
29. Longer, freer, more determined stride.
30. Snappy moving.
31. Powerful-fronted (strided).
32. Splay-fronted.
33. Splay-footed.

Femininity

1. Six well- (evenly) spaced prominent teats on each side of her underline.
2. A broody, roomy middle.
3. A gilt that has a more correct teat placement and a more prominent underline than any other animal in the class.
4. A larger number of evenly spaced nipples on her underline than No. 2 (12 nipples desirable).
5. Uneven in her teat spacing and lacking the prominence of underline displayed by the other animals in the class.
6. A gilt lacking in broodiness and capacity.
7. Not enough acceptable, normally developed nipples along her underline.
8. Lacking prominence of underline.
9. Blind, inverted or pin nipples.
10. Refined about the head, smooth in general appearance and possessing ample quality.
11. A breedy-headed, mild-dispositioned gilt.
12. A mellow, fine-haired, well-balanced gilt with tremendous style.
13. A clean-cut gilt that stands square and correct on her feet and legs.
14. Off on her feet and displaying a bad temperament.
15. Too coarse about the head and, in general, too rough throughout. Not enough refinement of the hide, hair coat and bone. Rather masculine-appearing.
16. Poorly spaced teats that are too few in number.
17. Two "blind" teats and a very flabby, loose, undesirable underline.
18. Larger vulva, indicating more reproductive maturity.
19. More prominent and evenly spaced underline.
20. More desirable underline in terms of greater teat distension from the milk wells.
21. More uniformity in teat distension and placement.
22. An underline that starts further forward.

Masculinity

1. A big, stout, rugged, heavy-boned, sound boar.
2. A vigorous constitution, a massive structure and an ambitious but controlled temperament.

3. A broad, nicely balanced head that shows strength and masculinity.
4. Well-developed testicles.
5. Only one testicle evident in the No. 1 boar, resulting in disqualification from competition.
6. Both testicles present definitely underdeveloped for an animal of this size and maturity.
7. Too vicious, easily irritated — the ranting kind; lacks constitution and strength of top.
8. A well-balanced, sound-footed boar that displays free and active movement at the walk.
9. A fine-boned, light-muscled, crooked-legged boar that travels too close behind.
10. A narrow-chested boar that lacks the constitution, structural correctness and substance of the other animals in the class.

Finish and Carcass

1. Should yield a higher percentage of lean cuts.
2. Should yield a higher percentage of ham and loin.
3. The No. 3 barrow will hang up a trimmer, meatier, longer, leaner, more shapely, more nicely balanced carcass than No. 4.
4. The No. 1 barrow will yield a wasty, light-muscled carcass, resulting in a lower percentage of lean cuts than any other pig in the class.
5. A short, fat carcass that will yield the lowest ham and loin percentage of any animal in the class.
6. Will hang a narrow, flat-muscled, meatless carcass.
7. A hard, trim, meaty barrow that is very correct and smooth in his finish.
8. Overfinished and soft. Lacks firmness in the flanks and behind the shoulders.
9. Cleaner in the jowl, trimmer and tighter in the middle and along the underline and firmer at the base of the ham than any other barrow in the class.
10. A rather soft, flabby finish. Uneven in his finish. A bit soft in his hams and along the sides.
11. Will hang a narrow, light-muscled, long-shanked carcass.
12. Will hang a carcass with more spread and thickness through the loin, combined with a longer, deeper, thicker ham than any other animal in the class.
13. Wraps more total red meat and muscle volume into one package.
14. When killed, carcassed and hung on the rail, should have a thicker, more desirable belly wall.
15. Should have less fat trim before it is placed in the self-service meat counter.
16. Should hang a carcass that could split a larger loin chop.
17. When killed and carcassed, should have more total muscle volume from rail to floor.
18. Should yield a carcass with a greater percent lean.
19. Should yield a higher cutability carcass, thus giving a greater return to both the packer and the producer.

20. Should yield a greater amount of retail lean per day of age.
21. Should require less fat trim before it enters merchandising channels.
22. Should yield higher retail lean per day of age.
23. A trimmer barrow in that he shows more shoulder-blade action while on the move.
24. Deeper-flanked, thus showing the potential to yield a carcass with a thicker, heavier belly wall.
25. Freer from waste from end to end (head to tail).

SAMPLE SET OF REASONS FOR
PLACING SWINE

The following is a sample set of reasons showing the correct use of judging terms and expressions. Note that the reasons compare the animals, not just describe them. These reasons are written about a class of Yorkshire breeding gilts.

"I placed this class of Yorkshire breeding gilts 1-2-3-4. I placed No. 1 at the top of this class and over 2, because she is a larger, longer, leaner, more level-designed gilt that is heavier-structured and freer-moving. She is a stouter-fronted, wider-chested gilt that is pulled apart more between the blades than 2, as well as a deeper, squarer-ribbed gilt with more depth of rib and looseness in her rear flank than 2. In addition, 1 is longer and more nearly level in her rump structure and possesses a higher tail setting than 2. She moves out squarer and truer behind on more rugged bone and has a more prominent underline. However, I'll grant that 2 is a longer-necked gilt that is more rectangular in her rib design, is looser and more expandable in her muscle make-up, has more give to her knees and cushion to her front pasterns and is more even in her teat spacing on the left side of her underline than 1.

"For my middle pair, I easily placed 2 over 3 because it more closely followed the type and pattern of the class-winning No. 1 gilt. No. 2 is a wider-skulled, stronger-jawed, more robust-fronted gilt that is wider-structured and stouter-featured than 3. She is deeper and squarer in her rib, longer-sided and looser spined than 3. In addition, she is longer from her hip to her hock and freer-moving, especially off her front end. No. 2 is leaner down her top and through the lower one-third of her body than 3 and possesses the most functional-appearing underline in the class. On the other hand, I'll admit that 3 has more Yorkshire breed character about her with a more erect set to her ear than 2. In addition, 3 is stronger in behind her shoulders; shows more muscle volume and development through her ham when viewed from the rear; and moves off wider at her hocks than 2.

"Coming to my bottom pair, I placed 3 over 4 because 3 appeared to be a larger-skeletoned, growthier gilt that was later-maturing and leaner-conditioned. No. 3 is wider-structured with more width through her chest chamber and all through the lower one-third of her body than 4. She is longer-rumped, higher in her tail setting and longer-muscled through her ham than No. 4. In addition, she is more mobile as she moves out, with a longer, freer-wheeling stride; stands down on a larger foot size; and appears to be a more desirable gilt than 4. However, I'll grant that 4 is a more level-designed gilt that is squarer-rumped and has an underline that starts farther forward and is freer of pin nipples than 3. I criticized 4 and placed her at the bottom of the class because she is an early-maturing, small-skeletoned, round, shallow-bodied, tight-flanked, tight-muscled gilt that is short-strided, gimpy-moving and lacking in the growability, durability and confinement adaptability of her contemporaries to merit a higher placing."

BREED REGISTRY ASSOCIATIONS

Breed	Association	Officer and Address
American Landrace	American Landrace Association, Inc.	Don Verhoff Executive Secretary P.O. Box 2340 West Lafayette, Indiana 47906
Berkshire	American Berkshire Association	Wesley Blanchard Secretary P.O. Box 2537 West Lafayette, Indiana 47906
Chester White	Chester White Swine Record Association	Daniel Parrish Executive Secretary 1803 West Detweiller Drive Peoria, Illinois 61615
Duroc	United Duroc Swine Registry	Gary Huffington Executive Secretary 1803 W. Detweiller Drive Peoria, Illinois 61615
Hampshire	Hampshire Swine Registry	Rick Maloney Executive Secretary 1111 Main Street Peoria, Illinois 61606
Hereford	National Hereford Hog Record Association	Ruby Schrecengost Route 1, Box 37 Flandreau, South Dakota 57028
Inbred Breeds	Inbred Livestock Registry Association	George W. Slater President Route 4, Box 207 Noblesville, Indiana 46060
OIC (Ohio Improved Chester)	OIC Swine Breeders' Association, Inc.	Thomas R. Hendricks Secretary-Treasurer P.O. Box 111 Greencastle, Indiana 46135

(Continued)

BREED REGISTRY ASSOCIATIONS (Continued)

Breed	Association	Officer and Address
Poland China	The Poland China Record Association, Inc.	Wesley Blanchard Secretary P.O. Box 2537 West Lafayette, Indiana 47906
Red Berkshire	Kentucky Red Berkshire Swine Record Association	Hogan Teater Secretary Lancaster, Kentucky 40444
Spotted	National Spotted Swine Record, Inc.	Larry Williams Executive Secretary P.O. Box 2807 West Lafayette, Indiana 47906
Tamworth	Tamworth Swine Association	Thomas Fenton, Jr. Secretary 2656 Horner Road Winchester, Ohio 45697
Wessex Saddleback	Wessex Saddleback Swine Association	A. M. McCracken Secretary 4010 Clinton Avenue Des Moines, Iowa 50310
Yorkshire	American Yorkshire Club, Inc.	Darrell D. Anderson Executive Secretary Box 2417 1769 U.S. 52 North West Lafayette, Indiana 47906

5

STOCK HORSES[1]

Stock horses have always been a very integral part of the range-cattle industry of the 17 western states. And as the industry expands into other areas where the grass-livestock economy is growing in importance, stock horses go with it.

Much has been written about the "wild" horses that roamed the cattle ranges many years ago. The conditions under which these horses developed were rugged. Nature selected them based on their ability to survive. Inasmuch as some of the best of these served as stock horses for the early cattle raisers and contributed to the foundation for the improved types, we should respect their ability to wear when we appraise the usefulness of the modern "cow ponies," now more appropriately referred to as stock horses.

Stock horses are versatile horses — the cattle industry cannot get along without them. There are roughly 500,000 to 600,000, or perhaps more of them, in service throughout the range area. Stock horses have been developed for hard work. Under a 225-pound weight, a stock horse must be able to overtake the most fleet-footed calf (Fig. 5-3) or anchor a steer approaching its own weight. Stock horses must be able to stop or start with unbelievable quickness and, under the most trying conditions, remain calm and controlled. They must be able to work long days, often traveling 25 miles or more over unbeaten paths with meager food and water. To be sure-footed, active, willing, even-tempered, strong and easily adjustable to all the varied chores of the cowhand is a must. Very few, if any horse types, are as versatile in the application of their qualities to heavy duty as are stock horses.

During the last quarter century, a rather uniform understanding of the general qualities that contribute to type has crystallized among stock-horse enthusiasts. Breeds have been developed and are continually being improved to comply with the generally accepted standards of quality that affect type and usefulness.

[1]Credit is acknowledged for valuable information on stock horses from the American Quarter Horse Association, the Palomino Horse Breeders of America, Inc., and the Appaloosa Horse Club, Inc.

STOCK-HORSE TYPE

"Type," in general, is that combination of characteristics that contributes to an animal's usefulness for the purpose at hand. When type, as it concerns usefulness, is confined to individuals within a breed, the agreement, or lack of agreement, between the animals of that breed is usually stressed. In effect, the individuals within a breed are judged on the basis of more or less fixed standards of quality that have been approved by the particular breed associations involved. Despite the relatively short time that breed associations have been active in formulating type for useful stock-horse performance, much has been accomplished in prescribing general standards that appraise useful values.

Type in stock horses involves an efficiently balanced combination of all physical characteristics and mental qualities that contribute to their usefulness as stock horses. Locale differences (environmental differences) affect type differences as they involve performance. The cattle-grazing ground in the United States differs from the near sea-level swamp lands in the Gulf coastal areas and the spacious southwestern plains with their modest evaluations to the varying elevations up to 8,000 feet or more in the forest- and brush-covered Rockies. Thus, stock-horse duty is strenuous in the rugged mountain areas of the West where brush cover; steep, rocky slopes; narrow trails — where there are not trails at all — and snowfalls that range from several inches to several feet are obstructions that necessitate sturdy characteristics not needed on the plains and in the lower altitude areas.

General Appearance

In general, the stock horse is medium-sized, though weights may vary considerably, from about 900 to 1,500 pounds. These extremes are not, as a rule, found in the same class. Handy-weight (average-weight) horses will stand from 14 to 16 hands and weigh from 1,000 to 1,300 pounds or more.

The body must be above average in total length, moderately long and strongly coupled in the back, broad in the loin, gently sloping and heavily muscled in the croup, neatly laid in the shoulders, medium in length of the neck and heavily muscled in the thighs and forequarters, with a roomy middle that is well-carried. Superior muscling must be evident throughout (see all pertinent figures).

Before you proceed with a detailed study of the various parts of the stock horse, you should refer to the points of the horse in Fig. 5-1. There must be unmistakable evidence of service in hard, flat, clean-cut bone, well-placed legs and well-shaped feet (Figs. 5-2 and 5-25). The feet should have a medium, rather round bearing surface, since feet of this shape distribute the concussion more evenly upon the hoofheads than do small, narrow feet. This is important insurance against fatigue. Good width at the heels and fullness of the frogs tend to prevent contracted heels.

The front of each hoof should have a 45- to 50-degree slope, which should continue up along the front of the pastern to the fetlock. The walls of each hoof

Fig. 5-1. Points of the horse.

1. Muzzle	15. Withers	29. Loin
2. Nostrils	16. Point of shoulder	30. Underline
3. Face	17. Breast	31. Rear flank
4. Eye	18. Arm	32. Foreflank
5. Forehead	19. Elbow	33. Coupling
6. Ear	20. Forearm	34. Croup
7. Poll	21. Knee	35. Tail
8. Cheek	22. Cannon	36. Point of buttock
9. Jaw	23. Fetlock point	37. Quarters
10. Throat	24. Pastern	38. Thigh
11. Neck	25. Foot	39. Sheath
12. Crest	26. Heart girth	40. Gaskin
13. Collarbed	27. Ribs	41. Hock
14. Shoulder	28. Back	42. Stifle

should have good depth and should spread somewhat toward the base. The sidewalls should not be vertical, as they do not allow the desired flexibility in the heel area when the foot makes contact with the ground.

The sole of the foot is somewhat concave rather than flat, and the size of the hoofhead must be proportional to the size of the foot with no undesirable fullness. The pastern should be set strongly into the middle of the hoof and should merge smoothly with the hoofhead.

The useful lifespan of stock horses is limited to the adequacy of their feet and legs (see Figs. 5-16, 5-17, 5-18 and 5-19 for diagrammatic positions). Note the good placing of legs in Fig. 5-35. In general, the correct placing of legs is reliable assurance that the action will be straight, smooth and controlled. Excessive width between the forelegs is often associated with lack of smooth shoulders and also lack of coordinated action. When desirable width is missing, the forequarters appear cramped.

The muscling of the forearms should be long, strong and pronounced (Figs. 5-2, 5-3, 5-4 and 5-23); the knees should be broad in front; and when viewed from the side, the knees should be clean-cut and strong. Any tendency toward "buck," or "calf," knees (Fig. 5-17) should be frowned upon. Excessive length in the **cannons** is undesirable, but adequate length is a necessity. The **tendons** should be set well back from the cannons and should merge smoothly into the knees to avoid a restriction there. When they are set well back, a groove will show between each cannon and its tendon, and the leg will appear flat (Fig. 5-2). **Fetlock** joints of ample size are favorably looked upon (Figs. 5-2, 5-24 and 5-25). The pasterns should be medium in length and set at a 45- to 50-degree angle, which then form a straight line with the foreparts of the hooves into which they should merge smoothly (Figs. 5-2 and 5-24). Since the forelegs carry the bulk of the weight, they should demonstrate unmistakable ability to absorb concussion and to wear well (Fig. 5-24).

Fig. 5-2. "The American Quarter Horse," by noted equine artist Orren Mixer. This painting of a modern American Quarter Horse stallion is recognized as the breed ideal for Quarter Horse breeders the world over. (Courtesy, American Quarter Horse Association)

Note the desirable set of the hind legs in Fig. 5-2. (See also Figs. 5-24 and 5-25.) As a source of power, the hind legs are more important that the front legs. Therefore, they must be as correct as possible for efficient power production, tremendous flexibility in action and long wear (Figs. 5-2 and 5-24). The **hocks** should not be set too far apart or too close together, as both of these extremes interfere with the free, straight and controlled action of the legs. The hocks should be set well under and directly below the point of the buttocks. This, in general, makes for a straight line of thrust. It also makes for a balanced distribution of concussion and wear on all moving parts of the legs.

When viewed from the front or the side, they should be wide. They should also be deep, sharply defined and free from what is usually termed "meatiness," "fullness," "puffiness" or "coarseness." Good size in hocks is desirable, as long as the hocks show pronounced wearing quality (Fig. 5-2). Small hocks are often weak and provide very little surface for the heavy wear to which they are subjected; thus, they are more susceptible to the development of unsoundnesses than are large, shapely hocks.

The tendons should be set well back so that the legs look flat in these areas. In order for the legs to be balanced in appearance and in wearing ability, the fetlock joints must be large, must be free from any thickness and must exhibit great strength (Fig. 5-2).

Head and Neck

The characteristic that is profoundly important in the head of the stock horses, and in which they differ rather widely, cannot be described in dimensions or in pounds. Dimensions and the most carefully selected adjectives fail to convey what the term "intelligence" means to a horse raiser. A horse with intelligence does not become greatly disturbed under the most trying duties and will remain calm yet alert to the rider's wishes at all times, diligently handling the job it has been taught. In other words, intelligence, occasionally referred to as "cow sense," conveys to the horse raiser unmistakable evidence that a horse so described has a good chance of easily learning to do well what its body is capable of doing. (Observe the illustrations throughout this discussion for head character and intelligence, or cow sense.)

A good, wide **forehead** and strong, deep **jaws** are desirable. The **jaws** should be rather wide apart at the angles to allow ample room for the windpipe that lies between them. The **throat** should be lean without any unnecessary filling between the jaws and throat to cause the horse to breathe heavily when the head is checked back. Sizeable **nostrils** that can expand readily when the horse is under pressure — yet not be too thin — and a relatively lean, shapely head with **ears** that are fine-textured, medium length, tapered and gracefully carried meet with much favor (Figs. 5-2 and 5-24).

The **neck** should be of medium length (Fig. 5-24). A short, stubby neck usually lacks the desired flexibility in head-neck action, which is significantly important; moreover, it hinders a horse from keeping its head up where it belongs. A smooth, deep attachment at the shoulders and a graceful taper toward the muscular attachment at the head are desirable. The crest should be small yet adequate to denote strength and style (Figs. 5-2 and 5-24). A heavy crest conflicts with flexibility. Exercise judgment in selection, even when you are choosing stallions.

Forequarters

The chest capacity of a horse determines, in large part, its endurance, as that is influenced by the free and unhampered use of its respiratory and circulatory centers. The modern stock horse should be heavily muscled through the shoulder, arm and forearm areas and should possess long, prominent, expressive muscling up in the "vee" and through the chest region. The fore-ribs should allow fullness and should be long, thus providing a full, deep heart girth. The shoulders, likewise, must be deep, muscular and smoothly laid. Wide shoulders are not unmistakable evidence of a large chest capacity; the chest is within the fore-ribs. When the shoulders are noticeably wide, they are frequently loosely laid and contribute general awkwardness to the action.

The slope of the shoulders usually corresponds to the slope of the pasterns, which is 45 to 50 degrees (Figs. 5-2 and 5-6). Note in Fig. 5-7 the reason for urging a good slope of the shoulders. If the shoulder blade (1-3) in Fig. 5-7 were straighter up and down, the inside angle between it and the humerus (14) immediately below would tend to straighten out, which would reduce the efficiency of that angle in the shoulder for absorbing concussion. Moreover, shoulders with adequate slope facilitate free, forward action of the front legs which is a must in stock horses.

The **withers** should be as high as, or preferably a little higher than, the croup (Fig. 5-2). The height of the withers in relation to the height of the croup can be influenced in a number of ways, any one or all of which are attributable to reduced length in the bones of the front legs or in the bones that influence the depth of the forequarters. Or, suppose the development of the rear quarters is stressed more than the forequarters. Such emphasis may result in some imbalance between them to the extent that the rear quarters have had an overall increase in size, including height, and may have outgrown the front quarters in height. Whatever the cause may be, low withers are annoying — the saddle may be difficult to keep in place, and in extreme cases, the rider may be uncomfortable, disliking the "downhill-ride feeling" when the horse is on level ground. Such a situation also crowds the saddle against the withers, thus causing saddle galls. Neatly laid withers of the desired height facilitate keeping the saddle in place and contribute to the desired wear balance in the forequarters, thus avoiding so-called saddle fatigue in the withers-shoulder area of the horse. Stock horses in high-show condition are at a disadvantage in showing the "working form" of their withers. You should observe this point carefully and strive to acquire a workable understanding of this situation. Low, flat, mutton-type withers are discriminated against when modern "working horses" are being selected either at the halter or under the saddle.

Back and Middle

A short, correctly turned, heavily muscled, well-carried back that is supported by a broad, strong loin is good assurance against undue back fatigue (Fig. 5-24). A liberal spring of ribs contributes to the width of the back and is generally associated with heavy muscling. Moreover, in contrast with ribs that drop away at the top suddenly (flat ribs), a neatly carried bottomline is associated with a good spring of ribs, which is important in heavy-duty stock horses. A long, clean underline (barrel) is indicative of capacity, and capacity is extremely important for endurance. A low-slung middle hinders mobility and controlled action and does not, as a rule, supply more food-storage space than a broad-ribbed, neatly carried middle.

Rear Quarters

The diversified activity of stock horses, involving sudden bursts of great speed, cutting and roping duties, coming to sudden stops at the end of a rope with a steer at the other end and climbing through rugged country calls for muscular, well-balanced hindquarters that are extremely powerful (Fig. 5-25). The rear quarters must be coordinated in functional harmony with all other parts. For example, a reasonably well-carried croup is preferred to a steep one, because the latter tends to push the hind legs under the horse, which in turn, interferes with the action of the hind legs, which is generally power and speed production (Figs. 5-3, 5-4, and 5-5).

Fig. 5-3. The American Quarter Horse at work. (Courtesy, American Quarter Horse Association)

Fig. 5-4. The American Quarter Horse in a sliding stop during reining competition. (Courtesy, American Quarter Horse Association)

Fig. 5-5. The American Quarter Horse performs during championship steer roping. (Courtesy, American Quarter Horse Association)

The hip bones should be wide and level, and the croup smooth and heavily muscled. Great power should characterize the **stifle,** with the muscles bulging at this point to the greatest width in the hindquarters (Fig. 5-25). The **thighs** should carry fullness to the bottom without unduly tapering off. The **gaskins** are broad from front to rear and very strong through both the inner and the outer areas (Fig. 5-25).

Action

A long, straight, determined, smooth stride at the walk is desired. The feet should be picked up freely, moved forward with good clearance and then placed squarely on the ground. Skimming the feet along the ground before they come to a stop (daisy-cutting) is objectionable and often causes stumbling. Stock horses spend more time at the walk than at any other gait. If they are right at the walk, they are not often disappointing at the trot or the gallop, though action should be checked carefully at all gaits.

Efficiency is the normal result of good balance and control. Action at the trot requires more effort, is usually more pronounced and is faster than at the walk. A well-trained horse should show a smooth, straight action with controlled speed, as that is more common than excessive speed at the trot. Objectionable features at the walk and at the trot are "paddling," swinging the feet too wide; "interfering," striking the supporting foot with the moving foot; "rope walking," swinging the striding leg around and in front of the supporting foot; and what is sometimes called "daisy-cutting," a skimming forward movement of the feet that does not allow enough ground clearance. Extremely high action wastes a stock-horse's energy.

When executing the gallop, the horse should show an easy, natural control and efficient coordination between the body and legs to make its going smooth (Fig. 5-16). A very restricted gallop requires the horse to go up and down too much for the distance traveled; it wastes energy and makes straight, smooth forward action difficult. Exaggerated speed at the gallop also requires more effort for the distance covered than a normal, natural gallop.

Breed Type

Except for color; breed-type differences are more easily seen than described. This is particularly true when two or more breeds have been developed for the same purpose, as very rarely do the sponsors of a breed depart much in their selection emphasis on traits that conflict with usefulness. There may be differences in size, body compactness, general body lines, style of head and neck and action, though you should exercise extreme caution in not overworking the use of breed type in any discussion, unless the differences involved are pronounced. Note the good mare in Fig. 5-29 that is registered in two breed associations.

The term *"**breed character**"* is often used to point out differences in the degree of excellence among individuals in well-established breeds, with reference to special display of style, head and neck carriage, action and coordinated control.

Balance

"Balance" refers to the harmonious development of all parts of the horse. This is important, because if any one part is deficient, then, obviously, full use cannot be made of the other parts of the body. "Balance" also describes the functionally efficient coordination of all the traits that make for straight, flexible, smooth action (Fig. 5-6).

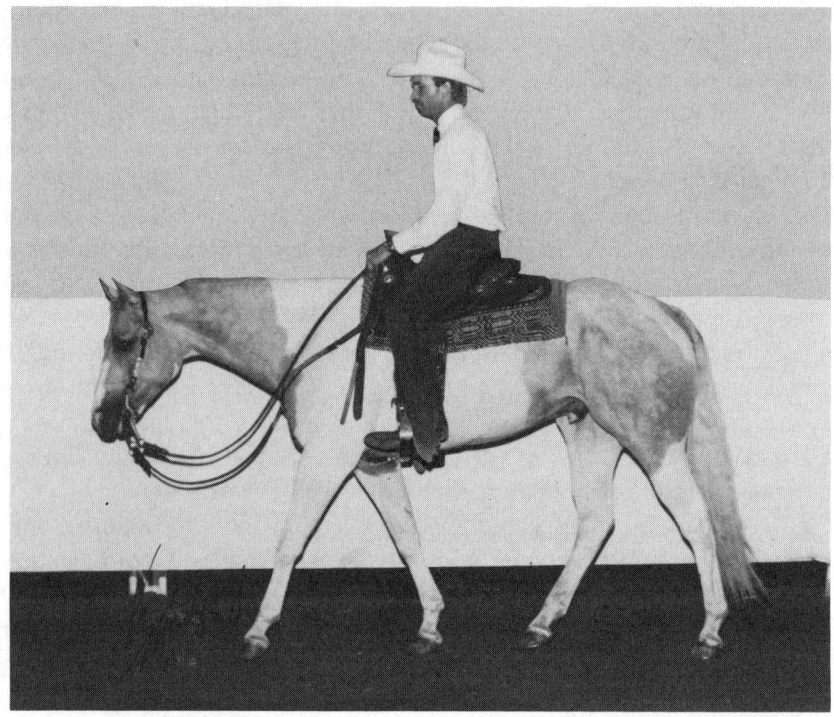

Fig. 5-6. Western Pleasure at its best. "Painted Snake" has earned a Superior in Western Pleasure. (Courtesy, American Paint Horse Association)

Constitution

Constitution in a stock horse involves those characteristics that reveal the ability to wear under hard work. A large chest capacity, a well-developed and neatly carried middle, ample muscling and strong, sound bone are evidence of constitution. (Observe the figures accompanying this discussion.[2])

---

[2]Specific figures that show particular positions and parts especially well are referred to often. Because all these pictures are of very good horses, you have a splendid opportunity to observe the results of much thoughtful effort in the development of these animals.

Size

The big, stout, rugged horses that stand 14.5 to 16.0 hands high and weigh 1,000 to 1,300 pounds each in working condition are in great demand today. These are the horses that have the strength, stamina and ability to put forth maximum effort for long periods of time.

Substance

Substance refers to bone size (ample bone) or sometimes to general ruggedness (Figs. 5-2 and 5-24).

Style

In general, style involves the natural expression of a controlled temperament, which the horse displays at rest or in action. Beautiful conformation, graceful action and general alertness all contribute to style. (See all the figures accompanying this discussion.)

Quality

"Quality" is generally used in describing sharply defined joints and bone — that is, ample in quantity and ruggedness to insure serviceability. Smooth, well-defined joints of adequate size over which the skin fits like a kid glove are usually defined as quality joints. Smoothness and good balance of body and style and a serviceable degree of fleshing contribute to quality. Excessive fleshing may conflict with the most desirable expression of quality in a stock horse. (Observe the superior quality that, in general, is in evidence in the pictures accompanying this discussion.)

The breeders of stock horses emphasize balance and usefulness. In placing a class of stock horses, first look for general appearance, type and balance (Fig. 5-25). Stock-horse type is characterized by size, scale, intelligent head, heavy muscling in both the forequarters and the hindquarters, moderately long back with strong coupling, ruggedness, flat bone, quality throughout and medium, well-rounded feet. All the characteristics of a stock horse should show unmistakable evidence of performance and endurance (Figs. 5-24 and 5-25).

THE SKELETON — ITS RELATIONSHIP TO THE APPEARANCE AND USEFULNESS OF A STOCK HORSE

Inasmuch as general type in stock horses is so basically influenced by the detailed structure of the skeleton, making a careful study of the skeletal variations, as they structurally influence body differences and efficiency in performance, is valuable. All the accompanying illustrations of the skeletal structure and the related discussion should be carefully studied, as they are designed to call attention to soundness, wearing qualities, flexibility in function and total usefulness.

The skeleton, or bone structure, is the framework of the horse (Fig. 5-7). It has a fundamental influence on the animal's appearance, general type and usefulness. The length, size, shape and position of the various bones have a direct relationship with the external development. In judging horses, you do not often refer directly to the skeletal details; nevertheless, it is fundamentally important to understand them well enough to appreciate the basic cause of many desirable as well as undesirable variations that occur in the details of conformation. The desirable as well as the undesirable variations in the bones are influenced by heritable factors, as are the details in the external development. Therefore, a comprehensive knowledge of the supporting structure, namely the skeleton, and its basic relationship to form and function of the stock horse, will be of service to you.

For convenience of study, the skeleton can be divided into a few groups, within which the bones are somewhat similar in form and function. The bones that have the most influence on the shape of the head are the *mandible,* or lower jaw; the *maxilla,* or upper jaw; the *nasal,* or face bone; the *frontal,* or bone of the forehead; and the *occipital,* or bone at the poll. There is often marked variation in the length of the lower and upper jawbones, as well as in their depth and width. The width between the mandibles may be too excessive, or it may be too narrow. If it is excessive, the head will appear coarse. If the area between the mandibles is too small, it does not allow enough space for the windpipe that lies between them. There are also marked variations in the angles at the back of the mandibles. This is apparent in the jaws of the horse, which may be slight or very full. The face bone is often not straight in profile, and the bone of the forehead may show a pronounced "dish," may indicate too marked a bulge or may, indeed, be too narrow. The occipital is often too small, and the horse will lack practical fullness at the poll. Inasmuch as there is very little tissue covering most of the head bones, it is apparent that head shape is influenced very directly by variations in the bones of the head.

You should observe that the variations in the length, the width, etc., of these bones are not always in the same proportions. Indeed, there may be such marked discrepancies in the variations that they constitute defects, as, for example, when the lower jaw is so much longer than the upper jaw that the front teeth do not meet normally. The angle at the back of the mandible may also influence the manner in which the upper and lower teeth come together. Such a defect is spoken of as an undershot jaw. The reverse of this, namely, too long an upper jaw or too short a lower jaw, whatever the case may be, constitutes an overshot jaw, in which the upper incisors (front teeth) do not come in direct contact with the lower incisors.

Both the overshot and the undershot defects may become so extreme that a horse affected with one or the other cannot eat grass efficiently. In less severe cases, the upper and lower teeth tend to half-lap rather than come squarely together, thus giving rise to unequal wear of the teeth surfaces, causing sharp points to develop on the teeth, which may injure the tongue. One sign of this is habitual slobbering. When these discrepancies affect the incisors, they also often affect, in like manner, the molars; hence the defect is very serious. The heritable force that stimulates

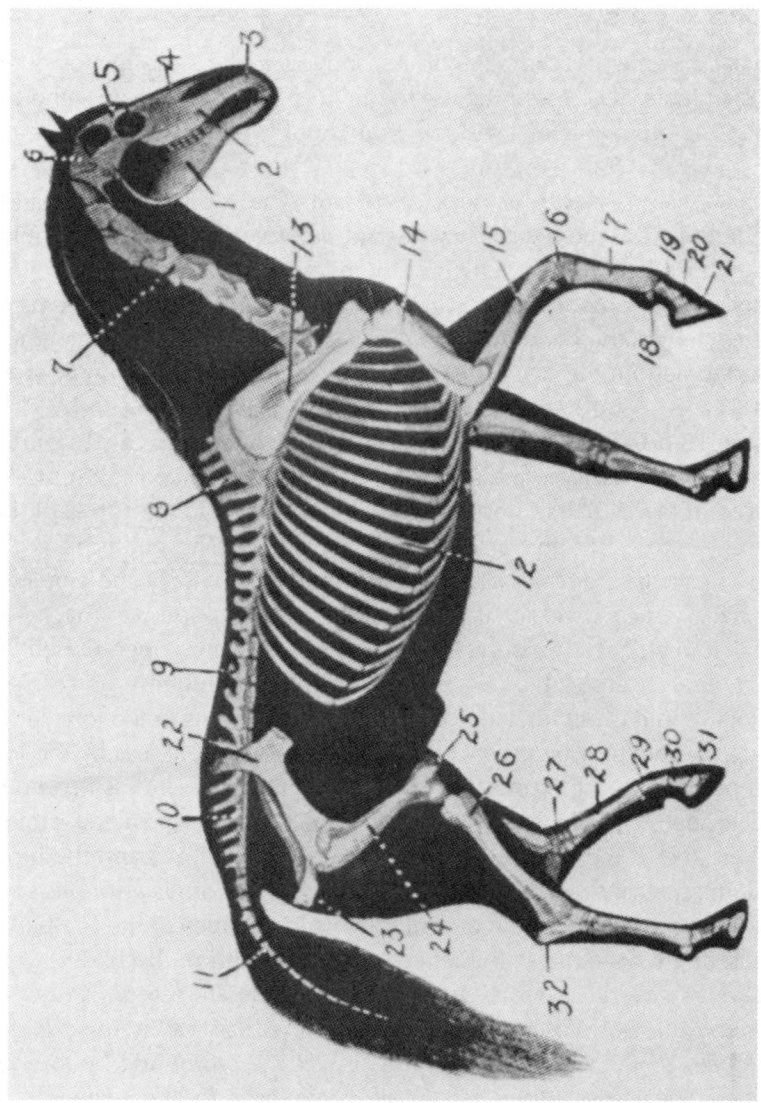

Fig. 5-7. Anatomy in its relation to the form of the horse. See Table 5-1 for the names of the bones. (Courtesy, Horse and Mule Association of America)

increased development in one bone does not always express itself to the same degree in another adjoining bone. Nor does increased development in the length of a bone necessarily imply that the bone will increase in width or diameter in precisely the same proportion.

Bones in the Skeleton

For your convenience, the parts of the body as well as the names of the bones are given (see Table 5-1). Inasmuch as the names of the bones, in general, are the same for the various types of livestock, you should learn them and the parts they involve.

There are seven *cervical,* or neck, vertebrae. These vary in length and directly influence the length of the neck. They are not, however, the only contributing factor to the variation in the apparent length of the neck in horses. The slope of the shoulder blades also has some influence. In addition, a lean, neat neck seems to have a more practical length than one that is thick, particularly at the throatlatch. In Fig. 5-7, the neck vertebrae join the back vertebrae at a point near the midpoint of the shoulder blades.

There are 18 *dorsal,* or back, vertebrae, namely, one for each pair of ribs. These have spinelike projections on the upper surface that give shape to the back and the withers. Note in Fig. 5-7 that those at the withers are the longest. It is probable that the latter may be too short in horses low at the withers. The variations in the length of the back are in part due to variations in the total length of the vertebrae of the back. The back appears shorter if the shoulder blades slope well back at the top. Long back vertebrae are, no doubt, closely associated with "open-ribbed" horses.

The *lumbar,* or loin, vertebrae are generally six in number in horses. The apparent length of the loin is influenced not only by the variations in the total length of the lumbar vertebrae but also by the lack of muscle development and by the location of the highest point of the hip bones. If the latter are tilted forward, as in a horse with a fairly level croup, the loin appears shorter than it does when the croup is very sloping. Then, too, if the distance between the last pair of ribs and the hip bones is small, the loin appears short. The apparent length of the loin is also influenced by its width. The length of the spines that project from both sides of the lumbar vertebrae and the muscling have a direct influence upon the width of the loin.

There are five *sacral* vertebrae in the croup. These are fused together to form a rigid structure. The slope of the croup is determined largely by the position of the sacral vertebrae, by the muscle development over the croup and by the position of the hip bones. The slope of the croup is also influenced by the angle which the hip bones make at the pelvic girdle, the point of attachment to the backbone.

The bones of the pelvic region that belong to the group of bones in the rear legs are the *ilium* (commonly referred to as the hip bone), the *ischium* (extending toward the rear of the croup, or buttock) and the *pubis* (lying between the ilium and the ischium and constituting the floor of the pelvic region). When these bones are fairly

TABLE 5-1. Bones in the Skeleton of a Horse[1]

Key Number in Fig. 5-7	Name	Number	Parts Involved
1	Mandible		Lower jaw
2	Maxilla		Upper jaw
3	Premaxilla		Nose
4	Nasal		Face
5	Frontal		Forehead
6	Occipital		Poll
7	Cervical vertebrae	7	Neck
8	Dorsal vertebrae	18	Back
9	Lumbar vertebrae	6	Loin
10	Sacral vertebrae	5	Croup
11	Caudal vertebrae	18–20	Tail
12	Ribs		Ribs
13	Scapula		Shoulder
14	Humerus		Arm
15	Radius and ulna		Forearm
16	Carpals	7–8	Knee
17	Metacarpals	3	Cannon
18	Sesamoid	2	Back of fetlock
19	First phalanx		Long pastern bone
20	Second phalanx		Short pastern bone
21	Third phalanx		Coffin bone—bone inside hoof
22	Ilium		These and the pubis lying
23	Ischium		between them—croup region
24	Femur		Thigh
25	Patella		Stifle
26	Tibia and fibula		Gaskin
27	Tarsals	6	Hock
28	Metatarsals	3	Cannon
29	First phalanges		Long pastern—pastern
30	Second phalanges		Short pastern
31	Third phalanges		Coffin bone—bone inside hoof
32	Tuber calcis, or fibular tarsal		Bone at point of hock

[1]See Fig. 5-7.

horizontal, there will be a tendency for the croup to be fairly level. On the other hand, if there is too large an angle between them and the sacral vertebrae, the croup will correspondingly slope.

The variation in the length of the legs is obviously due to the variation in the length of the bones of the legs. The length of these bones may not vary to the same degree. Often the pastern bones are too short in long-legged horses. The variations in the angles of the shoulder, elbow and pastern joints are fully as important as the variations in the length of the bones of the legs. The larger these inside angles are, the straighter the legs. Legs that are too straight are objectionable because they do not adequately absorb the concussion.

The slope of the **scapula,** or shoulder blade, has considerable influence on the slope of the shoulder. If it is firmly held, and not spreading at its base or at the shoulder joint, where it articulates with the **humerus,** the shoulder will appear neatly laid. This often occurs in horses that are too wide at this point. The muscle structure obviously influences the manner in which these parts are held together.

The depth of the body depends, in general, upon the length of the ribs. However, when you speak of a horse with a full rear flank as having ample "back rib," the rear ribs are not long enough to be the only influence on the rear flank (Fig. 5-7). A full rear flank is probably influenced as much by the muscle development as by the length of the rear ribs. It is generally associated with short-coupled horses.

Muscle development is important because it influences the usefulness as well as the beauty of horses. Horses in high condition are not necessarily heavily muscled. Muscle development can usually be determined with a fair degree of accuracy from observations of the stifle, gaskins, arms and forearms, the sizes of which are not greatly influenced by a high finish. Note carefully how the large, hard, clearly defined muscles in the horse shown in Fig. 5-2 contribute to beauty and correct stockhorse type.

BLEMISHES AND UNSOUNDNESSES

In horse judging, to become apt at identifying unsoundnesses and skilled at determining how much weight to place upon them requires experience and good judgment. While the responsibility of passing upon the soundness of a horse in the show ring is left to a veterinarian, especially at the larger shows, it, nevertheless, is important for a judge to be familiar with details regarding this matter. The rule governing this subject, as it appears in many state fair premium lists, is as follows: "The judges will discriminate severely against animals that have hereditary or transmissible defects or unsoundnesses, such as bog or bone spavin, ringbone, curb (when accompanied by curby hock), cataract, stringhalt and roaring. All questions concerning soundness of animals will be referred to a competent veterinarian."[3]

---

[3]It is the faulty conformation in the area involved, and not the defect itself as such, that is generally considered heritable.

Even though this assignment is left to a veterinarian, you should train yourself to the point where you not only can identify unsoundnesses but also can appraise the seriousness of these defects. It is important to distinguish between "discriminations" and "disqualifications." "Discrimination" implies that a horse can still be placed, but obviously down the line, while "disqualification" puts the horse at the bottom of the class. A horse is disqualified if it shows lame while in action, regardless of the cause of the lameness, or if it is blind. A stallion is disqualified if only one of his testicles is visible.

In the process of learning how much to discount horses for unsoundnesses, be guided not only by the magnitude of the defect but also by the apparent predisposing causes. For instance, if a horse that has short, straight pasterns, small feet and cramped heels also has sidebones, it is discounted more than a horse that has long, well-sloped pasterns, large feet and wide heels if it has sidebones. The sidebones in the former may be due to defective conformation, which in breeding animals would likely be transmitted to the offspring; whereas, in the latter, it may be due to an injury or excessive wear. A bone unsoundness, such as a ringbone, is more serious than a bog spavin, which involves fleshy tissue.

Defects such as scars or thickenings that result from injury sustained through contact with wire or any sharp object, rope burns, bruises, shoe boils, nail punctures or gravel injuries to the feet are usually termed **blemishes.** They are not looked upon favorably, even though they do not often seriously impair the usefulness of the horse. Mechanical injuries, however, often give rise to unsoundnesses. A kick on the hock, for instance, may be severe enough to cause the development of a bog spavin.

When blemishes are minor and do not cause lameness or do not interfere with the usefulness of the horse, they are discounted in the proportion in which they affect the looks and the quality of the horse. It is not easy to determine accurately the seriousness of a blemish, especially when it involves a joint. And, too, very often a horse with blemishes or unsoundnesses in the joints will not show lame after it has been warmed up.

Fig. 5-8. Location of some common unsoundnesses: (A) sidebone, (B) ringbone, (C) wind puffs, (D) splints, (E) bog spavin, (G) curb, (H) thoroughpin. Feet and legs of this conformation are subject to unsoundnesses. (See text for full discussion.)

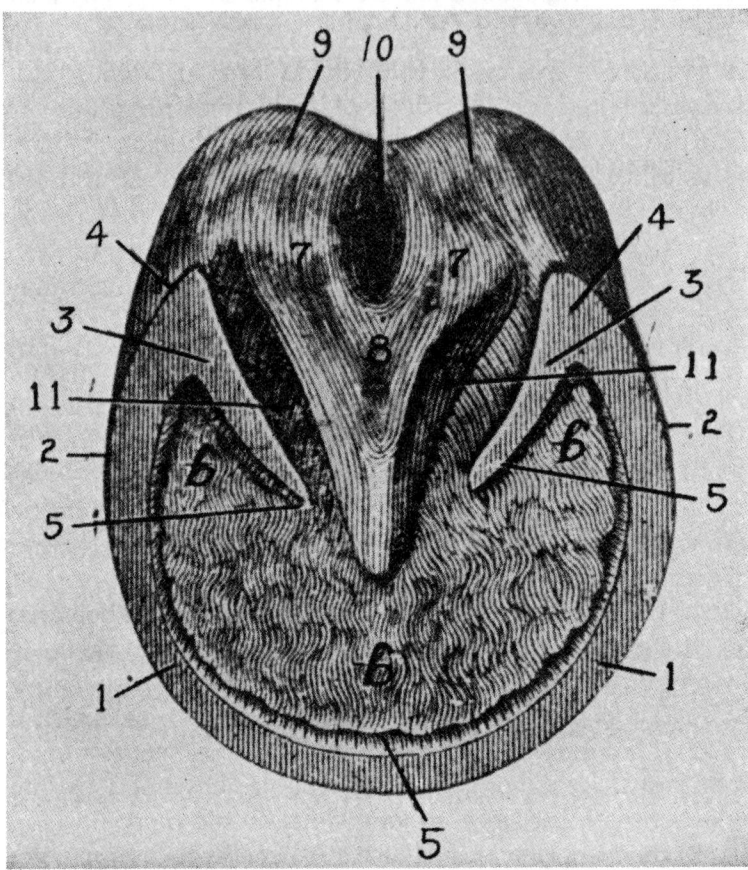

Fig. 5-9. Parts of the foot: (1) wall, (2) side of wall, (3) bars, (4) buttress, (5) leafy layer of toe and bars, (6) body of sole, (7) branches of frog, (8) body of frog, (9) bulbs of heel, (10) middle cleft of frog, (11) lateral clefts of frog. (Courtesy, University of Idaho)

There are a large number of unsoundness that interfere with the efficiency of a horse. As previously mentioned, some of these may develop from mechanical injury. They are also the result of faulty conformation in the tissues or parts involved. Too, they develop in horses that are constantly subject to overloads, or are otherwise abused, even though they may be physically well-formed with ample quality bone.

Unsoundnesses are often referred to as hereditary defects. For all practical purposes this statement is accurate. However, inasmuch as unsoundnesses are commonly the result of faulty conformation of the parts involved, it is the predisposition, such as faulty conformation, that is inherited, not the defect itself. It is important, therefore, for horses to be comparatively free from the structural and physical faults that make them subject to unsoundnesses. These faults are discussed under each unsoundness.

Sidebones

Sidebones are hard, bony projections located immediately above and toward the rear quarter of the hoofheads, as indicated in Fig. 5-8. They are perhaps the most common unsoundness of the feet of horses. Sidebones are found on the outside of the feet more often than on the inside, although they do occur on both sides of the same foot (B, Fig. 5-11). They very seldom occur on the hind feet, and when they do, they are of less importance. Sidebones are the result of the ossifying (developing into bone) of the lateral cartilages (Fig. 5-10). They prevent the normal expansion of the feet at the heel. This is very obvious when a normal coffin bone (A, Fig. 5-11) is compared to one that is badly affected with sidebones (B, Fig. 5-11). Sidebones cannot always be seen. Whenever present, although not visible, they can usually be detected by pressure from the thumb or other fingers on the suspected part of the foot. If that part is hard, this is evidence of a sidebone. Small sidebones that cannot be seen often cause more lameness than some that are very conspicuous. Skill and accuracy in identifying unsoundnesses will come only as a result of experience with them.

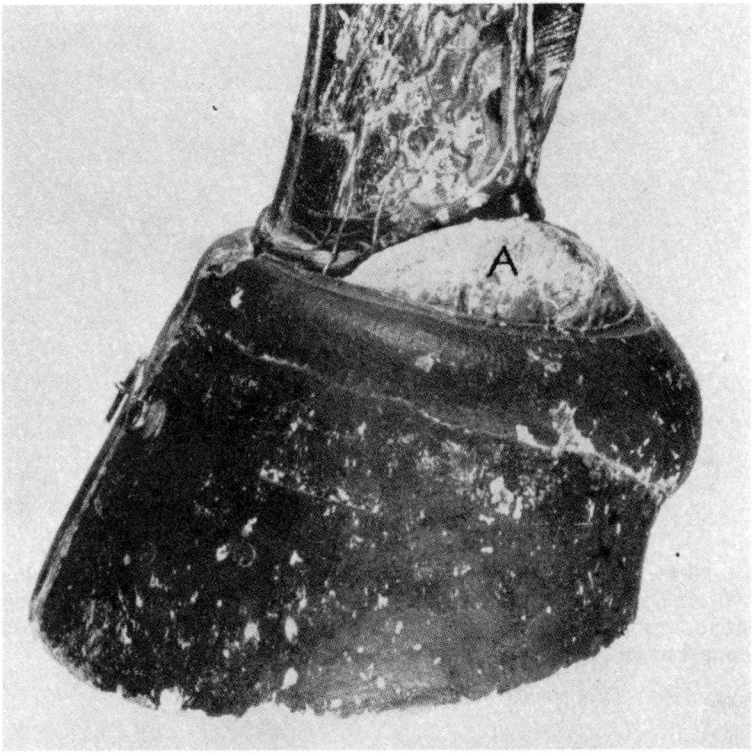

Fig. 5-10. A foot showing the location of the lateral cartilage, A. When this cartilage becomes injured, it ossifies and becomes sidebone. (See also Fig. 5-11.) (Courtesy, University of Idaho)

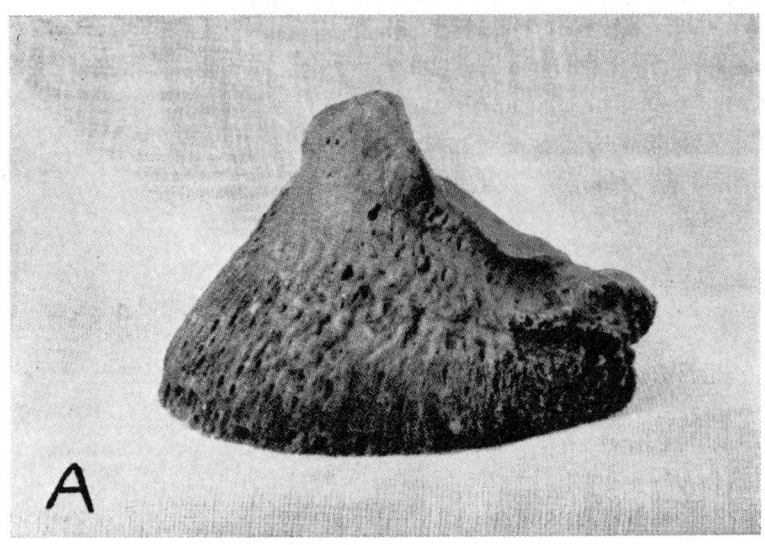

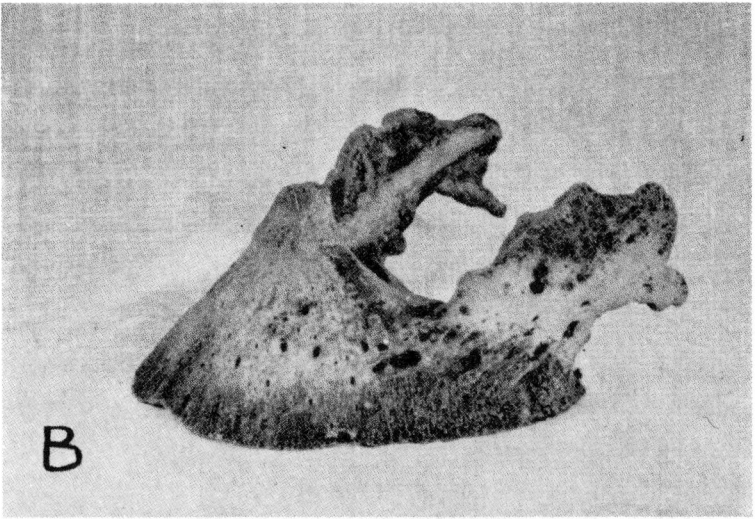

Fig. 5-11. *(Top)* A normal coffin bone. *(Bottom)* A coffin bone showing a very prominent sidebone on each side. It is impossible for the fleshy parts of the foot to spread properly inside this bony development when pressure is placed upon the foot. (Courtesy, University of Idaho)

A horse with narrow, or cramped, heels, with hooves that are straight up and down at the heels, or quarters, and with thick, short, straight pasterns is likely to have sidebones. With such a conformation, being able to distribute the pressure uniformly on all parts of the foot is impossible. Sidebones do not always make a horse go lame. Although, as a rule, in a large percentage of the cases, impaired action is evident in a short stride.

Splints

Splints are abnormal bony growths found on cannon bones, usually on the inside surface, but occasionally on the outside (Fig. 5-8). They are most common on the front legs. When found on the hind cannons, they are generally on the outside. Splints may enlarge and interfere with a ligament, thus causing irritation and lameness. Their presence detracts from the appearance of the animal, even when there is no lameness. They often disappear on young horses.

Wind Puffs, or Windgalls

Wind puffs, or windgalls, are enlargements of the fluid sac (bursa) located immediately above the pastern joints on the forelegs and rear legs. They reflect too much hard work, especially on pavement, and a deficiency in the shock-absorbing ability of the parts involved (Fig. 5-8). They detract from the clean-cut appearance of the leg, although they may not cause lameness. Horses with a small amount of bone and small joints usually develop wind puffs sooner than horses with bone of ample size and quality. Wind puffs are generally recognized as a blemish; if they cause lameness, they are considered as an unsoundness.

Ringbones

Ringbones are bony enlargements that involve the pasterns (Fig. 5-12). They interfere with the action of the joints and tendons, thus causing lameness accompanied by stiff ankles. They are located just above the hooves in front of and on the sides of the pasterns on the forefeet (Fig. 5-8). They are not often found on the hind feet. Lameness may be evident, even though the ringbones are not large enough to be seen. If they are sufficiently developed, you can feel them by examining the lower front part of the pasterns. However, they do not always develop entirely around the lateral ringbones. They may be caused by bruises, strains, overexertion, improper shoeing or faulty conformation, such as short, straight pasterns, cramped hoofheads or feet that are too small for the weight they carry, which puts excessive strain on all parts of the feet.

Horses with moderately long, sloping, clean-cut pasterns, large hoofheads and shapely, well-developed hooves are not susceptible to this unsoundness.

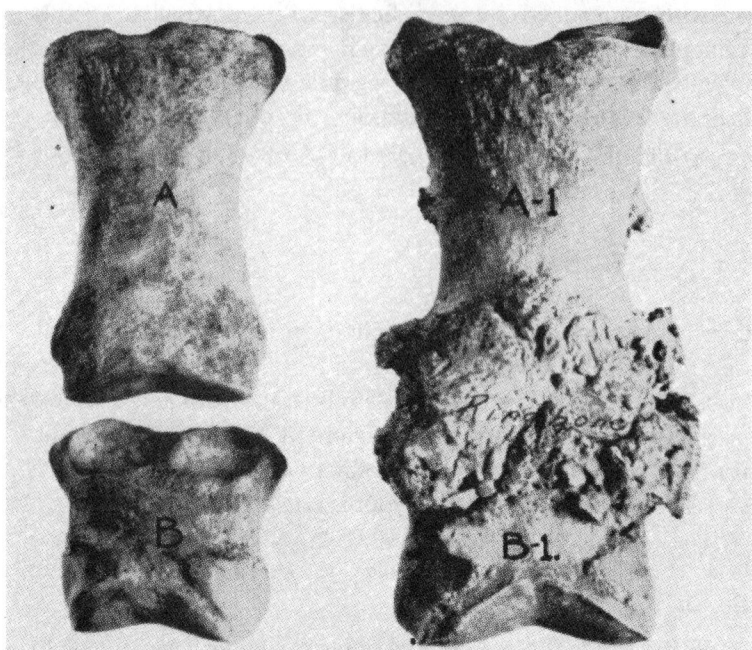

Fig. 5-12. Ringbone. *A* and *B* are normal long and short pastern bones. *A-1* and *B-1* are fused, and the result is an unsoundness, namely, a ringbone. The joint is stiff and obviously affects the action and movement of the horse.

Bone Spavins

Bone spavins, sometimes referred to as "Jack" spavins, are bony developments in the hocks. They interfere with action in the hocks in the same way that ringbones interfere with action in the pasterns. They are located on the inside of the hocks toward the front at the point where the base of the hocks taper into the cannon part of the legs (Fig. 5-8). The two hocks should be compared. A prominent development of bone at a hock joint usually indicates a bone spavin. A horse with small, crooked, meaty hocks that are not properly supported with wide cannons is usually predisposed to this defect.

A bone spavin may involve injury to one or more of the bones of a hock. Sprains, overexertion and slipping are among the causes. Bone spavins are one of the most destructive conditions affecting the usefulness of a horse because they usually cause a very painful lameness. The lameness is most evident when the animal is used following rest. Bone spavins are a heritable weakness.

Bog Spavins

Bog spavins are fillings of the natural depressions, particularly on the inside and front of the hocks (Fig. 5-8). They are usually associated with thick, coarse hocks. The hocks should be clean-cut, with every normal depression showing distinctly. This defect does not always cause lameness, but it detracts from the appearance of the hocks. Bog spavins are much larger than blood spavins.

Blood Spavins

Blood spavins are varicose vein developments located immediately above bog spavins (Fig. 5-8). These depressions in sound hocks are well-defined. When advanced, they may cause lameness. They are often associated with well-developed bog spavins.

Scratches

Scratches refers to a traverse cracking of the skin in the hollow of the heels. Lack of exercise, mud left to dry on the legs, too much grain as feed, especially when the horse is not hard at work, are some of the predisposing causes. Scratches is unsightly and affects the quality of the heels.

Grease

Grease may involve any part of the legs from the hocks down. It does not often occur in the front legs. It is most frequently found in horses that have a heavy covering of long, stiff hair in this area. Horses that are coarse in the feet and legs are also subject to this trouble. Grease is usually accompanied by itching, thickened skin, erect hair on the affected parts and sores. The discharge from these sores causes the affected parts to appear greasy, hence, the name.

Corns

Bruises of the soft tissue underlying the horny soles of the feet, which are manifested as reddish discolorations of the soles immediately below the affected areas, are called corns. Fast work on hard, rough roads, flat soles, weakened bars and poor shoeing may also cause corns.

Capped Elbow, or Shoe Boil

When the horse is lying down, an irritation at the point of the elbow (capped elbow) from the shoe may cause sufficient disturbance to develop a tumorous growth that may become a running sore (shoe boil).

Thrush

Thrush is a disease of the frog. It is characterized by a very offensive odor and often a discharge of pus from the cleft of the frog. It is most commonly found in the hind feet and is caused by unsanitary conditions in the animal's stall. Most cases will respond if the affected frog is trimmed away, sanitation is improved and an antiseptic is used.

Founder, or Laminitis

Founder, or laminitis, is a serious ailment of the fleshy laminae. It is characterized by the sole of the foot dropping, the toes turning up and small ridges appearing around the foot parallel to the hoofhead. It is particularly apparent in the front feet.

Founder may occur when horses overeat, when they are overworked, when they drink too much cold water when they are hot or when mares foal and the uterus becomes inflamed. There is usually severe pain accompanying founder. However, proper treatment by a competent veterinarian will usually prevent permanent injury.

Curbs

Curbs are located at the rear of the legs and below the hock joints (Fig. 5-8). They are due to thickening of the ligaments or tendons about 5 or 6 inches below the hock points. In well-defined cases, lameness results. Horses with crooked, or "sickle," hocks are subject to curb. While generally of less consequence than bony unsoundnesses, curbs are usually recognized as an unsoundness.

Quarter Cracks, or Sand Cracks, and Toe Cracks

Vertical splits in the horny wall of the inside of each hoof (in the region of the quarter), which extends from the coronet, or hoofhead, downward, are known as "quarter cracks," or "sand cracks." They are seldom found in the hind legs. When the cracks are on the forepart of the toes, they are termed "toe cracks." This condition usually results when the hooves are allowed to become too dry and brittle or when they are improperly shoed.

Capped Hocks

Bruising the hocks may cause them to swell (capped hocks) and become unsightly. This constitutes a blemish.

Stringhalt

Stringhalt is indicated by excessive flexing of one or both of the rear legs. It is very conspicuous and is looked upon with much disfavor. It is especially noticeable when you are backing up a horse.

Cocked Ankles

Cocked ankles, a condition usually limited to the hind feet, is found in horses that stand bent forward on their fetlocks in a cocked position. This condition can be corrected by proper foot trimming.

Roaring

The breathing of a horse should be normal. The term "roaring" refers to a defect in the air passage of a horse, which causes it to roar or whistle when it forces respiration. If this defect is suspected, it is good practice to exercise the horse at the trot sufficiently to determine if it prevails. This is one reason why judges require horses to repeat their actions at the trot. Roaring can be corrected by surgery.

Broken Wind, or Thick Wind

Broken wind is another form of difficult breathing. A horse with this defect inhales normally but exhales with considerable effort. Exhaling is characterized by double effort. Therefore, the breathing rhythm is not normal. When the horse is exercised, the defect becomes more apparent.

Blindness

Partial or complete loss of vision is known as blindness. Either eye or both eyes may be affected. A blind horse usually has very erect ears and a hesitant gait. Frequently, blindness is also reflected by discoloration of the eye(s). Moving a hand gently back and forth in front of the eye(s) can confirm or negate any suspicions.

Moon Blindness, or Periodic Ophthalmia

Moon blindness, or periodic ophthalmia, is a cloudy or inflamed condition of the eye that disappears and returns in cycles that are often completed in about a month. In the past, many people believed these cycles were related to changes of the moon; thus, this condition was given the name "moon blindness." It is a nutritional deficiency disease, in some cases at least, caused by a lack of riboflavin, a B vitamin. Riboflavin is abundant in green forages of all kinds — green hays and green pastures; consequently, supplying ample quantities of these feeds may prevent the condition.

Cataract

When the lens of the eye is clouded with a light grey film, this condition is commonly identified as a cataract.

Bowed Tendons

Bowed tendons are enlarged tendons behind the cannon bones, in both the front and the hind legs. Descriptive terms of "high" and "low" bow are used by horse raisers to denote the location of the injury, the high bow is just under the knee and the low bow is just above the fetlock. This condition is often caused by severe strain, such as heavy training or racing. When bowed tendons are pronounced, more or less swelling, soreness and lameness are present.

Thoroughpin

Thoroughpin is puffiness in the web of a hock (Fig. 5-8). It can be determined by movement of the puff, when pressed to the opposite side of the leg.

Stifled

The stifle corresponds to the knee in humans. A horse is said to be stifled when the patella of the stifle joint has been displaced. Occasionally, the patella can be put back into its normal position, but more often, the affected horse is rendered useless.

Contracted Feet

Contracted feet is characterized by a drawing-in, or contracting, at the heels. It most often occurs in the forefeet. A tendency toward contracted feet may be inherited, but improper shoeing usually aggravates the condition.

Quittors

Quittors are deep-seated running sores at the coronets, or hoofheads. They cause severe lameness. The infection may arise from a puncture wound or from corns and sand cracks, or it may be carried in the bloodstream. Quittors are usually confined to the forefeet, but they sometimes occur on the hind feet.

Osselets

"Osselets" is a rather inclusive term that refers to a number of inflammatory conditions around the ankle joints. Generally, it denotes swellings that are fairly well-defined and located slightly above or below the actual center of the joints and, ordinarily, a little to the inside or outside of the exact front of the legs. When touched, it feels like putty or mush, and it may be warm or hot. The pain will correspond to the degree of inflammation, as evidenced by swelling and fever. An afflicted horse will travel with a short, choppy stride and will show evidence of pain when it flexes the inflamed ankle.

Knee-sprung

The condition of over in the knees or with the knees protruding too far forward is known as knee-sprung, or buck-kneed.

Calf-kneed

Standing with the knees too far back, directly opposite to buck-kneed, or knee-sprung, is called "calf-kneed."

Sweeney

Sweeney is a depression in a shoulder due to atrophied muscles. Sweeney is caused by nerve injury.

Fistula

Fistulous withers is an inflamed condition in the region of the withers (Note Fig. 5-1, no. 15), commonly thought to be caused by bruising. Fistula and poll evil are very similar except for location.

Poll Evil

Poll evil is an inflamed condition in the region of the poll (the area on top of the neck and immediately behind the ears). It is usually caused by bruising on the top of the head. The swelling, which may be on one or both sides, usually contains pus or a straw-colored fluid. At first the affected area is hot and painful, but later the acute symptoms of inflammation subside. Poll evil is slow to yield to treatment, and it may break out again after the external symptoms have disappeared.

Parrot-Mouth and Undershot Jaw

Both parrot-mouth and undershot jaw are hereditary imperfections in the way in which the teeth come together. In parrot-mouth, or overshot jaw, the lower jaw is shorter than the upper jaw. The reverse condition is known as undershot jaw.

Hernia, or Rupture

A hernia, or rupture, is a protrusion of any internal organ through the wall of its containing cavity; but it usually refers to the pushing out of a portion of the intestine through an opening, or tear, in the abdominal muscle.

Heaves

Heaves is difficulty in forcing air out of the lungs. It is characterized by jerking of the flanks (double-flank action) and coughing after the horse has drunk cold water. There is no satisfactory treatment, although affected animals are less bothered if they are turned to pasture, if they are used for only light work, if their hay is sprinkled lightly with water at the time of feeding or if their entire ration is pelleted.

COMMON DEFECTS IN WAY OF GOING

The feet of a horse should move straight ahead and parallel to a center line drawn in the direction of travel. Any deviations from this way of going constitute defects, some of which are:

1. **Cross-firing.** Cross-firing, which is generally confined to pacers, consists of a "scuffing" on the inside of the diagonal forefeet and hind feet.

2. **Dwelling.** Most noticeable in trick-trained horses, dwelling consists of a noticeable pause in the flight of the foot, as though the stride was completed before the foot reached the ground.

3. **Forging.** Forging is the striking of the forefoot by the toe of the hind foot.

4. **Interfering.** Interfering refers to the striking of the fetlock, or cannon, by the opposite foot that is in motion. This condition is predisposed in horses with base-narrow, toe-wide or splay-footed standing positions.

5. **Lameness.** Lameness can be detected when the horse is standing, for it will favor the affected foot. In action, the load on the ailing foot eases and the head bobs as the affected foot strikes the ground.

6. **Paddling.** Paddling refers to throwing the front feet outward as they are picked up. Horses with toe-narrow or pigeon-toed standing position are predisposed to this condition.

7. **Pointing.** Pointing refers to the perceptible extensions of the stride with little flexion.

8. **Pounding.** Pounding is a condition in which there is heavy contact with the ground in contrast to the desired light, springy movement.

9. **Rolling.** Excessive lateral shoulder motion, characteristic of horses with protruding shoulders, is known as rolling.

10. **Scalping.** Scalping occurs when the hairline at the top of the hind foot hits the toe of the forefoot as it breaks over.

11. **Speedy cutting.** Speedy cutting is a condition in which the inside of the diagonal forepastern and the hind pastern make contact. It is sometimes found in fast trotting horses.

12. **Stringhalt.** Stringhalt is characterized by excessive flexing of the hind legs. It is most easily detected when the horse is backed up.

13. **Trappy.** A short, quick, choppy stride is known as trappy. Horses with short, straight pasterns and straight shoulders are predisposed to this condition.

DETERMINING THE AGE OF A HORSE

The age of a horse is determined by the condition of its front, or incisor, teeth, of which there are six in the upper and six in the lower jaw. When the foal is born, it may have two temporary teeth in the middle front of each jaw. If they are not present then, they will appear within 7 to 10 days. When the colt is about five weeks of age, one tooth will appear on either side and close by the first pair in the upper and lower jaws (Fig. 5-13). These are the temporary intermediates. The corner pain in each jaw appear when the horse is from 6 to 10 months of age. This completes the temporary, or mild, teeth.

When the horse is one year old, the center pairs show wear. When it is 1¹/₂ years old, the temporary intermediates show wear. When it is two years old, the temporary, or milk, teeth show wear.

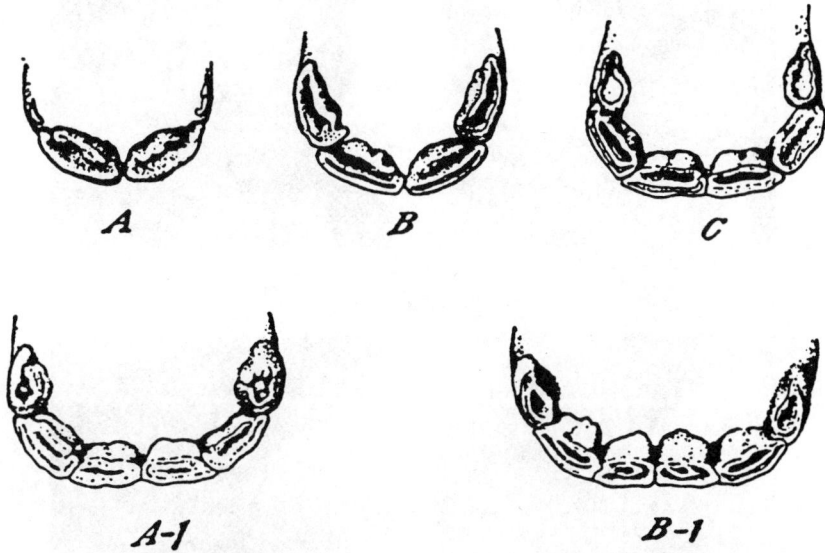

Fig. 5-13. Appearance of the incisors of a colt: (A) immediately after colt's birth (see text), (B) at 6 weeks, (C) from 6 to 10 months after birth, (A-1) at 12 months, (B-1) at 18 months. (Courtesy, USDA)

The next major change takes place when the horse is 2¹/₂ years of age. The central pair in the upper and lower jaws begin shedding (Fig. 5-14). These are replaced by the permanent central pairs when the horse is three years old. In a four-year-old horse's mouth, the temporary intermediates have been replaced by the permanent intermediates. The two corner temporary pairs have been replaced by permanent corner teeth by the time the horse reaches five years of age. A six-year-old horse will have the permanent corner teeth up to a point level with the intermediate and central pairs. The central pair will show considerable wear and will be smooth, with the cup, or hollow, in the center of the teeth disappearing.

In the seven-year-old horse's mouth, the intermediates show wear. Upon careful inspection of the corner pair in the upper jaw, you will find a projection in the posterior edge of these two teeth. This is caused by unequal wearing of the corresponding lower teeth, which are somewhat smaller than the uppers. All three pairs show wear in the mouth of the eight-year-old — the cup has worn down in the central

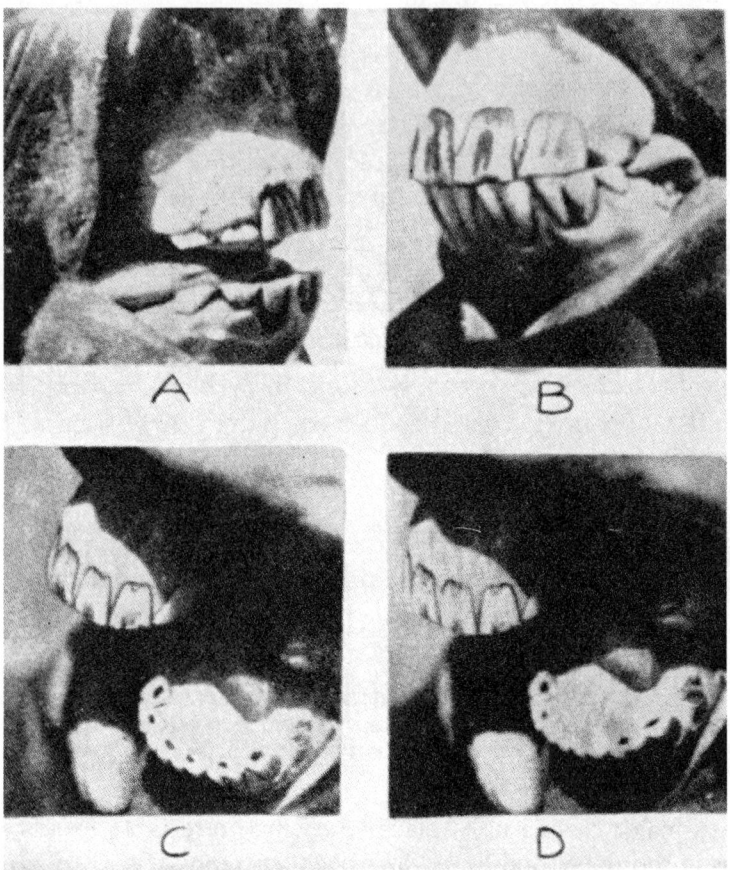

Fig. 5-14. (A) The mouth of a three-year-old horse in which the middle pair of incisors above and below have replaced the middle pair of milk teeth; (B) the mouth of a four-year-old horse, showing two pairs of permanent incisors; (C) the mouth of a five-year-old horse, with all the permanent incisors in place but with the corners not yet worn; (D) the mouth of a six-year old horse, in which the cups in the center pair of incisors have been worn down. In a seven-year-old horse, the cups of the intermediate teeth will be gone, while in an eight-year-old horse, the cups will be worn out of all the lower incisors. (Courtesy, USDA)

pairs, shows slightly in the middle pair and is still quite prominent in the corner pairs.

After the horse has reached its eighth year, it is difficult to determine the animal's age with much accuracy (Fig. 5-15).

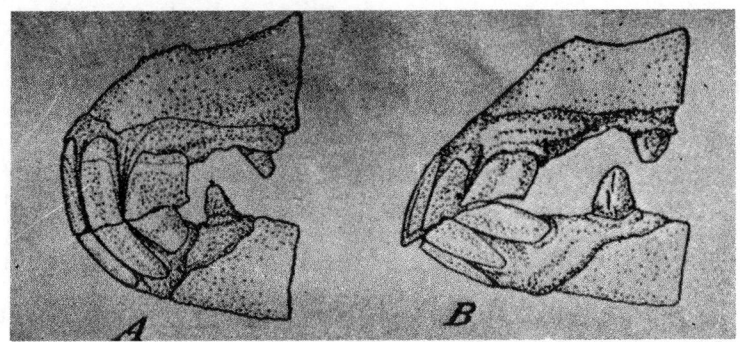

Fig. 5-15. The teeth tend to straighten out as the horse advances in age. (*A*) 6-year-old teeth; (*B*) 20-year-old teeth. (Courtesy, USDA)

ALIGNMENT OF THE LEGS AND FEET

Inadequacies in the alignment of the legs and feet, as illustrated in Figs. 5-16, 5-17, 5-18 and 5-19, contribute to inefficiency in action and to unsoundness in the legs and feet, as well as reducing the functional efficiency and length of usefulness in horses. These diagrams demonstrate how and why discrepancies in the proper stance are major causes of general inefficiency. The feet and legs should move in a straight line. Any variation from that contributes to wasted effort. Moreover, if the weight on the legs is not supported by proper alignment of the bones involved, fatigue and uneven wear on the articulating joints will usually result in unsoundness and general loss of efficiency.

Any variation from free and agile action is burdensome for the horse and takes its toll increasingly as severity of the imperfections conflict with the action. The importance of making observations from both the front and the rear, as well as from the sides, is self-evident in Figs. 5-16, 5-17, 5-18 and 5-19.

FORM IN RELATION TO FUNCTION

Figs. 5-22 and 5-25 of horses at halter and Figs. 5-3, 5-4, 5-5 and 5-20 of horses in action provide ample opportunity for observing the actual form, or type, in detail, similar to that described under "The Quarter Horse." Observe the perfection of form in these pictures, just as you would in a judging class. Looking for imperfections is

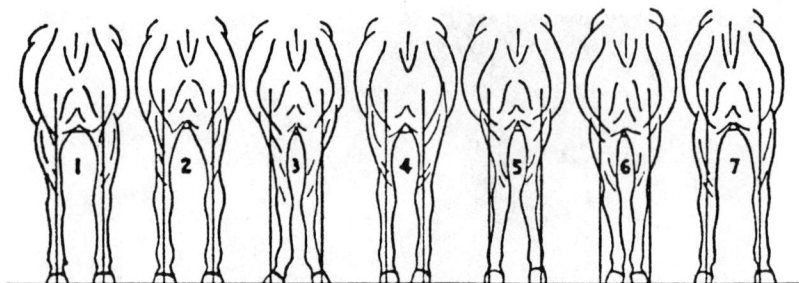

Fig. 5-16. Front view of forelimbs. A perpendicular line drawn downward from the point of the shoulder should fall upon the center of the knee, cannon, pastern and foot. (1) Correct conformation; (2) slightly bow-legged; (3) close at knees and toed-out (splay-footed); (4) toed-in (pigeon-toed); (5) knock-kneed; (6) base narrow; (7) base wide. (Courtesy, Horse and Mule Association of America)

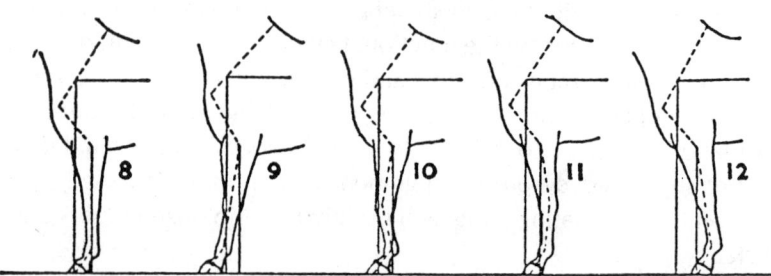

Fig. 5-17. Side view of forelimbs. A perpendicular line drawn downward from the center of the elbow point should fall upon the center of the knee and pastern and back of the foot. A perpendicular line drawn downward from the middle of the arm should fall upon the center of the foot. (8) Correct conformation; (9) leg too far forward; (10) knee sprung; (11) knees set back; (12) foot and leg too far back. (Courtesy, Horse and Mule Association of America)

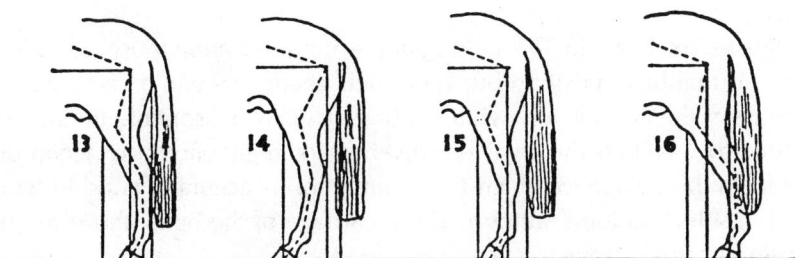

Fig. 5-18. Side view of hind limbs. A perpendicular line drawn downward from the hip point should fall upon the center of the foot and divide the gaskin in the middle. A perpendicular line drawn downward from the point of the buttock should just touch the upper rear point of the hock and fall barely behind the rear line of the cannon and fetlock. (13) Correct position; (14) leg too far forward (sickle-hocked); (15) entire leg too far under; (16) entire leg too far back. (Courtesy, Horse and Mule Association of America)

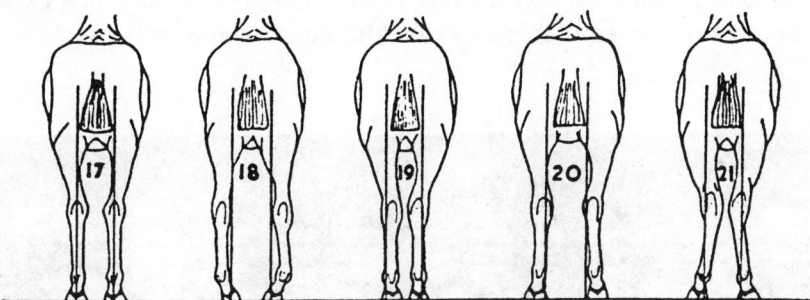

Fig. 5-19. Rear view of hind limbs. A perpendicular line drawn downward from the point of the buttock should fall in line with the center of the hocks, cannons, pasterns and feet. (17) Correct position; (18) too wide at hocks; (19) base too narrow and hocks too close; (20) base wide and hocks too far apart; (21) hocks too close together and toes too wide apart (cow-hocked). (Courtesy, Horse and Mule Association of America)

also challenging and stimulating, as breeders have made measurable progress in their efforts to combine, from various sources, the characteristics and qualities that reflect the requirements in the form and function of stock horses.

Totalness in function is clearly in evidence in the action pictures. Every one of the horses in action knows its assignment and loses no opportunity in the execution of it. Ability, willingness and agility all contribute to the accomplishment of the difficult assignment.

The horses involved in the cutting-horse demonstration pose at halter in an apparently normal head position but keep their heads low while "cutting." Since all the horses carry their heads low while cutting, it seems reasonable to conclude that they do this to facilitate the total effectiveness of their efforts to accomplish the assignment. To the degree to which this assumption is accurate, good judgment has been used in selecting for a rather medium carriage of the head above the backline in the horse's normal stance.

THE QUARTER HORSE

The Quarter Horse was originally developed for quick get-away and for great speed at short distances — the quarter mile. Such speed, coupled with the ability to turn quickly, has made the Quarter Horse an ideal horse from which to rope, or "cut," cattle, and as a result, it has become prominent as a "cow horse." It has a quiet disposition and unusual nervous stability, even in the presence of much excitement. The Quarter Horse, in general, is described in the following sections.

Fig. 5-20. "Jet View" winning the El Primero del Ano Derby at the Los Alamitos Race Course. (Courtesy, American Quarter Horse Association)

Head

The head of a Quarter Horse is beautiful because of its prominent and wide-set eyes, clean-cut appearance and pronounced expression of intelligence (Figs. 5-2 and 5-26). The head is broad, relatively short and topped by small ears, which are, in some specimens, referred to as "fox ears." The forehead is not dished. The jaws are especially wide and deep but not excessively so, and the nostrils are large and prominent. Well-developed jaws and a short muzzle give the impression of power and strength. There should be smooth balance of all parts so that there is harmony between the head, neck and shoulders.

Neck

The neck of a Quarter Horse should join the head at a 45-degree angle and merge smoothly into the shoulders (Figs. 5-2 and 5-21). The neck crest should not be prominent. There is usually a distinct space between the jawbone and the neck muscle so that there is no interference with breathing when the head is checked back. A medium-length neck is favored over an extremely short one because it facilitates freedom of head-neck action.

Shoulders

In general, the shoulders of a Quarter Horse should be shapely, should add to the balance and beauty and, at the withers, should provide a good place for the saddle (Figs. 5-2 and 5-21). The correct slope of the shoulder bed is about 45 degrees. Well-placed shoulders are fundamental to a true straight action and make for ease in the riding gait. Horses with straight, steep shoulders are usually short and stubby in their gait and are rough to ride. The withers are medium in height and relatively sharp, extending well back and merging smoothly with deep, well-laid shoulders.

Chest

The chest should be wide, broad and deep, with a heavy-muscled "vee" between the legs (Fig. 5-22). This type of chest indicates superior constitution and provides plenty of space for the lungs and heart (Fig. 5-23). The ribs should be well-sprung, carrying down into a deep, full heart girth. Flat-sided ribs are objectionable because they reduce the space for the heart and lungs.

Back and Barrel

Particularly unusual characteristics of the Quarter Horse are the short, strong saddle back, close coupling and powerful loin (Figs. 5-2, 5-24 and 5-26). A short back and strong loin are essential in providing a riding horse with the ability to endure long hours of supporting the weight of a rider and saddle, ranging in most cases from 150 to 250 pounds. The last rib should be rather close to the hips, thus

Fig. 5-21. "Super Impressive" holds 10 grand championships and 2 reserve championships. This splendid stallion is owned by Dr. and Mrs. Derwood L. Ashworth, Hickory, North Carolina.

Fig. 5-22. "Skipster 249119," 15.3 hands, chestnut stallion, AQHA Champion, with a total of 223 foals on the ground. His get have accumulated 1,025 Performance points and 841 Halter points. He was Honor Roll Stallion in Indiana six out of eight years and won Get of Sire seven out of eight years. He is owned by Clayton Benker, Lafayette, Indiana. (Courtesy, Clayton Benker, Lafayette, Indiana; photo by Carter Allen)

Fig. 5-23. "The Ole Man" has proven his versatility and superiority as a sire. In one year he was listed on six AQHA leading sire lists and was the only sire to be included on all these lists. He was also the Leading Sire of Race Winners (most wins), of Race Winners (most winners), of Money Earners, of Performance Class Winners, of Halter Class Winners and of Two-Year-Old Qualifiers for the Register of Merit (race). Owned by Roy Browning Ranches, Ada, Oklahoma.

Fig. 5-24. "Hot Scotch Man," two-time World Champion at halter. (Courtesy, American Paint Horse Association)

eliminating any appearance of a "hollowed out," or gaunt, look in this region. The ideal shape of the middle, or barrel, consists of deep, well-sprung ribs extending back to the hip joints while the underline comes back straight to the flanks. An ideal cow horse should not be too paunchy, or heavy, in the middle because this hampers its ability to move freely.

Rear Quarters

A "must" of the Quarter Horse, as well as being one of its most distinguishing features, is the heavy, deeply muscled hindquarters (Fig. 5-23). From either the side or the rear view, the rear quarters should be broad, deep with bulging muscles, which are carried down full into the stifle and gaskins. The Quarter Horse should have exceptionally good width through the stifle and plenty of length in the hips. Short hips and light quarters are very undesirable because they denote lack of power. By observing a stock horse from the rear, you can obtain a mental picture (Figs. 5-2 and 5-23). Note that the legs are squarely set under the body, the gaskins are muscled inside and out, the hocks are clean and wide and the cannons are moderate in length.

Feet and Legs

"No feet, no horse" is a common statement among judges of horses. The feet should be medium-sized and round, with especially good depth and width at the heels. The pasterns should be moderate in length and should carry some slope; but, if the pastern slope is excessive and too long, the horse is usually more suspectible to leg injury and is not as sure-footed. Too steep pasterns result in a hard ride and an awkward gait. The cannon bones should be moderate in length and clean-cut, and the tendons should be well-attached and strong. The groove between the bone and the tendon in both cannon regions should be deep and clean. All leg joints should provide evidence of strength and freedom from unsoundnesses. (Refer to the material under the heading "Blemishes and Unsoundnesses" for a discussion of this subject.) The legs should set well under (Figs. 5-2, 5-21 and 5-26).

The hocks should be set wide. Good length above and below the hocks gives the legs desirable leverage, thus adding to the power and quick get-away of the Quarter Horse. Try to form a mental picture of the ideal setting of the hocks and rear legs, as shown in Figs. 5-2 and 5-26. (See also Figs. 5-22, 5-23, 5-24 and 5-25.)

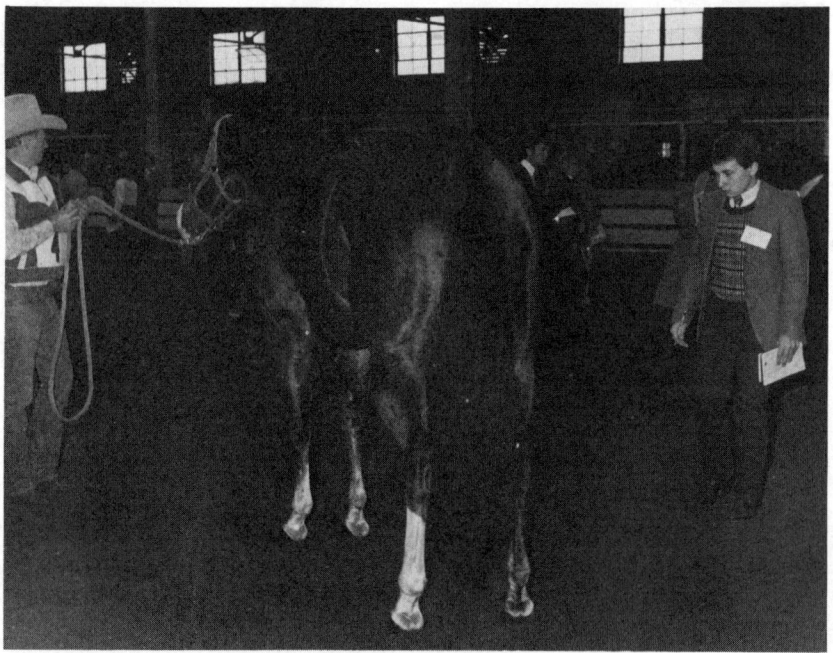

Fig. 5-25. Action shot during intercollegiate Quarter Horse judging contest. Note close observation of the feet and legs by the contestant. (Courtesy, American Quarter Horse Association)

Action

The best way to judge the action of a Quarter Horse is to see that animal perform under the saddle with a rider. The rider should be requested to demonstrate the horse's ability at the walk, trot and canter, or "restrained gallop." It is also of importance to observe its ability to start quickly, stop quickly, back up and "turn on a dime" (Figs. 5-3, 5-4 and 5-5). A well-trained Quarter Horse responds readily to the rein and bit without any tugging or pulling by the rider.

Many times it is not possible to have the horse's action shown with a rider. When this is the case, the action should be observed at the walk or the trot. The action of the Quarter Horse at any speed is very collected at all times, indicating perfect control. The stride at both the walk and the trot is straight. When standing, the Quarter Horse is perfectly at ease with legs well under. This allows the animal to move quickly in any direction (Figs. 5-2, 5-21, 5-22, 5-26 and 5-27).

Fig. 5-26. "Watch Joe Jack," chestnut stallion sired by "Two-eyed Jack," and "Out of Watch Joe Moore," is an AQHA Champion, an AQHA High-Point Western Pleasure Stallion. He has won 83 Grand and 34 Reserve championships, 218 Halter points and 148 Western Pleasure points. Owned by Howard Pitzer Ranch, Ericson, Nebraska.

Fig. 5-27. "Debonaire Lady," AQHA Superior Halter Horse. Owned by Larry B. Sullivant, Gainesville, Texas.

Color

The color of the Quarter Horse varies. Bay, chestnut, black and brown are the most prevalent. Palomino, grey and a few lighter shades also occur.

Height and Weight

The Quarter Horse stands at 14 to 17 hands and weighs from 900 to 1,300 pounds.

THE PALOMINO

As with other western stock horses, the Palomino in its early history felt the influence of the Arabian, Barb, Thoroughbred, Standardbred, Morgan and other saddle stock horses. In general, these influences, in large part, were rather indirect and obtained essentially from the breeds that contributed to the Palomino in more recent years. During the foundation days of the Palomino on the southwestern plains, when the other breeds were combined directly and indirectly with the environmentally well-adapted western cow ponies, the foundation was laid for types that combined the physical form, strength, maneuverability and intelligence that so prominently characterize the useful qualities of the current stock-horse type.

After many years of experience in the rugged cattle-producing areas of the West and Southwest, that include the mountain and plateau grazing lands west of the Mississippi, between the Mexican and Canadian borders, ranchers now recognize a large number of characteristics that are of special service for stock-horse performance. Lines of breeding have been developed, and are being improved, to comply with the standards of serviceable qualities that are generally accepted by stock producers as contributing to overall stock-horse usefulness.

Illustrations of stock horses at halter, and also in action, have been used rather liberally throughout Part 5 so that you will have the opportunity to compare the important type characteristics and to evaluate the duties that the stock horse in action is called upon to perform. To appreciate fully the significance of the judgment that the stock-horse breeders have used to develop a type that qualifies for the difficult assignments, it is important for you to learn the names and to understand the functions of the different parts of the horse. This is also helpful as you try to increase your ability to judge form and function with greater accuracy. (See the detailed discussion under "Stock-Horse Type.")

In general appearance, the Palomino stock-horse type is medium in size, though weights may vary considerably, from about 900 to 1,300 pounds. These extreme weights, however, are as a rule not found in horses under heavy-duty assignments. A division in show classes is usually made between light and heavy classes at 1,100 pounds. Handy-weight Palomino stock horses will stand from 14^1/$_2$ to 16 hands and weigh from about 1,000 to 1,200 pounds. These weights apply to horses that are in strong condition. Horses in working condition vary in weight.

In general, the **body** is medium in length, moderately long and strongly coupled in the back, broad in the loin, gently sloping and heavily muscled in the croup, neatly laid in the shoulders, medium in the length of neck and heavily muscled in the thighs and forequarters, with a roomy, "useful" middle that is neatly carried. Superior muscling throughout should be evident. (Study all pertinent figures.)

There must be unmistakable evidence of clean-cut, well-placed **legs** and well-shaped **feet** (Figs. 5-28 and 5-31). Before you proceed with a detailed study of the various parts of a stock horse, you should review the "Points of a Horse" (Fig. 5-1), "The Skeleton — Its Relationship to the Appearance and Usefulness of a Stock Horse" and "Blemishes and Unsoundnesses," because those materials, which deal with the usefulness of the stock horse, are basic to a comprehensive understanding of the "mechanics" involved, as they relate to the sum total of the body functions.

When breeds are developed for the same general purpose, to function in similar

Fig. 5-28. "Mary J. Sweet PHBA PB-20134 AQHA P-73795," Palomino Quarter Horse filly. Owned by Horse Patch Farms, Jack Janowitz, Littleton, Colorado. (Courtesy, Jack Janowitz, Littleton, Colorado)

environments, invariably, more, or less, duplication of the type sought among the
related groups occurs, which, by the same token, leads to parallelism in the terminology used to describe their respective type characteristics. To an appreciable degree, this duplication is unavoidable, inasmuch as the characteristics involved in
function under comparable conditions are basically the same. These similarities reflect the general agreement among the sponsors that there is a specific relationship
between form and function in stock horses.

Many of the horses that contributed substantially to the stock-horse qualities of
the Palomino, were of Quarter Horse breeding. Some of these have been recognized
as major contributors, not only to type but also to the golden color of the Palomino,
which is summed up in Palomino Horses, the official publication of the Palomino
Horse Breeders of America, Inc., as follows: "Because of the fact that the Palomino
color was first perpetuated on horses in the Southwest, it is a natural assumption that
the American Quarter Horse has contributed more extensively to this registry than
any other type."[4]

---

[4]*Palomino Horses*, July 1947.

The discussion of the Palomino in this text is limited to its use as a stock-horse. The pictures illustrating the Palomino are of winners in stock-horse classes, essentially double-registry Palominos — namely, those that have been registered in both the American Quarter Horse Association and the Palomino Horse Breeders of America, Inc.

The distinctive color of the Palomino, as described by the Palomino Horse Breeders of America, Inc.,[5] must be, as nearly as possible, that of a newly minted gold coin and must not vary significantly from that, either lighter or darker. The mane and tail must be white, with not more than 15 percent or chestnut hair in either. The Palomino may have white markings on the face or white stockings or socks below the knees and hocks. It must be free of white spots on the body, except for those caused by saddle rubbing or accidents.

The skin of the Palomino must be basically dark (a horse is considered basically dark-skinned when the eyes are dark and the color of the skin around the eyes and nose is dark or black). The eyes must be dark brown, hazel or black, and both must be the same color.

A Picture Study of Palominos

Placing a class of four stock horses from pictures, as with other classes of livestock is difficult because action is such a vital part of the total usefulness of these horses. Too, it is just as difficult to procure equally suitable pictures, with acceptable photographic technique and proper stance of the individuals involved. Consequently, an attempted placing would not be feasible or meaningful. Therefore, the mares in Figs. 5-29, 5-30 and 5-31 are discussed individually. In general, the photographic technique in each of these pictures is the same, and essentially the same qualities have been emphasized in each stance. The mares involved seem completely relaxed, and their stance is normal.

The Palomino mare in Fig. 5-29 is a good example of photographic skill in "totalness" of study that can be accomplished in one view — in this case the broadside view. The mare appears comfortable and is under no immediate discipline — as shown by the slack lead rope. Thus, you do not need to estimate adjustments that you might otherwise have had to make.

The broadside stance in Fig. 5-29 allows you to observe the coordination of all the parts of the body, except those that the front and rear views would contribute. However, in this picture the broadside view of the front and rear quarters leaves little doubt of the adequate development of these two parts. The uniformity in the development of excellence of all parts of the body, head and neck, feet and legs contributes to the general all-over appraisal of correctness and balance.

Fig. 5-30, which is a slightly to-the-rear broadside view, reveals all-over development and symmetry in body conformation, feet and legs, stock-horse alertness and

---

[5]PHBA Registration Rules.

Fig. 5-29. "Josie Mark PB-13341 AQHA 40175," Grand Champion Palomino mare and the dam of grand champions, two of which are shown in Figs. 5-30 and 5-31. Owned by Bent Arrow Ranch, Broken Arrow, Oklahoma. (Photo by Strayhorn)

Fig. 5-30. "Sugar Bar Maid PB-15881 AQHA P-97777," Grand Champion Palomino mare foaled by "Josie Mark" in Fig. 5-29. Owned by Bent Arrow Ranch, Broken Arrow, Oklahoma. (Photo by Strayhorn)

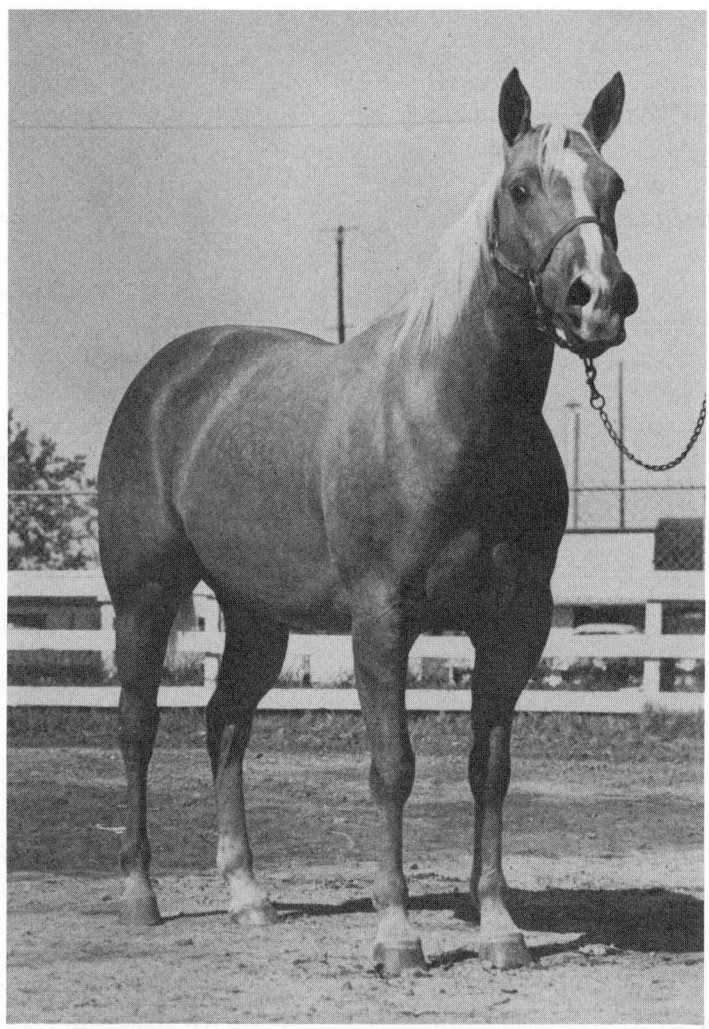

Fig. 5-31. "Hustle PB-14766 AQHA P-14766," Grand Champion Palomino mare foaled by "Josie Mark" in Fig. 5-29. Owned by M. T. McCormick. (Photo by Strayhorn)

disposition control. The ready-for-service shows to advantage the overall qualities of substance in body balance, adequacy in bone and shapely pasterns and feet. The heavy-muscled development in the arms and forearm is evident. The slope of the shoulders merges smoothly into the withers — thus providing a satisfactory placing for the saddle — and the back is well-carried, denoting strength and smoothness in the loin-croup coupling. Heavy muscling in the croup that continues well down into the gaskins, is clearly evident.

Each hock is wide from front to rear, deep and strongly supported by a flat cannon, with the tendon set well back. (For further discussion on the influence of the shape and setting of the feet and legs in the stock horse in relation to appearance and usefulness, refer to Figs. 5-16, 5-17, 5-18 and 5-19).

The side-front view of "Hustle," Fig. 5-31, provides a very good opportunity for you to study the setting and carriage of the head and neck, the front quarters, the conformation and setting of the legs and feet and the general relationships of these parts to the apparently equally well-developed middle and rear quarter characteristics. This mare is apparently well-supplied with power delivery facilities, as well as with the qualities that make it possible for her to use that power for the assigned purpose in stock-horse performance. Substance, muscling in the chest and "vee," quality, all-over correctness in conformation, even temperament and alertness deserve attention.

THE APPALOOSA

In general, published information on the Appaloosa was not readily available until 1938 when the Appaloosa Horse Club, Inc., was organized.[6] Since that time, however, active research on the history of the breed has produced much information relative to its history and characteristics.

Appaloosas have often been referred to as the "rough-country" stock horses. For about 200 years, the Nez Percé tribe subjected them to service in the rugged mountains and precipitous canyons in the drainage areas of the Columbia River and its tributaries. During this period, these horses were primarily used in warfare and in the search for food. Efficiency in performance and ability to survive undoubtedly were the bases for selection. Stallions that did not meet the desired requirements were castrated. Thus, conformation was probably a minor factor in formal selection, having much less influence than performance in the service areas essentially needed for stock-horse use.

Appaloosas seem to have an innate ability to respond to a broad category of disciplines. Some horse breeders characterize this ability as "intelligence," which cannot be described in pounds or inches.

In a survey conducted by the Appaloosa Horse Club, Inc., regarding the varied uses of Appaloosas, the first choice of two-thirds of the members (every member

---

[6]Much of the information on the Appaloosa, both past and present, and many of the pictures are courtesy, Appaloosa Horse Club, Inc., Moscow, Idaho.

voted) was as stock horses. This result is significant. Doubtless even though Appaloosas are ridden in parades, in rodeos, in high schools, for drills, for jumping, for pleasures, they are still preferred for stock-horse service.

Head

The head of the Appaloosa, in general, is medium in length, broad across the forehead and conservative in the depth and fullness of the jaw (Figs. 5-32, 5-33 and 5-34). Excessive development in the jaw contributes to coarseness. A Roman profile, when it does occur, is frowned upon. A slight dish in the forehead is acceptable, though a straight faceline is preferred. The general lines of the head should suggest leanness and quality. Prominent eyes well set apart and shapely, well-placed, medium-length, pointed ears contribute favorably to an alert appearance (Fig. 5-36).

Contributing to the general quality appearance of the head are the following: (1) jaws that are medium-sized, muscular and spreading well at the throatlatch to facilitate breathing when the head is checked back and in the few cases when the head is carried too high; (2) nostrils that are good-sized, set well apart and relatively free from thickness in the walls, which facilitates dilation in breathing (Figs. 5-35 and 5-36); (3) muscular lips that are well-carried and functionably sizeable; and (4) contour of the head and manner in which the head is carried, which are indicative of alertness, even temperament and ambition (Figs. 5-35 and 5-36).

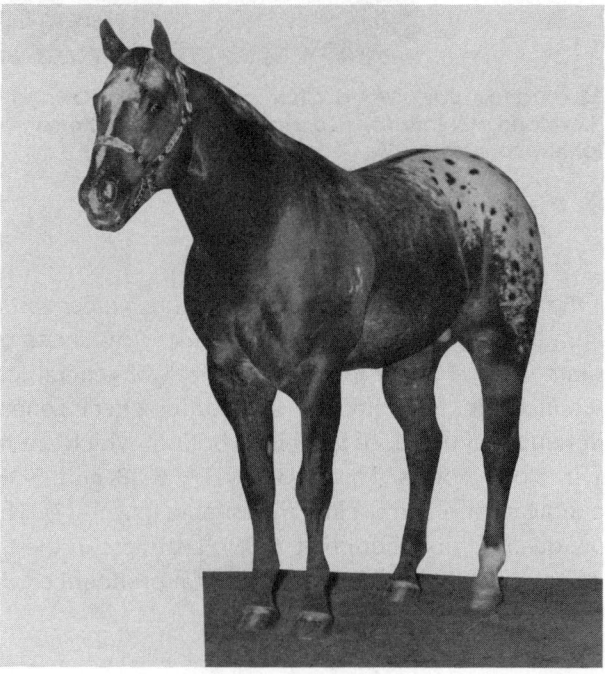

Fig. 5-32. "Breezing Bandit," World Champion Aged Stallion. Owned by White Sulphur Springs Farm, Georgetown, Kentucky. (Courtesy, Appaloosa Horse Club, Inc.; photo by Johnny Johnston)

Fig. 5-33. "Darla's Boy," World Champion Gelding. Owned by Steve Bateman, Loveland, Colorado. (Courtesy, Appaloosa Horse Club, Inc.; photo by Johnny Johnston)

Neck

The length of the neck, in general, is in harmonious agreement with the length of the body. Medium length is preferred. In the relatively moderate number of cases when the neck is either too short or too long, the lack of general length balance in the horse is easily noticeable. In addition, a short or long neck generally contributes to awkwardness in reining and lack of flexibility, both of which are important in the head-neck action in stock horses (Figs. 5-36, 5-37, 5-38 and 5-39). Just enough fullness of crest to indicate strong muscling is desirable (Fig. 5-37). The neck must be clean-cut and show quality throughout (Fig. 5-39). Leanness in the throat area, with special attention to the junction at the jaw, should be pronounced (Fig. 5-39).

Shoulders

The depth, width from front to rear, muscling and slope of the shoulders are all important, so much so that you should not become impatient if the owner of a horse

Fig. 5-34. "First Lady Reed," Grand Champion Mare at a National Appaloosa Horse Show, Santa Rosa, California. Owned by American Horsemen, Inc., Virden, Illinois, and shown by Tommy Manion. (Courtesy, Appaloosa Horse Club, Inc.; photo by Johnny Johnston)

insists that the shoulders are "the balance wheel of the entire structure." Virtually all the action that is communicated to the saddle area from the ground is routed through the shoulder area and determines the character of the action in terms of a "smooth" or "rough" ride, which is highly significant to the person whose assignment it is to get the best service out of the horse with minimum wear on both the rider and the animal (Figs. 5-32, 5-33, 5-34, 5-35 and 5-36).

For a basic understanding of the shoulder structure and slope, observe the bone arrangement in Fig. 5-7. Note in particular the two bones marked 13 (shoulder blade, or scapula) and 14 (arm bone, or humerus). The inside angle — that is, the angle between these two bones over the rib area — determines the degree of smoothness of impact that is transmitted through the shoulders from the legs. Thus, as the top of the shoulder blades tilt increasingly forward, the shoulder-neck attachment becomes decreasingly less sloping, the concussion with the ground increases and the wear on the horse, as well as on the rider, is correspondingly emphasized.

Not only do steep shoulders contribute to rough riding but they also predispose the feet and legs to unsoundness. The shoulders should be neatly laid in at the top to facilitate smooth and prominent withers that are extremely important for a proper "sit" to the saddle (Figs. 5-33, 5-34, 5-35, 5-36, 5-37 and 5-38).

Chest

While, in general, the chest area varies with the depth and width of the shoulders, excessive width should not be favored in this area in stock horses, as it is usually associated with the width between the front legs. To the degree to which this relationship is applicable, emphasizing chest capacity through depth rather than through excessive width would appear advisable. Horses with extreme width between the front legs generally travel with a "rolling" action, which is more or less associated with awkwardness in leg action and short, insecure stride. Appaloosas are noticeably free from this tendency (Figs. 5-32 and 5-38).

Back and Middle

In an acceptable topline, the back is short and well-supported (Figs. 5-36, 5-37 and 5-38). The loin is wide and strong, filling the space between the ribs and the hips smoothly (Fig. 5-38). A strong spread of the ribs is usually good assurance that the weight of the middle is supported in a well-carried underline. The flank should be well let down. A long, relatively level underline is favorable to freeness in body action (Figs. 5-36, 5-37, 5-38 and 5-39). Sluggishness is associated with a low-slung middle.

Rear Quarters

The rear quarters of a stock horse operating in rough country should be heavily muscled and well-balanced within the entire body. Considerable length, good width, moderate slope of the croup and heavy muscling running well down over the thighs through the stifle and into the gaskins assure strength in this area. A moderate slope of the croup is usually associated with a well-supported setting of the hind legs. Observe the setting of the hind legs in Fig. 5-39.

Feet and Legs

The forearms are broad at the base of the shoulders, strongly muscled and tapered smoothly into broad, deep, clean-cut knees that are supported by short, quality cannons. The attachment of the tendon at the back of each cannon bone should not be restricted; instead it should be set somewhat apart from the cannon to give the area a flat appearance with a broad attachment at the base of the knee. Thus, the groove between the tendon and the cannon bone should clearly be evident and free

Fig. 5-35. Champion Appaloosa Halter Horse. (Courtesy, Appaloosa Horse Club, Inc.)

Fig. 5-36. Champion Appaloosa Halter Horse. (Courtesy, Appaloosa Horse Club, Inc.)

Fig. 5-37. "Paul's Silver Fox," an Appaloosa mare, has fared well in the show ring, having been Reserve Grand Champion Mare at the Cow Palance and Reserve Grand Champion Mare at a Denver Livestock Show. Owned by Kennedy Farms, Inc., Tracy, California.

Fig. 5-38. This outstanding Appaloosa, "Goer," was First Stallion and Grand Champion Stallion in Fort Worth, San Antonio and Houston. Owned by Simmons Ranch, Penyn, California.

from unnecessary filling. The fetlock should be strong and lean in appearance and well-supported by not less than a medium-length pastern set at a slope somewhat near a 45-degree angle. The slope of the pasterns is a critical point at which, in large part, the hard-riding qualities originate. Thus, the slope of the pasterns and the slope of the shoulders, as well as to the wearing ability of the parts involved in the fore-quarters, (Figs. 5-33, 5-36 and 5-37), are important contributors to riding comfort.

Quality is one of the very important characteristics to look for in the rear legs. Note in Fig. 5-34 the pronounced evidence of quality, as well as the moderately long cannons and their attachments at the knees and hocks, both of which show considerable strength, desirable leanness and acceptable size. Quality is also reflected in the pasterns and pastern joints, hooves and hoofheads.

Action

A long, straight, determined, smooth stride at the walk wins favor. The feet should be picked up freely, moved forward with good clearance and then placed squarely on the ground (Fig. 5-39). Skimming the feet along the ground before they come to a stop is objectionable, as this often may cause stumbling. The stock horse spends more time at the walk than at any other gait. Though correct at the walk, it may be disappointing at the trot or the gallop — each an important gait; hence, action should be checked at all standard gaits.

Fig. 5-39. Champion Western Pleasure Appaloosa. (Courtesy, Appaloosa Horse Club, Inc.)

Efficiency in action is the normal result of good balance and control. Action at the trot requires more effort, is usually more pronounced and is faster than at the walk. A well-trained horse should exhibit smooth, straight action with controlled speed, as that is more common than excessive speed at the trot. Extremely high action uses up energy. The direction of hoof movement should be straight forward. Variations from this contribute to wasteful action.

When executing the gallop, the horse should show an easy, natural control and an efficient coordination between the body and legs to make the action smooth (Fig. 5-40). A very restricted gallop requires the horse to go up and down too much for the distance it travels. This gait drains the horse's energy, makes straight and smooth forward action difficult and has little, if any, application in stock-horse service. Exaggerated speed at the gallop also requires more effort for the distance covered than a normal, natural gallop.

Fig. 5-40. Appaloosa in action, Camas Prairie Stump Race. (Courtesy, Appaloosa Horse Club, Inc.)

Weight and Height

While 1,100 to 1,200 pounds is preferred to 900 to 1,000, the most popular weight is from 1,000 to 1,100 pounds. Although there is more preference for 15.2 to 16.0 hands than for 14.2 to 15.0, the most popular height is between 15.0 and 15.2 hands.

Color

The Appaloosa is characterized by various color markings that definitely distinguish it from other breeds, as well as distinguishing the individuals within the breed from each other. The breed is unique in this respect. The Appaloosa Horse Club Stud Book and Registry, page 57, "does not prefer any particular pattern of markings and does not place any weight on marking in judging Appaloosas."

However, inasmuch as the color is so strikingly unusual, the following discussion concerning various color patterns is quoted from the Stud Book.

> A horse having dark roan or solid color foreparts, white with dark spots over the loins and hips. In the Palouse country, Appaloosas with this pattern were commonly said to have "Squaw" spots. With few exceptions, Appaloosas with this pattern show it from the birth. This pattern is one of the most common in the breed [Figs. 5-32, 5-33, 5-34, and 5-36]. Old timers claim the dark blue roan, and white with black spots over the loin and hips to have been the most popular with the Nez Percé. The dark spots on Appaloosas appear in several shapes, such as round, oval, pointed or leaf shapes and diamond shapes.
>
> A white horse with spots over the entire body. One type of this spotting will show spots very close together on the head and neck, sometimes giving an almost solid appearance. The spots will become farther separated toward the loin and hips but will remain quite uniform in size. In the other type, spots will appear much larger over the loin and hips, becoming smaller and farther apart toward the head.
>
> A horse having roan or solid colored foreparts with white over the loin and hips. This pattern is quite common among Appaloosas.
>
> A horse having dark base color with various sizes of white spots or specks over the body.
>
> A horse having mottling of dark and white covering the body. This color sometimes resembles an ordinary roan except for the mottling, colored skin and other characteristics.
>
> Old timers speak of Appaloosas as being either a "red" or a "blue" Appaloosa — the red applying to the chestnut, bay and red roans, and the blue to the black and blue roans. The terms "red" and "blue" with reference to Appaloosa are very common in the Palouse country. Duns, buckskins and Palominos with Appaloosa markings were not known to the early pioneers. They are the results of crossing an Appaloosa with a dun or Palomino. The last two of the color patterns listed are usually foaled solid color and change with age. Often a horse of these patterns will first become covered with white specks as a yearling or two-year-old and become very mottled at three or four. Some Appaloosas turn white with old age. Too, the coat will often change color considerably with seasons, such as light in summer and dark in winter. Although Appaloosas of the last two patterns are not as colorful as some discussed earlier, they very often produce foals of the other more colorful patterns. There are many examples of a rather plain-colored mare producing a very flashy marked foal even from a solid color stallion. An Appaloosa stallion will often sire foals of several different color patterns.[7]

---

[7]*The Appaloosa Horse Club Stud Book and Registry,* p. 56.

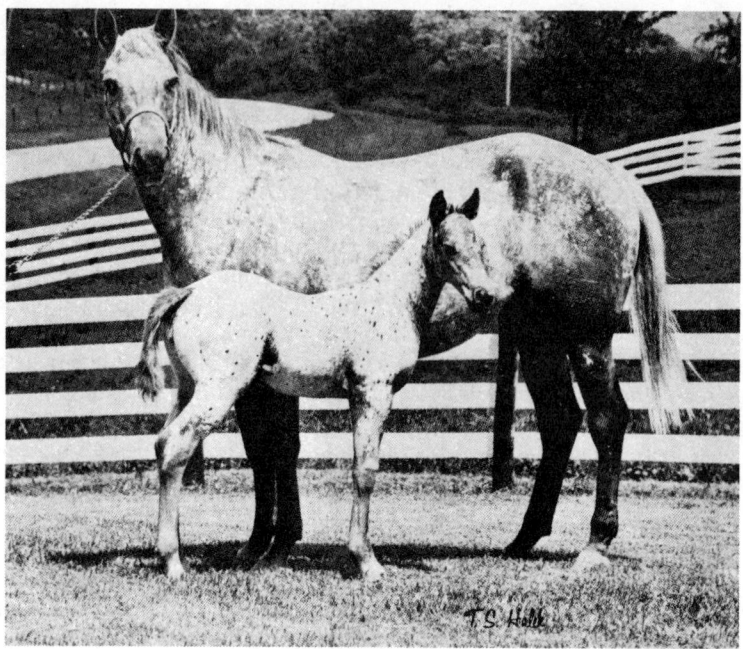

Fig. 5-41. This attractive Appaloosa mare and her foal are at home at Money Creek Ranches, Houston, Minnesota.

HORSE JUDGING TERMINOLOGY

Horse shows have influenced the types of certain breeds of light horses. Other breeds have been affected through selections based on their performance on the race tracks. The vast majority of horses are evaluated by horse owners who take pride in owning good horses and who do their own buying and selling.

Judging horses, like all livestock judging, is an art, the rudiments of which must be obtained through patient study and long practice. Successful horse producers are usually competent judges. Likewise, shrewd traders are usually masters of the art, even to the point of deception.

Accomplished stock producers generally agree that horses are the most difficult to judge of all classes of livestock. In addition to conformation, which is the main criterion in judging other livestock, action and numerous unsoundnesses are of paramount importance.

There are four methods of selection: (1) selection based on type or individuality, (2) selection based on pedigree, (3) selection based on show winning or other performance and (4) selection based on production testing. Only the first one — judging — is expanded upon in this text.

Successful horse judges must know what to look for. That is, they must have in mind an ideal or a standard of perfection.

Good horse judges possess the ability to observe both good conformation and quality performance, as well as defects, and the ability to weigh and evaluate the relative importance of the various good and bad features.

There is always great danger of beginners making too close an inspection by getting "so close to the trees that they fail to see the forest."

Good judging procedure consists of the following three steps: (1) observing at a distance to get a panoramic view of several horses, (2) seeing the animals in action and (3) inspecting them close up. Also, in viewing a horse from all directions, you should use a logical method, as for example, (1) front view, (2) rear view, (3) side view and (4) action and soundness. Thus, you avoid overlooking anything and to a great extent you are able to retain the observations that you made.

The following extensive list of terms and expressions will be of service to you in your attempt to describe horses with more precision and to speak with greater ease and familiarity about the qualities in horses. This list should aid you in making a careful examination of horses and in describing your observations with well-selected terms that not only specifically indicate the problems at hand but also do it in terms that are familiar to horse producers. As you acquire experience with horses, you should seek to enlarge your vocabulary of descriptive terms, particularly so that you can acquire an appreciation of desirable group expressions that you can freely use to describe various combinations of characteristics, as for instance, "A horse with quality, substance, constitution and balance" or "A feminine-headed, high-quality filly with a long barrel that is more correct in her feet, legs and action than any other individual in the class." Expressions of this nature should be used only after you have become fully familiar with the characteristics that contribute to the qualities that are involved. Likewise, you should refrain from using the term "quality" in a general way unless quality is apparent in bone, feet, hair coat, body and muscling.

You should have at your command a workable vocabulary consisting of a large number of expressions, not memorized from any specified list, but rather those that through use have become a natural part of your ability to relate your understanding of horses. These expressions should be carefully selected and specifically applied to the problems at hand so that there is little if any confusion regarding their significance.

The following expressions are examples of different combinations that can be made of various descriptive terms. You should not attempt to memorize the combinations as they appear in this list, because if you do, you will be looking for a horse which fits these memorized expressions; whereas, you should be learning how to formulate expressions that adequately and specifically describe the horse at hand. However, you should carefully study the various terms that are used and the manner in which they are combined in summarizing the qualities of a horse.

General Expressions

1. High head carriage, alert ears, excellent disposition and beautiful conformation, (Fig. 5-4).

2. All parts well-developed and nicely blended together.
3. Lacking style and beauty.
4. Lacking balance and symmetry.
5. More Quarter Horse breed type and sex character about the head, ears and neck then any other mare in the class.
6. More femininity and Quarter Horse character than No. 3.
7. A stallion with more masculinity and ruggedness, possessing more desirable Quarter Horse breed type than No. 2 (Fig. 5-2).
8. More quality and refinement about the head, ears, hide, bone and hair coat than any other individual in the class.
9. A bigger, stouter, longer-muscled stud that is fundamentally more correct than any other animal in the class (Fig. 5-21).
10. A nicer-balanced, stronger-coupled, heavier-muscled mare that is structurally more correct than No. 2.
11. Showing plenty of quality, which is indicated by clean, flat bone, well-defined joints and tendons, refined head and ears and fine skin and hair coat (Fig. 5-2).
12. Excellent breed type (size, color, shape of body and head and action true to the breed represented).
13. A mare that is flat in her withers, weak-topped, short and steep in her croup, light-muscled and lacking the quality and substance to place any higher in this class today.
14. A big, stout, rugged stallion that shows more muscle development through the arms and forearms and in the gaskin (inner and outer) and stifle region than any other animal in the class (Fig. 5-26).
15. A feminine-headed, high-quality filly with a long barrel that is more correct in her feet, legs and action than any other individual in the class.
16. The most correct gelding in the class that moves out straighter and truer than No. 3.
17. A masculine-headed, strong-jawed stallion that is cleaner through the throat-latch, stronger-coupled, heavier-muscled through the stifle and gaskins and that moves out in a freer, more collected manner (well-coordinated action) than No. 2.
18. A strong-fronted mare that is well "veed" up in the chest, long in the underline but wide at the hocks and sluggish in action at both the walk and the trot.
19. A low-quality, plain, light-muscled, crooked-legged filly with the least desirable action of any individual in the class.
20. A flat-withered, long-backed, shallow-middled stallion that is off on his feet and legs — the poorest-moving animal in the class.

Feet and Legs

1. Straight, true and squarely set on his feet, and legs (front, rear or all feet and legs) (Figs. 5-35 and 5-36).

2. Standing more correct on his feet and legs with a more desirable slope to the pasterns than any other animal in the class (Figs. 5-25 and 5-35).
3. Ample bone for her size.
4. Standing on more substance of bone and cleaner and neater in the joints than No. 2.
5. Standing on more good, clean, flat, high-quality bone than any other mare in the class.
6. Narrow and shallow at the heels and somewhat thick and puffy about the hocks.
7. Sound and free from blemishes.
8. Toed-in behind and standing too wide at the hocks.
9. Toed-in front (pigeon-toed).
10. Toed-out behind and close at the hocks (cow-hocked).
11. Toed-out in front (splay-footed).
12. Sickle-hocked or camped under (too much set to rear legs).
13. Buck-kneed.
14. A bit back at the knees.
15. Pasterns too short and straight.
16. Long, gently sloped pasterns (about 45-degree angle) (Fig. 5-6).
17. Straight, strong legs with ample bone.
18. Knock-kneed.
19. Close at the ground (usually accompanies a narrow chest, a narrow chest floor, lack of constitution).
20. Crooked and off on his front feet (rear feet, all feet) and legs.
21. Possessing larger, denser hooves with more width and depth of heel and more correct set to the feet and legs than No. 2.
22. A small-footed, shelly-hooved mare that is contracted at the heels and too straight in her pasterns to stand any higher in this class today.
23. Clean, dense, quality bone.
24. Correct in set of legs. Well-placed underpinning (Fig. 5-21).
25. Wide, flat cannons, and smooth joints (Fig. 5-39).
26. Lacking somewhat in substance of bone, especially below the knees and hocks.
27. Clean-cut knees of good size. Strong, well-supported knees.
28. Somewhat longer pasterns set at a bit more angle.
29. Pasterns too short and straight.
30. Somewhat weak in the pasterns.
31. "Cocked" in the ankles.
32. Large, clean-cut, wide, deep hocks (Fig. 5-5).
33. Large, round, well-balanced feet.
34. Clean hoofheads that show quality (Fig. 5-33).
35. Possessing hoofheads that lack quality. They show too much thickness.
36. A bit back on the heels.
37. Shapely feet that are indicative of long wear (Fig. 5-40).

38. Coon-footed.
39. Thick, puffy, meaty, unsightly hocks.
40. Tending to travel tender on left front foot.

Head, Neck and Shoulders

1. Well-proportioned head compared to rest of body (Figs. 5-21 and 5-35).
2. Possessing a more desirable chiseled appearance about the head, stronger jaws and shorter, more attractive, correctly set ears than No. 2 (Fig. 5-32).
3. Refined, clean-cut head (Fig. 5-36).
4. Full forehead with great width between the eyes (Fig. 5-32).
5. Broad and strongly muscled jaws (Fig 5-32).
6. Medium-sized, well-carried, attractive ears.
7. A bold, masculine-headed stud that is stronger-jawed and cleaner in the throat-latch than any other animal in the class (Fig. 5-29).
8. A mare with more refinement and femininity about the head and neck than No. 3 (Fig. 5-34).
9. A plain-headed, weak-jawed stallion that lacks the masculinity of the other individuals in the class.
10. Roman-nosed (Fig. 5-39).
11. Parrot-mouthed (overshot jaw).
12. Pig-eyed.
13. Undershot jaw.
14. Fairly long neck, carried high (Fig. 5-27).
15. Clean-cut about the throatlatch (Figs. 5-21 and 5-24).
16. Having head and neck well set on shoulders.
17. A short, thick, cresty-necked mare.
18. Ewe-necked.
19. Desirable slope to the shoulders (Figs. 5-21, 5-24 and 5-26).
20. A thin-necked, straight-shouldered stud.
21. Sloping shoulders (about a 45-degree angle).
22. A stud that is heavier-muscled through his arms and forearms, showing more prominence of muscling up in the "vee" and through the chest than any other animal in the class (Fig. 5-26).
23. A flat-muscled, narrow-chested mare with very little evidence of muscling up in the "vee" through the arm and forearm region.
24. Refined features about the head. Clean-out, well-defined lines. Excellent quality in the head and neck.
25. Showing head that lacks the feminine qualities desired in a brood mare.
26. Well-shaped, neatly carried ears.
27. Short, foxy ears (Fig. 5-26).
28. Plain head, lacking quality and alertness.
29. Having eyes that are a bit small and dull. Large, bright, prominent eyes that are well-balanced with the general quality of the head.

30. Having neck that has good length and balance and head that is functionally well-carried. The head is carried too low, making the horse appear sluggish. The head is alertly carried.
31. Well-placed crest that shows ample fullness (Fig. 5-32). Too full or heavy crest. Inadequately developed crest.
32. Neat, clean-cut, well-up head and neck that show excellent breed type and character (Fig. 5-2).
33. Too short and stubby neck.
34. Smooth shoulders, set well back on the top and neatly laid in.
35. Very acceptable shoulder slope, affording ample room for action in the front feet as well as a suitable place for the saddle.
36. Deep, muscular shoulders that are snugly laid in.
37. Low and flat at the withers (mutton withers).
38. Shallow-shouldered and light-muscled through the shoulder, arm and forearm region.
39. Excellent balance in the deep, heavily muscled, neatly laid shoulders and the refined, breedy, well-carried head and neck (Fig. 5-34).
40. Prominent withers that will "sit" a good saddle.
41. Poor withers.
42. Light-muscled through the arms and forearms.
43. A deep, wide-chested mare that is muscled up better in the "vee" than No. 2.
44. A narrow-chested, light-muscled individual.
45. Thick and heavy (full) in the throatlatch (Fig. 5-29).

Topline and Rear Quarters

1. Clearly defined withers on the same height as the high point of the croup (Figs. 5-21, 5-24 and 5-26).
2. Short coupling, as denoted by the last rib being close to the hip (Fig. 5-21).
3. A strong-backed, shorter-coupled mare, showing more strength of underline (long-barreled) than No. 2.
4. A sway-backed mare that is long in the coupling and steep over the croup.
5. A mare that is shorter and stronger in the back and loin, with a longer, more nicely turned, heavier-muscled croup (Fig. 5-34).
6. Longer and more level over the croup with a high, well-set tail.
7. Ample middle due to long, well-sprung ribs (Fig. 5-37).
8. Deep in the barrel (heart, heart girth) and well let down in the rear flank.
9. Lacking middle development capacity (both length and depth).
10. A deep-ribbed, long-middled mare.
11. A shallow-ribbed, short-middled stud.
12. High-cut rear flank, or "wasp-waist."
13. A stronger-topped, wider-loined, neater-coupled mare that is longer and more level over the croup and has more muscle development through the gaskin and stifle area than No. 3.

14. A weak-topped, long-coupled, short-rumped, light-muscled stallion.
15. Wide and muscular over the croup and through the rear quarters.
16. Lacking width and length over the croup and muscling through the rear quarters.
17. Short, very well-developed, closely coupled loin.
18. Neatly turned over the loin. Strong loin. Powerful loin.
19. Powerful back and loin, extremely muscular in this area (Fig. 5-26).
20. A bit slack in the back. Low in the back. Easy in the back. Dipping a bit in the back, sagging behind the withers.
21. A short-backed, weak-topped individual.
22. A long, level croup.
23. Extremely thick and muscular through the thighs, gaskins and stifle (Fig. 5-25).
24. Showing much more prominence of muscling in the outer gaskin than through the inner gaskin (Fig. 5-23).
25. Lacking development of inner (outer, both) gaskin muscle.
26. Well-balanced, shapely thighs.
27. "Pear-hipped."
28. Tapering from top to croup to lower thigh area (wider at top than at bottom — lacking uniformity of width or balance).
29. Light-muscled through the thighs, gaskins and stifle.
30. Loaded with hard, firm muscle (Fig. 5-23).
31. Lacking the muscling over the top and in the rear quarters that horse producers prefer.
32. Neat-hipped.
33. Possessing topline and quarters that are evenly balanced with a deep, long, neat underline.
34. A well-balanced, strong-topped, long-quartered stallion.
35. Strong in the coupling and showing unusual power through the rear end (quarter, thigh and stifle areas) (Fig. 5-23).
36. Enough spring and depth of fore-rib to "sit" a good saddle.
37. A stronger-loined, longer-rumped, heavier-muscled mare through the gaskin and stifle area than No. 2.
38. Too long in the back and lacking in middle and capacity, compared to the other fillies in the class.

Action

1. Moving straight and correct at both the walk and the trot (Fig. 5-6).
2. Carrying hocks close and high and clearing the ground handily.
3. Correctly set hocks that are under control.
4. Too much movement of the hocks at the walk or the trot.
5. Long, straight, determined stride.
6. Straight, smooth, aggressive, long, free action.
7. Flashy way of going but lacking substance and ruggedness of bone.
8. Collected manner of moving. Well-coordinated action.

9. Steady, true, practical action.
10. Regular action in feet and legs that is well-coordinated with the movement of the knees and hocks.
11. Long, bold stride.
12. Well-balanced, straight mover but lacking a bit in the size of the feet.
13. Moving with plenty of ease but traveling wide at the hocks (or close at the hocks).
14. A bit short in stride. Sluggish in action.
15. Fairly good at reaching out, but with the toes striking the ground too soon.
16. Traveling with too much motion at the shoulders (rolling action).
17. Too slow at the walk and awkward and clumsy at the trot. A bit draggy. Stale in action.
18. Paddling. Swinging out. "Toe-narrow" (toed-out, or splay-footed) or "base-narrow." Winging somewhat. "Toe-wide" (toed-in, or pigeon-toed) or "base-wide."
19. Short and stubby or stilted way of going. A little tender on left front foot. Goes too close. "Rope-walker."
20. Easy, prompt, balanced movement.
21. A long step, with each foot carried forward in a straight line.
22. Stepping with feet lifted off the ground.
23. Rapid, straight, elastic trot with joints well-flexed.
24. Short step with feet not lifted off the ground (daisy-cutter).
25. Winging, forging, interfering.
26. Slow, collected canter, which is readily executed on either lead.
27. Fast, extended canter.
28. Short, choppy stride.
29. Moving close (narrow) in front (behind).
30. Moving wide in front (behind).

Quality, Style and Bloom

1. Excellent quality, as indicated by a well-balanced combinations of smoothness, refinement in the bone, hair and skin and clean-cut features (Fig. 5-24).
2. Excellent quality, but too much refinement in the bone and general lack of desirable ruggedness.
3. Lacking quality in the joints and especially in the hocks. Coarse, thick, meaty, poorly set hocks.
4. Carrying head up high with unusual style (Fig. 5-21).
5. Flexible way of going and well-coordinated body movements.
6. Excellent bloom, correct fleshing and physically very fit.
7. Lacking a bit in bloom and action to be distinctive in style.
8. Lacking necessary bloom. Stale. Overdone. Sluggish.
9. Excellent quality, as denoted by clean, flat bone, well-defined joints and tendons (Fig. 5-26).

10. Refined about the head and ears and possessing fine hair and a thin, pliable hide (Fig. 5-26).
11. Lacking quality and refinement (coarse, low quality).
12. Possessing more style, quality and animation than any other individual in the class (Fig. 5-36).
13. A stylish filly that has more flash and coordination of movement than No. 2.
14. Plain, off-type individual.
15. A smooth, high-quality filly that is quick to catch the judge's eye at either the walk or the trot.

In general, "quality" refers to the degree of excellence in refinement as it contributes to working qualities, "style" involves the manner in which the horse displays itself and "bloom" indicates the appearance as it is influenced by finish and physical fitness.

REASONS FOR PLACING A CLASS OF QUARTER HORSE MARES

"I placed this class of Quarter Horse mares 1-3-4-2. In my top pair, I placed 1 over 3 in a close placing, as 1 is a growthier, more correctly balanced, broodier mare that possesses a longer, slimmer neck than 3. She is "veed" up higher in the chest and tied down farther on the forearm than 3. In addition, she has a more desirable slope to the shoulders and a more correct angle to her pasterns. She is also longer and more leveled out over the croup and shows a longer, cleaner barrel and underline. However, I grant that 3 has more Quarter Horse character about the head and ears, is cleaner in the throatlatch, is shorter and stronger in the back and stands on higher-quality, cleaner, flatter bone.

"In coming to the middle pair and most logical decision in the class, I placed 3 over 4 because 3 is a more feminine, cleaner cut, more symmetrical mare than 4. She has more character and quality about the head, ears, throatlatch and neck. She also is shorter and stronger in the back and has a more desirable turn out over the croup. She has a more desirable angle to both the front and rear pasterns, which should allow for a more pleasureful ride and greater durability under working conditions. I admit that 4 is a larger, growthier, more rugged mare with a longer underline.

"For the bottom pair, 4 placed over 2, as 4 is a longer, growthier, broodier, higher-capacity mare. No. 4 is wider and deeper through the rib, giving her more capacity and constitution. She is more correct in her topline and stands down straighter and squarer on her feet and legs than 2. I will grant that 2 shows more quality and refinement about the head, ears, bone and joints but I fault her and leave her at the bottom of the class for being a small, unbalanced, lower-capacity mare that lacks the muscling and skeletal correctness to place any higher in this class today."

Fig. 5-42a. Rear view of Quarter Horse mare No. 1.

Fig. 5-42b. Front view of Quarter Horse mare No. 1.

Fig. 5-42c. Side view of Quarter Horse mare No. 1.

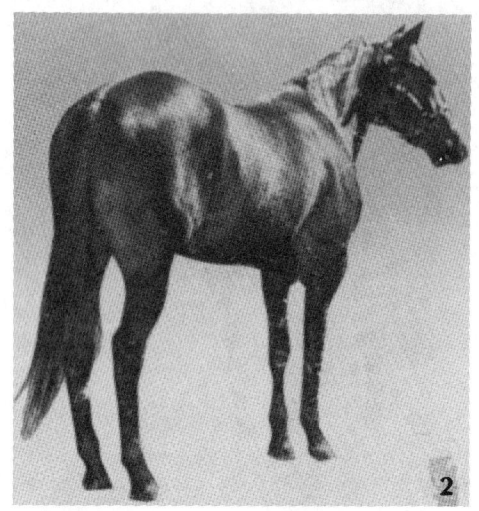

Fig. 5-43a. Rear view of Quarter Horse mare No. 2.

Fig. 5-43b. Front view of Quarter Horse mare No. 2.

Fig. 5-43c. Side view of Quarter Horse mare No. 2.

Fig. 5-44a. Rear view of Quarter Horse mare No. 3.

Fig. 5-44b. Front view of Quarter Horse mare No. 3.

Fig. 5-44c. Side view of Quarter Horse mare No. 3.

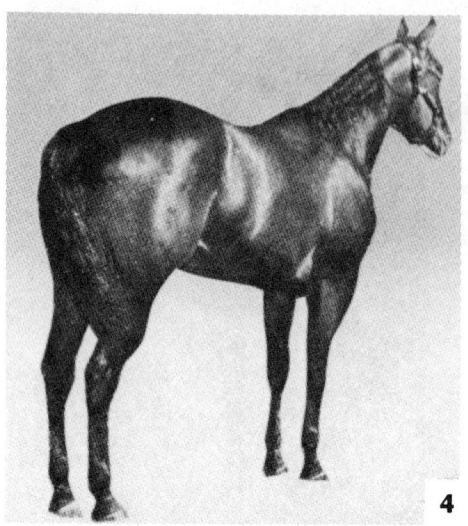

Fig. 5-45a. Rear view of Quarter Horse mare No. 4.

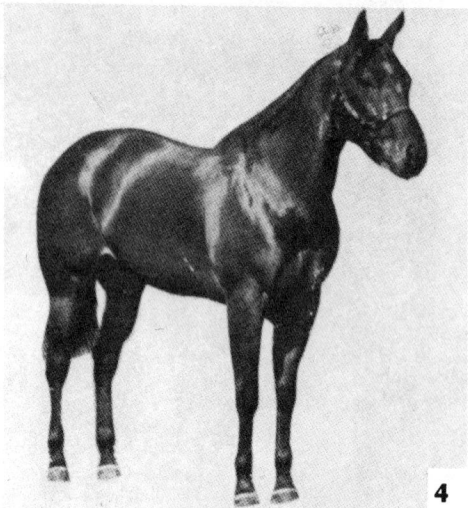

Fig. 5-45b. Front view of Quarter Horse mare No. 4.

Fig. 5-45c. Side view of Quarter Horse mare No. 4.

REASONS FOR PLACING A CLASS
OF APPALOOSA GELDINGS

"I placed this class of Appaloosa geldings 2-1-4-3, feeling the class broke down into two distinct pairs. I placed 2 over 1 because 2 is a larger, growthier, heavier-muscled gelding than 1. He shows more character and cleanness in the throatlatch and neck. He also is longer from hip to hock and shows more muscle through the stifle, gaskin and forearm. In addition, he has longer, sharper, more prominent withers. I grant that 1 is stronger-topped and shorter-coupled and possesses a straighter underline.

"In coming to my middle pair, I placed 1 over 4 as 1 is a more athletic-appearing, longer-muscled gelding than 4. He has a more intelligent-appearing head and ears, is longer and slimmer-necked and has a longer, more desirable slope to the shoulders. He is stronger-backed and shorter-coupled and has a more desirable turn out over the croup. He is longer and cleaner in the underline and has more angle to his pasterns than 4. I admit that 4 has more depth of heart and possesses sharper, more prominent withers than 1.

"In my bottom pair, I placed 4 over 3 in a very difficult decision. He shows greater over-all balance and is more correct in his feet and leg structure. He has a slimmer, leaner neck, is sharper and more prominent in his withers and shows more length from his hips to hocks. He has a more desirable slope to his shoulders and a more correct angle to his pasterns, which should allow for more give and cushion under saddle. I grant that 3 has more muscle expression through the arms, forearms and stifle, but I fault 3 and place him last because he is an unbalanced, thick-necked, sluggish-appearing individual standing on the least desirable set of feet and legs in the class."

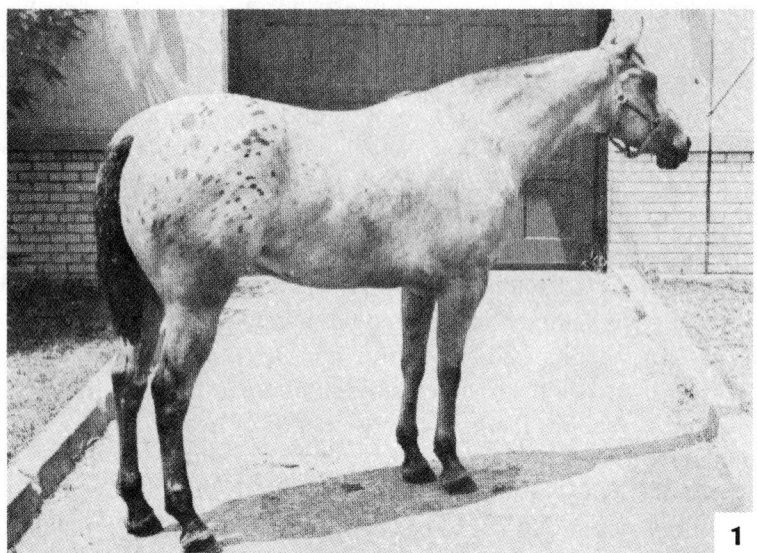

Fig. 5-46. Side view of Appaloosa gelding No. 1.

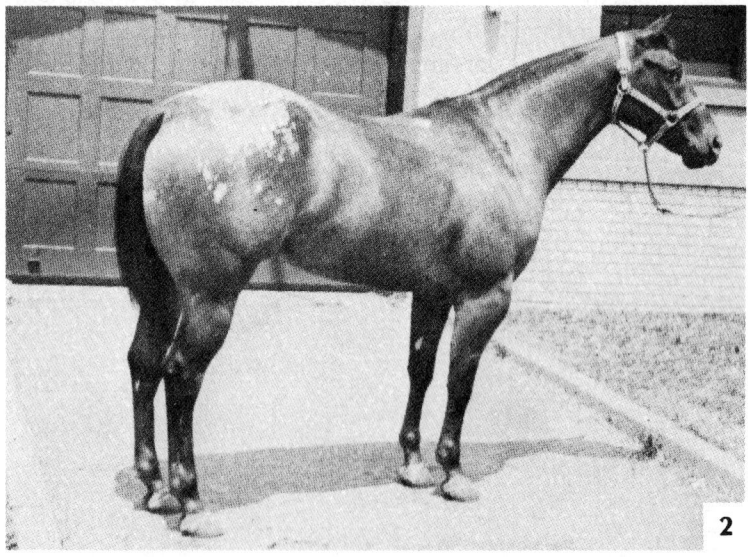

Fig. 5-47. Side view of Appaloosa gelding No. 2.

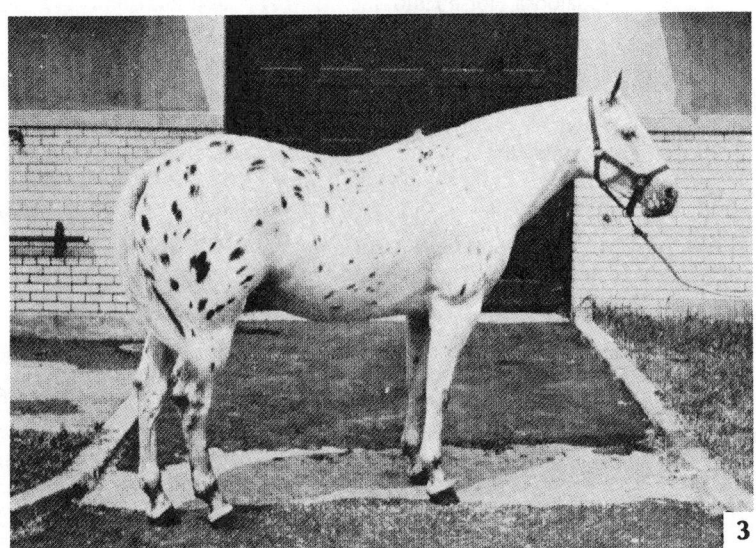

Fig. 5-48. Side view of Appaloosa gelding No. 3.

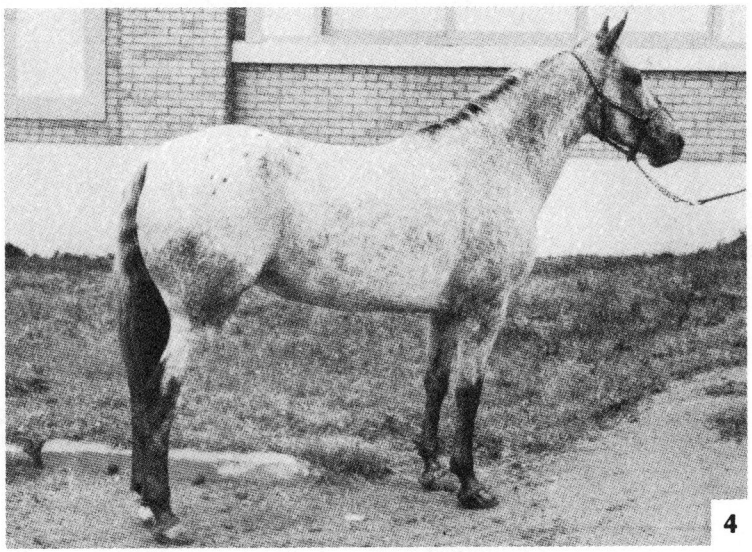

Fig. 5-49. Side view of Appaloosa gelding No. 4.

BREED REGISTRY ASSOCIATIONS

Breed	Association	Officer and Address
Appaloosa	Appaloosa Horse Club, Inc.	Ron Robison Executive Secretary Box 8403 Moscow, Idaho 83843
Paint	American Paint Horse Association	Ed Roberts Executive Secretary P.O. Box 18519 Fort Worth, Texas 76118
Palomino	The Palomino Horse Association, Inc.	P.O. Box 324 Jefferson City, Missouri 65101
	Palomino Horse Breeders of America, Inc.	Melba Lee Spivey Secretary Box 249 Mineral Wells, Texas 76067
Quarter Horse	American Quarter Horse Association	Ronald Blackwell Executive Vice President P.O. Box 200 2701 I-40 East Amarillo, Texas 79168

Index